Surveys and Tutorials in the Applied Mathematical Sciences

Volume 10

Featuring short books of approximately 80-200pp, Surveys and Tutorials in the Applied Mathematical Sciences (STAMS) focuses on emerging topics, with an emphasis on emerging mathematical and computational techniques that are proving relevant in the physical, biological sciences and social sciences. STAMS also includes expository texts describing innovative applications or recent developments in more classical mathematical and computational methods.

This series is aimed at graduate students and researchers across the mathematical sciences. Contributions are intended to be accessible to a broad audience, featuring clear exposition, a lively tutorial style, and pointers to the literature for further study. In some cases a volume can serve as a preliminary version of a fuller and more comprehensive book.

Hengguang Li

Graded Finite Element Methods for Elliptic Problems in Nonsmooth Domains

Springer

Hengguang Li
Department of Mathematics
Wayne State University
Detroit, MI, USA

ISSN 2199-4765 ISSN 2199-4773 (electronic)
Surveys and Tutorials in the Applied Mathematical Sciences
ISBN 978-3-031-05820-2 ISBN 978-3-031-05821-9 (eBook)
https://doi.org/10.1007/978-3-031-05821-9

Mathematics Subject Classification: 65N30, 35B65, 65N50, 65N12, 65N15, 35J25

This Springer imprint is published by the registered company Springer Nature Switzerland AG
The registered company address is: Gewerbestrasse 11, 6330 Cham, Switzerland

To my family.

Preface

Elliptic boundary value problems are essential models in many scientific disciplines. The development of effective finite element methods to solve these equations is a central focus in computational mathematics. The performance of the numerical algorithm especially relies on the smoothness of the solution. The standard finite element method can approximate smooth functions well, while for singular solutions (solutions with radical changes in their derivatives), it can have seriously deteriorated convergence and sometimes yield false approximations. This is largely because the uniform construction of the approximation space is inadequate for capturing the multiscale structure of the singularity. The graded finite element method, allowing mesh size variations in different regions of the computational domain, is a widely used technique for effective model simulation and analysis in the presence of singularities. In elliptic problems, the solution can possess singularities near the nonsmooth points of the domain and near the points where the boundary condition changes, even when the given data is smooth. These singular solutions, depending on the domain geometry and the boundary condition, represent some of the most common singularities in practice.

This book presents graded finite element methods for elliptic problems in nonsmooth domains with mixed boundary conditions. Meant to be introductory and self-contained, it covers basic mathematical theory and numerical tools that address major challenges imposed by the singular solution. Since Kondratiev's 1967 seminal paper [71], there has been significant progress on the mathematical analysis of singular problems. One purpose of this book is to synthesize the relevant literature, and introduce important theoretical results and their applications to a broader audience in the computation community. With a systematic approach, this book demonstrates the development of graded mesh finite element techniques for singular solutions, in which concepts in several areas (e.g., functional analysis, Sobolev space theory, numerical analysis, and scientific computing) play a critical role.

This book is intended primarily for mathematics graduate students and researchers. It can be used for a graduate topics course for students who have had a first course in analysis and in finite elements. This book can also serve as a reference for applied and computational mathematicians, scientists, and engineers working on numerical methods who may encounter singular problems.

Chapters 1–4 contain the essential material for a course. Chapter 1 provides a brief review of the finite element method. Chapter 2 collects key concepts in Sobolev space theory and basic finite element error analysis results. Chapter 3 derives representations of the singular solution in two- and three-dimensional domains, and formulates the principle of the graded mesh construction. Chapter 4 targets two-dimensional singular problems. In particular, it develops regularity analysis in weighted Sobolev spaces and lays out the explicit graded mesh algorithm along with the associated finite element error estimates. The material in Chaps. 1–4 gradually builds up and becomes more involved. Chapters 1 and 2 are mostly independent of each other, both providing the basis for the exposition in Chaps. 3 and 4. Some of the topics in Chaps. 1 and 2 can be skipped if covered in the prerequisite course. Chapters 5 and 6 are devoted to graded finite element methods for three-dimensional singular problems that are more technically demanding yet less explored. These two chapters summarize recent theoretical and numerical advances for these problems, which can be used as additional reading material for students and researchers who are interested in three-dimensional anisotropic algorithms.

There is a vast literature on finite element methods for singular problems. By focusing on graded finite element techniques and the associated analytical tools, this book largely omits many other active developments in this area. For example, one notable omission is the adaptive finite element method (AFEM) for singular problems. Compared with the graded finite element method that is based on a-priori analysis of the equation, the AFEM is usually built upon a-posteriori estimations. The AFEM has been investigated in various research books, articles, and conference proceedings, some of which are listed in the References. See [3, 15, 30, 40, 48, 96, 115, 117], for example. Moreover, for the mathematical analysis of singular problems, this book considers second-order elliptic equations with constant coefficients, and generally avoids high-order problems, time-dependent problems, and problems with variable coefficients. This allows one to derive more explicit results with a smaller set of technical terms and reasonably limited space. Readers who are interested in extending the application of the analysis to these problems are referred to [49, 61, 71, 93, 101] and references therein.

The author is indebted to many people who have helped at various stages and contributed to the research program over the years. The author expresses his warmest thanks to I. Babuška, C. Bacuta, U. Banerjee, S. Brenner, L. Chen, Y.-J. Lee, S. Nicaise, V. Nistor, J. Ovall, L. Scott, L.-Y. Sung, Q. Zhang, Z. Zhang, and L. Zikatanov for exciting and fruitful discussions on singular problems. The author is particularly grateful to D. Drucker, P. Yin, and the anonymous reviewers for reading the manuscript and providing helpful suggestions.

Detroit, MI, USA
March 2022

Hengguang Li

Contents

Chapter 1
The Finite Element Method

We describe the basic components of the finite element algorithm, using the Poisson equation in a bounded d-dimensional domain for $d = 1, 2,$ and 3 as a model problem. In addition, we give detailed examples of the finite element method in different dimensions, illustrating concepts important for implementation, such as the mesh, the basis functions, and the assembling of the stiffness matrix. This chapter is suitable for readers who are interested in a concise review of the finite element method.

1.1 The Finite Element Algorithm

Let $\Omega \subset \mathbb{R}^d$, $d = 1, 2, 3$, be a bounded d-dimensional domain with a reasonably smooth boundary $\partial\Omega$. Consider the Poisson equation with the Dirichlet boundary condition as the model problem

$$-\Delta u := \sum_{i=1}^{d} \partial_{x_i}^2 u = f \quad \text{in} \quad \Omega \qquad \text{and} \qquad u = 0 \quad \text{on} \quad \partial\Omega. \tag{1.1}$$

The finite element method, in its simplest form, is a process of producing numerical solutions from a finite-dimensional space to approximate the differential equation. It transforms the original equation that is in an infinite-dimensional space to a finite-dimensional problem that the computer is able to compute. The finite element method consists of two main components: the variational formulation and the finite element space. We shall describe the finite element algorithm by applying it to solve (1.1).

H. Li, *Graded Finite Element Methods for Elliptic Problems in Nonsmooth Domains*, Surveys and Tutorials in the Applied Mathematical Sciences 10,
https://doi.org/10.1007/978-3-031-05821-9_1

1.1.1 The Variational Formulation

Suppose the solution u in (1.1) is sufficiently smooth. Let v be a smooth function such that $v = 0$ on $\partial\Omega$. Multiply both sides of (1.1) by v. Then integration by parts yields

$$\int_\Omega -v\Delta u \, dx = \int_\Omega f v \, dx$$
$$\Rightarrow \int_\Omega \nabla u \cdot \nabla v \, dx - \int_{\partial\Omega} \nabla u \cdot n v \, ds = \int_\Omega f v \, dx. \tag{1.2}$$

Here, $\nabla = (\partial_{x_1}, \ldots, \partial_{x_d})^T$ is the gradient operator and

$$n(x) := \big(n_1(x), n_2(x), \ldots, n_d(x)\big)^T$$

is the outward normal vector at $x \in \partial\Omega$. Meanwhile,

$$\int_\Omega \nabla u \cdot \nabla v \, dx = \int_\Omega \sum_{i=1}^{d} (\partial_{x_i} u \partial_{x_i} v) \, dx$$

and

$$\int_{\partial\Omega} \nabla u \cdot n v \, ds = \int_{\partial\Omega} v \sum_{i=1}^{d} (n_i \partial_{x_i} u) \, ds.$$

Recall $v = 0$ on $\partial\Omega$ for (1.1), and therefore, $\int_{\partial\Omega} \nabla u \cdot n v \, ds = 0$. Set

$$a(u, v) := \int_\Omega \nabla u \cdot \nabla v \, dx \qquad \text{and} \qquad (f, v) := \int_\Omega f v \, dx. \tag{1.3}$$

Note that $a(\cdot, \cdot)$ is a symmetric bilinear form. We call (1.2) or simply

$$a(u, v) = (f, v) \tag{1.4}$$

the *variational formulation* of (1.1). To derive the finite element approximation, we also need the finite element space.

1.1.2 The Finite Element Space

We start with some necessary notation. For each integer $m \geq 0$, we denote by P_m the space of all polynomials of degree $\leq m$ in the variables $x_1, \ldots, x_d$. Note that the number of basis functions in P_m depends on m and d.

Proposition 1.1 *For $1 \leq d \leq 3$, the dimension of the space P_m satisfies*

$$\dim(P_m) = \binom{m+d}{m} = \frac{(m+d)\cdots(m+1)}{d!}. \tag{1.5}$$

Proof We show (1.5) for different values of d. First, consider the one-dimensional ($d = 1$) case. Let x be the variable. It is clear that the basis $\{1, x^1, \ldots, x^m\}$ of P_m consists of $m + 1$ basis functions, which leads to (1.5) for $d = 1$.

For the two-dimensional ($d = 2$) case, let x and y be the variables. For $i, j \geq 0$, the set $\{x^i y^j\}_{i+j \leq m}$ is a basis of P_m. Note that for each fixed value of i, with the constraint $i + j \leq m$, there are $m + 1 - i$ basis functions of the form $x^i y^j$. Therefore, the dimension of P_m is

$$\dim(P_m) = \sum_{i=0}^{m} (m + 1 - i) = \frac{(m + 2)(m + 1)}{2}, \tag{1.6}$$

which verifies (1.5) for $d = 2$.

For the three-dimensional ($d = 3$) case, let the variables be x, y, and z. For $i, j, k \geq 0$, consider the basis $\{x^i y^j z^k\}_{i+j+k \leq m}$ of P_m. Then for each fixed value of i, the number of basis functions of the form $x^i y^j z^k$ is $(m + 2 - i)(m + 1 - i)/2$. This is because with i fixed in $x^i y^j z^k$, $y^j z^k$ is a polynomial of degree $\leq m - i$ with two variables. Based on (1.6), the number of such terms (i.e., $y^j z^k$) is $(m+2-i)(m+1-i)/2$. Therefore, the dimension of P_m for $d = 3$ is

$$\dim(P_m) = \sum_{i=0}^{m} \frac{(m + 2 - i)(m + 1 - i)}{2} = \frac{(m + 3)(m + 2)(m + 1)}{3!}.$$

This completes the proof. □

For a subset $G \subset \mathbb{R}^d$, we denote by $P_m(G)$ the space which is the restriction of P_m to the set G,

$$P_m(G) = \left\{p|_G : p \in P_m\right\}.$$

The dimension of $P_m(G)$ is also given by (1.5) if the interior of G is not empty.

Let $a_i = (a_{1,i}, \ldots, a_{d,i})^T$, $1 \leq i \leq d + 1$, be $d + 1$ points in $\mathbb{R}^d$. Assume these points are linearly independent, which is equivalent to the matrix

$$\begin{pmatrix} a_{1,1} & a_{1,2} & \ldots & a_{1,d+1} \\ a_{2,1} & a_{2,2} & \ldots & a_{2,d+1} \\ \vdots & \vdots & \ddots & \vdots \\ a_{d,1} & a_{d,2} & \ldots & a_{d,d+1} \\ 1 & 1 & \ldots & 1 \end{pmatrix}$$

being nonsingular. Namely, the $d + 1$ points a_i are not contained in the same hyperplane. Then a *d-simplex* is the convex hall U of these $d + 1$ points,

$$U = \left\{x = \sum_{i=1}^{d+1} \lambda_i a_i : 0 \leq \lambda_i \leq 1, \sum_{i=1}^{d+1} \lambda_i = 1\right\}.$$

The points $a_1, \ldots, a_{d+1}$ are the vertices of the d-simplex. Note that a 1-simplex is an interval, a 2-simplex is a triangle, and a 3-simplex is a tetrahedron. For an integer

$0 \le i \le d-1$, an i-dimensional face of U is any i-simplex formed by $i+1$ vertices of U. For example, a zero-dimensional face is a vertex, a one-dimensional face is an edge of U, and a two-dimensional face is a triangular face of the tetrahedron.

For any point $x = (x_1, \ldots, x_d) \in \mathbb{R}^d$, the solutions $(\lambda_1, \ldots, \lambda_{d+1})^T$ of the linear equations

$$\begin{pmatrix} a_{1,1} & a_{1,2} & \ldots & a_{1,d+1} \\ a_{2,1} & a_{2,2} & \ldots & a_{2,d+1} \\ \vdots & \vdots & \ddots & \vdots \\ a_{d,1} & a_{d,2} & \ldots & a_{d,d+1} \\ 1 & 1 & \ldots & 1 \end{pmatrix} \begin{pmatrix} \lambda_1 \\ \lambda_2 \\ \vdots \\ \lambda_d \\ \lambda_{d+1} \end{pmatrix} = \begin{pmatrix} x_1 \\ x_2 \\ \vdots \\ x_d \\ 1 \end{pmatrix}$$

are the *barycentric coordinates* of x with respect to the $d+1$ vertices a_i. These barycentric coordinates are unique since the matrix, as mentioned above, is nonsingular. The barycentric coordinates play an important role in identifying points in $\mathbb{R}^d$ in relation to the vertices of the d-simplex. For example, for $1 \le i, j \le d+1$, the barycenter (center of gravity) of U is the point x such that $\lambda_i(x) = 1/(d+1)$. Also $\lambda_i(a_j) = 1$ if $i = j$ and $\lambda_i(a_j) = 0$ otherwise. Now we outline the two ingredients of the finite element space: domain decomposition and the basis functions.

Domain decomposition refers to techniques that divide the domain into smaller regions, upon which a finite-dimensional approximation can be built. In the finite element context, domain decomposition is characterized by the establishment of a *triangulation*.

Definition 1.1 (Triangulation) A triangulation $\mathcal{T}$ of the domain is a collection of subsets (elements) T of Ω. These elements T are often the d-simplexes that satisfy the following properties:

- $\bar{\Omega} = \cup_{T \in \mathcal{T}} T$.
- Each element T is closed and the interior $\mathring{T}$ of T is not empty.
- Different elements $T_1, T_2 \in \mathcal{T}$ do not overlap, i.e., $\mathring{T}_1 \cap \mathring{T}_2 = \emptyset$.
- The boundary of each element $T \in \mathcal{T}$ is Lipschitz-continuous.
- A face of an element $T_1 \in \mathcal{T}$ is either a subset of the domain boundary $\partial\Omega$ or a face of an adjacent element $T_2 \in \mathcal{T}$.

Such a triangulation is also called a *conforming mesh*. Let $\mathcal{T}$ be a triangulation such that for $0 < h_{\mathcal{T}} \le 1$,

$$\max_{T \in \mathcal{T}} \{\mathrm{diam}(T)\} \le h_{\mathcal{T}} \mathrm{diam}(\Omega),$$

where $\mathrm{diam}(\cdot)$ is the diameter of the region under consideration. For any $T \in \mathcal{T}$, let B_T be the largest ball contained in T. Then $\mathcal{T}$ is *shape regular*, if there exists a constant $c_0 > 0$ such that for all $T \in \mathcal{T}$

$$\mathrm{diam}(B_T) \ge c_0 \mathrm{diam}(T).$$

The mesh is said to be *quasi-uniform* if there exists $c_1 > 0$ such that

$$\min_{T \in \mathcal{T}} \{\mathrm{diam}(B_T)\} \ge c_1 h_{\mathcal{T}} \mathrm{diam}(\Omega).$$

Thus, a quasi-uniform mesh is a shape regular mesh but not conversely. For a quasi-uniform mesh $\mathcal{T}$, the *mesh size* is denoted by $h := \max_{T\in\mathcal{T}}\{\mathrm{diam}(T)\}$.

The basis functions of the finite element space can be defined using the elements in the triangulation. The finite element space consists of piecewise polynomial functions associated with the underlying triangulation with certain continuity constraints across the faces of the elements. An important requirement is that each basis function has compact support that is the union of several adjacent elements. We describe one of the most widely used finite elements—the *Lagrange element* associated with a simplicial triangulation (a triangulation comprised of simplexes). For more comprehensive discussions on the finite element method, we refer interested readers to [20, 35, 43, 55, 69, 108, 110, 121].

Definition 1.2 (Lagrange Element) Given a triangulation $\mathcal{T}$ with d-simplexes ($d = 1, 2, 3$), let $T \in \mathcal{T}$ be an element. Recall the barycentric coordinates $(\lambda_i)_{i=1}^{d+1}$ with respect to the vertices a_i of T. For a given integer $m \geq 1$, define the set of *nodes* (or nodal points) on T by

$$L_m(T) = \left\{x = \sum_{i=1}^{d+1} \lambda_i a_i : \sum_{i=1}^{d+1} \lambda_i = 1,\ \lambda_i \in \left\{0, \frac{1}{m}, \dots, \frac{m-1}{m}, 1\right\}\right\}.$$

$L_m(T)$ may include nodes on the boundary ∂T and nodes in the interior of T, depending on the dimension d and the value of m. See Figs. 1.1 and 1.2. In addition, when $T_1, T_2 \in \mathcal{T}$ share the same face (resp. vertex), the nodes on the common face (resp. vertex) coincide. The overlapping nodes from adjacent simplexes are identified with one single node in the triangulation $\mathcal{T}$. Let N be the number of nodes in $\mathcal{T}$. For the ith node, $1 \leq i \leq N$, define the associated *nodal basis function* ϕ_i such that

$$\phi_i|_T \in P_m(T), \quad \forall\, T \in \mathcal{T} \quad \text{and} \quad \phi_i(x_j) = \delta_{ij} := \begin{cases} 0 & \text{if } i \neq j, \\ 1 & \text{if } i = j, \end{cases} \tag{1.7}$$

where x_j is the jth node in $\mathcal{T}$, $1 \leq j \leq N$. Then the *Lagrange finite element space* of degree m associated with $\mathcal{T}$ is the linear span of the basis functions:

$$S_L := \left\{v = \sum_{1\leq i\leq N} c_i\phi_i : c_i \in \mathbb{R}\right\} = \{v \in C(\bar{\Omega}) : v|_T = P_m(T),\ \forall\, T \in \mathcal{T}\}.$$

See Fig. 1.3 for illustrations of linear ($m = 1$) Lagrange basis functions.

Remark 1.1 A polynomial of degree $\leq m$ on T is uniquely determined by its values at the nodes on T. Therefore, the basis function is well defined via the conditions in (1.7) and is continuous over the domain Ω [35, 43]. It is also clear that ϕ_i has compact support, namely $\phi_i = 0$ outside of the union of simplexes sharing the ith

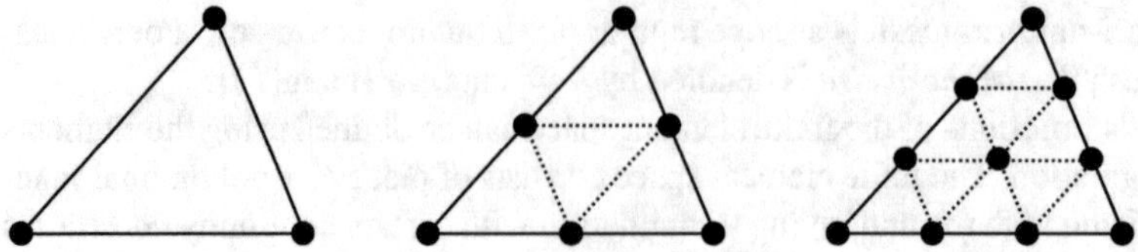

Fig. 1.1 Lagrange nodal points on a triangle (left–right): $m = 1$, $m = 2$, $m = 3$.

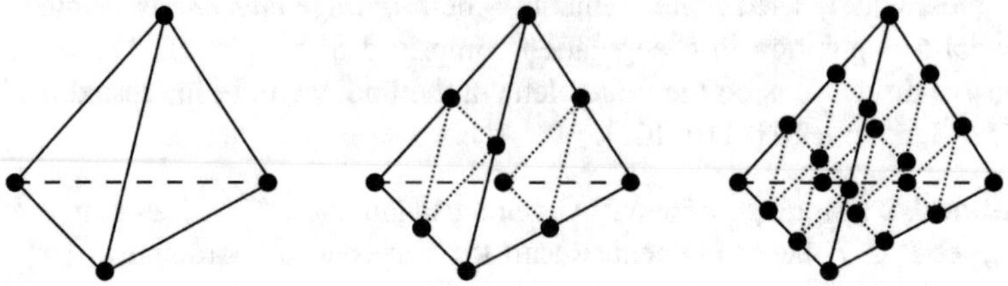

Fig. 1.2 Lagrange nodal points on a tetrahedron (left–right): $m = 1$, $m = 2$, $m = 3$.

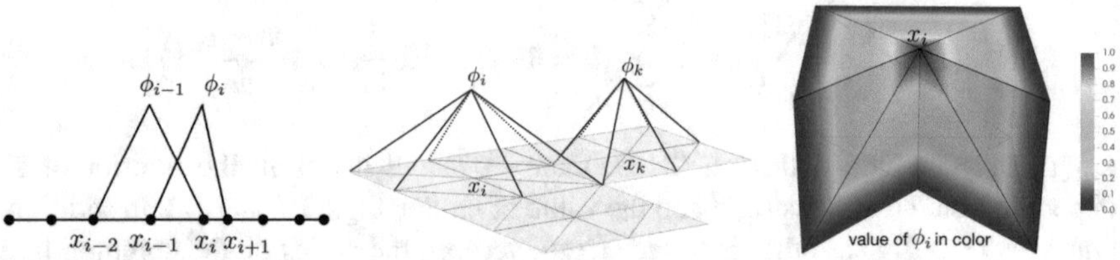

Fig. 1.3 Linear Lagrange basis functions in d-dimensional domains, $d = 1, 2, 3$.

node. To solve (1.1), the Dirichlet boundary condition can be incorporated by using a slightly different version of the Lagrange finite element space

$$S = S_L \cap \{v|_{\partial\Omega} = 0\}. \tag{1.8}$$

S is the linear span of all basis functions associated with the nodes in the interior of $\bar{\Omega}$. In what follows, by abuse of notation, we shall still use N to denote the number of basis functions in S.

Then the finite element solution $u_h \in S$ of (1.1) is defined by the equation

$$a(u_h, v_h) = (f, v_h), \quad \forall\, v_h \in S, \tag{1.9}$$

where $a(u_h, v_h) = \int_\Omega \nabla u_h \cdot \nabla v_h \, dx$ and $(f, v_h) = \int_\Omega f v_h \, dx$ are the same forms as in (1.3). Represent u_h as a linear combination of the basis functions by writing $u_h = \sum_{i=1}^N u_i \phi_i$, where u_i's are unknown real coefficients. Choosing $v_h = \phi_j$ for $j = 1, \ldots, N$, we see that (1.9) is equivalent to a system of N linear equations

$$\int_\Omega \sum_{i=1}^N u_i \nabla\phi_i \cdot \nabla\phi_j \, dx = \int_\Omega f \phi_j \, dx, \qquad j = 1, \ldots, N,$$

or its matrix form

$$\mathbf{A}\boldsymbol{u} = \boldsymbol{b}, \tag{1.10}$$

where $\boldsymbol{u} = (u_1, \ldots, u_N)^T$, $\boldsymbol{b} = (\int_\Omega f\phi_1\,dx, \ldots, \int_\Omega f\phi_N\,dx)^T$, and

$$\mathbf{A} = \begin{pmatrix} \int_\Omega \nabla\phi_1 \cdot \nabla\phi_1\,dx & \ldots & \int_\Omega \nabla\phi_N \cdot \nabla\phi_1\,dx \\ \vdots & \ddots & \vdots \\ \int_\Omega \nabla\phi_1 \cdot \nabla\phi_N\,dx & \ldots & \int_\Omega \nabla\phi_N \cdot \nabla\phi_N\,dx \end{pmatrix}. \tag{1.11}$$

Here $\mathbf{A}$ is the stiffness matrix, which is symmetric and positive definite. According to (1.7), $\int_\Omega \nabla\phi_i \cdot \nabla\phi_j\,dx \neq 0$ only when their supports overlap. Given a shape regular mesh $\mathcal{T}$, the number of nonzero entries in each row of the stiffness matrix is bounded, independent of the dimension N of the finite element space. Thus, $\mathbf{A}$ is also sparse.

1.2 Examples

We present examples of the *linear* Lagrange finite element method (1.9) solving (1.1) with domains in different dimensions.

Let $(e_i)_{i=1}^d$ be the canonical basis of the space $\mathbb{R}^d$. Namely, $e_i = (0, \ldots, 1, 0, \ldots)^T$, where the only nonzero value (i.e., 1) occurs at the ith entry. Let e_0 be the zero vector in $\mathbb{R}^d$. Denote by $\hat{T}$ the standard d-simplex with the vertices $e_0, \ldots, e_d$. Let T be a d-simplex with vertices $a_1, \ldots, a_{d+1}$. Then there exists an affine mapping $F : \hat{T} \to T$ that is also a bijection, such that

$$F(e_\ell) = a_{\ell+1}, \qquad 0 \le \ell \le d. \tag{1.12}$$

We call $\hat{T}$ the *reference element* of T. See Fig. 1.4 for illustrations of the reference element in different dimensions. The affine mapping F is important in both implementation and theoretical analysis of the finite element algorithm. For instance, using the chain rule, one can calculate the stiffness matrix $\mathbf{A}$ and the vector $\boldsymbol{b}$ in (1.10) using F and the standard basis functions on the reference element $\hat{T}$.

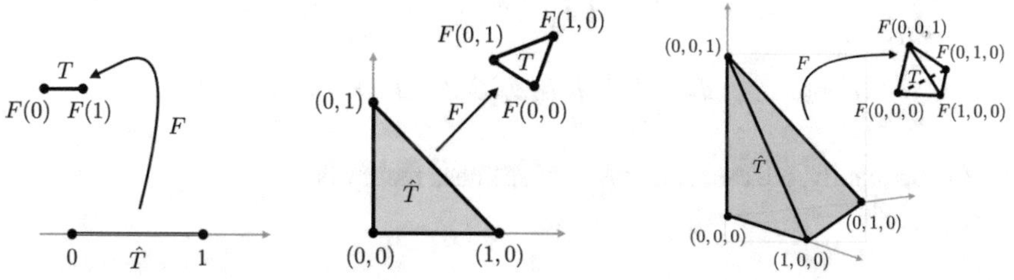

Fig. 1.4 Reference elements (d-simplexes) $\hat{T}$ and the affine mappings, $d = 1, 2, 3$.

1.2.1 A One-Dimensional Example

Consider the Poisson problem in a one-dimensional (1D) domain:

$$-\partial_x^2 u = 1 \quad \text{in} \quad \Omega = (0,1) \qquad \text{and} \qquad u(0) = u(1) = 0. \tag{1.13}$$

We use the uniform simplicial mesh $\mathcal{T}_h$ for (1.13), where $0 < h < 1$ is the element size such that h^{-1} is an integer. Then the dimension of the linear Lagrange finite element space is $N = h^{-1} - 1$. The domain Ω is divided into $N + 1$ subintervals by $N + 2$ equidistant nodes x_i, $0 \le i \le N + 1$, such that $x_0 = 0$ and $x_{N+1} = 1$. Let $T_i = [x_{i-1}, x_i] \in \mathcal{T}_h$, $1 \le i \le N + 1$, be the ith subinterval. Recall that each linear basis function ϕ_i is associated with an interior node x_i, $1 \le i \le N$. Thus, the integral $\int_\Omega \partial_x\phi_i \partial_x\phi_j \, dx$ for $1 \le i, j \le N$ is nonzero only when $|j - i| \le 1$. Consequently, in the ith row of the stiffness matrix (1.11), there are at most three nonzero entries:

$$\int_{T_i} \partial_x\phi_{i-1}\partial_x\phi_i \, dx, \quad \int_{T_i\cup T_{i+1}} \partial_x\phi_i\partial_x\phi_i \, dx, \quad \text{and} \quad \int_{T_{i+1}} \partial_x\phi_{i+1}\partial_x\phi_i \, dx. \tag{1.14}$$

Let $\hat{T} = [0, 1]$ be the reference element and let $F_i : \hat{T} \to T_i$ be the aforementioned affine mapping. See the 1D case in Fig. 1.4. Then with F_i, the integral on T_i can be transformed into an integral on $\hat{T}$. Through this scaling process, the evaluation of integrals on the subintervals to assemble the stiffness matrix $\mathbf{A}$ and the vector $\boldsymbol{b}$ in (1.10) is replaced by the evaluation of integrals on the reference element. We now demonstrate this process in calculating the integrals in (1.14). For any $x \in T_i$, $\hat{x} := F_i^{-1}(x)$ is its image under the mapping F_i^{-1} and $\hat{x} \in \hat{T}$. For a function $v(x)$, define $\hat{v}(\hat{x})$ such that $\hat{v}(\hat{x}) = v(x)$. It is clear that F_i^{-1} involves scaling and shifting of the element. In fact,

$$F_i^{-1}(x) = h^{-1}(x - x_{i-1}) \quad \text{and} \quad F_i(\hat{x}) = h\hat{x} + x_{i-1}.$$

Let $\hat{\phi}_{e_0} = 1 - \hat{x}$ and $\hat{\phi}_{e_1} = \hat{x}$ be the standard basis functions on $\hat{T}$ that are associated with $e_0 = 0$ and $e_1 = 1$, respectively. Then using the chain rule, one obtains

$$\begin{aligned}
\int_{T_i} \partial_x\phi_{i-1}\partial_x\phi_i \, dx &= h^{-1} \int_{\hat{T}} \partial_{\hat{x}}\hat{\phi}_{e_0}\partial_{\hat{x}}\hat{\phi}_{e_1} \, d\hat{x} = -h^{-1}, \\
\int_{T_i\cup T_{i+1}} \partial_x\phi_i\partial_x\phi_i \, dx &= h^{-1}\left(\int_{\hat{T}} (\partial_{\hat{x}}\hat{\phi}_{e_1})^2 \, d\hat{x} + \int_{\hat{T}} (\partial_{\hat{x}}\hat{\phi}_{e_0})^2 \, d\hat{x} \right) = 2h^{-1}, \\
\int_{T_{i+1}} \partial_x\phi_{i+1}\partial_x\phi_i \, dx &= h^{-1} \int_{\hat{T}} \partial_{\hat{x}}\hat{\phi}_{e_1}\partial_{\hat{x}}\hat{\phi}_{e_0} \, d\hat{x} = -h^{-1}.
\end{aligned}$$

Consequently, the associated $N \times N$ stiffness matrix is

$$\mathbf{A} = h^{-1} \begin{pmatrix} 2 & -1 & 0 & 0 & 0 & \dots \\ -1 & 2 & -1 & 0 & 0 & \dots \\ 0 & -1 & 2 & -1 & 0 & \dots \\ \vdots & \ddots & \ddots & \ddots & \ddots & \ddots \\ \vdots & \dots & 0 & -1 & 2 & -1 \\ 0 & \dots & \dots & 0 & -1 & 2 \end{pmatrix}. \tag{1.15}$$

Hence, according to (1.9) and (1.10), the finite element approximation $u_h = \sum_{i=1}^{N} u_i \phi_i$ of (1.13) can be obtained by solving the system of linear equations

$$\mathbf{A}\boldsymbol{u} = \boldsymbol{b},$$

where the ith entries of $\boldsymbol{u}$ and $\boldsymbol{b}$ are u_i and $b_i = (1, \phi_i)$, respectively.

The actual solution $u = -\frac{1}{2}(x - \frac{1}{2})^2 + \frac{1}{8}$ of (1.13) and the numerical solutions corresponding to different mesh sizes are illustrated in Fig. 1.5. One observes that the numerical approximation becomes more accurate on a finer mesh (smaller h). For $h = 0.02$, u and u_h are almost indistinguishable given the current image resolution! Note that (1.13) is a rare example in which the solution can be expressed explicitly using known functions. This is not the case for most partial differential equations, which is the main reason we need numerical methods to solve these problems.

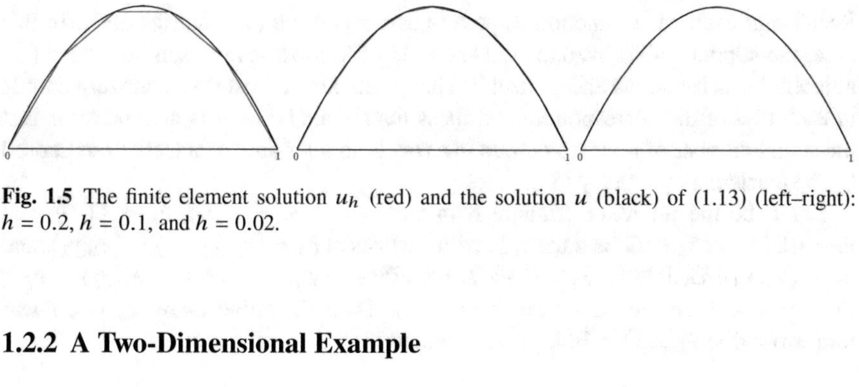

Fig. 1.5 The finite element solution u_h (red) and the solution u (black) of (1.13) (left–right): $h = 0.2$, $h = 0.1$, and $h = 0.02$.

1.2.2 A Two-Dimensional Example

Consider the Poisson problem in a two-dimensional (2D) domain:

$$-\Delta u = 1 \quad \text{in} \quad \Omega = (0, 1)^2 \qquad \text{and} \qquad u = 0 \quad \text{on} \quad \partial\Omega. \tag{1.16}$$

We use a quasi-uniform simplicial triangulation $\mathcal{T}_h$ in the linear finite element approximation of (1.16), where $0 < h < 1$ is the mesh size. For a given mesh $\mathcal{T}_h$, we follow the uniform *midpoint decomposition* to generate a finer mesh $\mathcal{T}_{h/2}$: divide each triangle $T \in \mathcal{T}_h$ into four equal sub-triangles by connecting each pair of midpoints on the edges of T. Implementing this procedure repeatedly, one obtains a sequence of *nested* triangulations, in which the nodes in the coarser mesh are also nodes in the finer mesh. Consequently, the associated finite element spaces are also *nested*: the finite element space on the coarser mesh is a subspace of the finite element space on the finer mesh! See Fig. 1.6 for meshes from consecutive refinements.

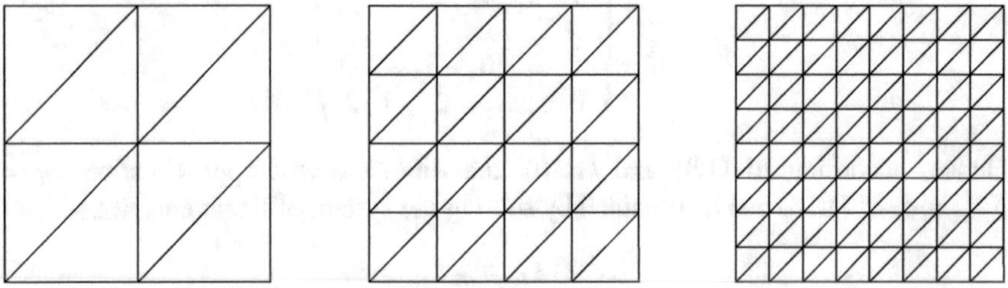

Fig. 1.6 Midpoint decompositions (left–right): initial mesh, mesh after one refinement, mesh after two refinements.

Let $\mathcal{T}_h$ be a triangulation obtained by repeated midpoint decompositions starting with the initial mesh in Fig. 1.6. Recall the associated linear finite element space S in (1.8) for which $m = 1$. Then the finite element solution $u_h \in S$ of (1.16) is defined by (1.9): for any $v_h \in S$,

$$\int_\Omega \nabla u_h \cdot \nabla v_h \, dx = \int_\Omega v_h \, dx.$$

Recall that each basis function $\phi_i \in S$ is associated with an interior node. In this case, the support of ϕ_i overlaps the supports of at most seven basis functions (six adjacent basis functions and ϕ_i itself). Thus, there are at most seven nonzero entries in each row of the corresponding stiffness matrix $\mathbf{A}$ (1.11). It is also possible that the actual number of nonzero entries in a row is smaller due to cancellations caused by the particular symmetry of the mesh.

Let $\hat{T}$ be the reference triangle with vertices at $e_0 = (0, 0)$, $e_1 = (1, 0)$, and $e_2 = (0, 1)$. Let $T_i \in \mathcal{T}_h$ be a triangle with vertices at $a_1 = (x_1, y_1)$, $a_2 = (x_2, y_2)$, and $a_3 = (x_3, y_3)$. Denote by $F_i : \hat{T} \to T_i$ the affine mapping such that $F_i(e_\ell) = a_{\ell+1}$ for $\ell = 0, 1, 2$ (see the 2D case in Fig. 1.4). Then the affine mapping is a linear transformation $F_i(\hat{x}, \hat{y}) = \mathbf{B}(\hat{x}, \hat{y})^T + \boldsymbol{c}$, where

$$\mathbf{B} = \begin{pmatrix} x_2 - x_1 & x_3 - x_1 \\ y_2 - y_1 & y_3 - y_1 \end{pmatrix} \quad \text{and} \quad \boldsymbol{c} = \begin{pmatrix} x_1 \\ y_1 \end{pmatrix}. \tag{1.17}$$

For $(x, y) \in T_i$, let $(\hat{x}, \hat{y}) := F_i^{-1}(x, y)$ be its image on $\hat{T}$ under F_i^{-1}. For a function $v(x, y)$ in T_i, define $\hat{v}(\hat{x}, \hat{y})$ in $\hat{T}$, such that $\hat{v}(\hat{x}, \hat{y}) = v(x, y)$. Note that the standard linear basis functions on $\hat{T}$ are

$$\hat{\phi}_{e_0}(\hat{x}, \hat{y}) = 1 - \hat{x} - \hat{y}, \quad \hat{\phi}_{e_1}(\hat{x}, \hat{y}) = \hat{x}, \quad \hat{\phi}_{e_2}(\hat{x}, \hat{y}) = \hat{y}.$$

As in the 1D case, one can use the affine mapping to transform the integrals in the stiffness matrix and the vector on the right hand side of (1.10) to integrals in terms of the standard basis functions on $\hat{T}$. Consequently, the evaluation of the integral (numerical quadrature) is carried out merely on the reference element.

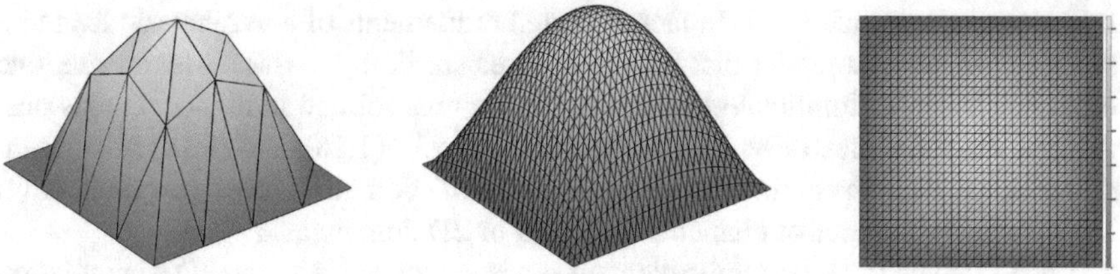

Fig. 1.7 Numerical solutions of (1.16) on quasi-uniform meshes (left–right): mesh after one refinement, mesh after four refinements, top view of the solution (mesh after four refinements).

In Fig. 1.7, we display the linear finite element solutions of (1.16) associated with different triangulations. These meshes are derived from the initial mesh in Fig. 1.6 using repeated midpoint decompositions. One observes that the numerical solution represents a better approximation on the finer mesh.

1.2.3 A Three-Dimensional Example

Consider the Poisson problem in a three-dimensional (3D) domain:

$$-\Delta u = 1 \quad \text{in} \quad \Omega = (0,1)^3 \qquad \text{and} \qquad u = 0 \quad \text{on} \quad \partial\Omega. \tag{1.18}$$

We use a quasi-uniform simplicial triangulation $\mathcal{T}_h$ in the linear finite element approximation of (1.18), where $0 < h < 1$ is again the mesh size. For a given mesh $\mathcal{T}_h$, we follow the *midpoint decomposition* in [29] to produce a finer mesh $\mathcal{T}_{h/2}$: divide each tetrahedron $T \in \mathcal{T}_h$ into eight sub-tetrahedra by connecting midpoints on the edges of T. This 3D version of the midpoint refinement involves more complex procedures than the midpoint decomposition for triangles in the 2D case. We shall elaborate on the detailed steps of the refinement in Chap. 3. See Fig. 1.8 for the first few triangulations of the domain (unit cube).

The complexity of the problem can grow rapidly as the domain dimension increases. Unlike the 2D case, the eight sub-tetrahedra in the initial element after one

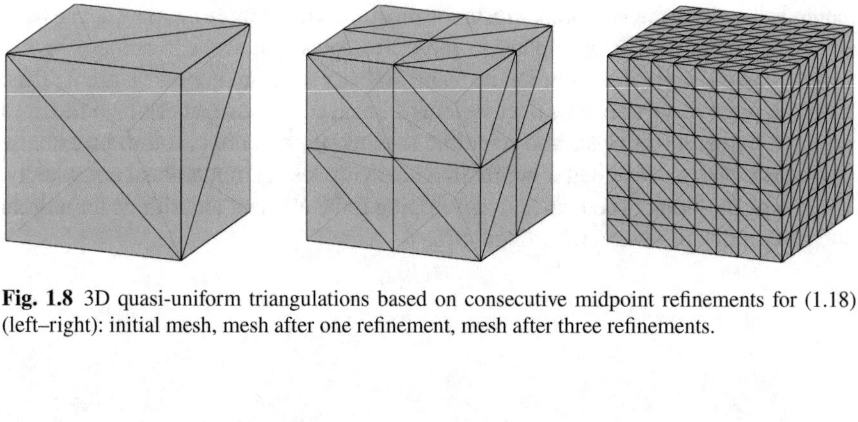

Fig. 1.8 3D quasi-uniform triangulations based on consecutive midpoint refinements for (1.18) (left–right): initial mesh, mesh after one refinement, mesh after three refinements.

refinement are not identical. In fact, repeated refinements of a tetrahedron lead to a sequence of sub-tetrahedra that belong to three similarity classes. Meanwhile, one often needs more simplicial elements to fill the unit volume in higher dimensions. Assume the same mesh size h for the problems (1.13), (1.16), and (1.18) in different dimensions. Then the number of elements in the 3D domain is much (exponentially) larger than the number of elements in the 1D or 2D domain as $h \to 0$.

Let S (1.8) be the linear finite element space associated with the 3D triangulation $\mathcal{T}_h$. Despite the aforementioned geometric and computational complexity, the finite element solution $u_h \in S$ for (1.18) is defined in the same variational formulation (1.9): for any $v_h \in S$,

$$a(u_h, v_h) = (1, v_h),$$

which results in the system of linear equations $\mathbf{A}\boldsymbol{u} = \boldsymbol{b}$ (1.10).

Let $\hat{T}$ be the reference tetrahedron with vertices $e_0 = (0, 0, 0)$, $e_1 = (1, 0, 0)$, $e_2 = (0, 1, 0)$, and $e_3 = (0, 0, 1)$. Let $T_i \in \mathcal{T}_h$ be a tetrahedron with vertices $a_1 = (x_1, y_1, z_1)$, $a_2 = (x_2, y_2, z_2)$, $a_3 = (x_3, y_3, z_3)$, and $a_4 = (x_4, y_4, z_4)$. Denote by $F_i : \hat{T} \to T_i$ the affine mapping such that $F_i(e_\ell) = a_{\ell+1}$ for $\ell = 0, 1, 2, 3$ (Fig. 1.4). With a straightforward calculation, one obtains $F_i(\hat{x}, \hat{y}, \hat{z}) = \mathbf{B}(\hat{x}, \hat{y}, \hat{z})^T + \boldsymbol{c}$, where

$$\mathbf{B} = \begin{pmatrix} x_2 - x_1 & x_3 - x_1 & x_4 - x_1 \\ y_2 - y_1 & y_3 - y_1 & y_4 - y_1 \\ z_2 - z_1 & z_3 - z_1 & z_4 - z_1 \end{pmatrix} \quad \text{and} \quad \boldsymbol{c} = \begin{pmatrix} x_1 \\ y_1 \\ z_1 \end{pmatrix}. \tag{1.19}$$

Consequently, as in the 1D and 2D cases, one can evaluate the integrals in $\mathbf{A}$ and in $\boldsymbol{b}$ of (1.10) by using numerical quadrature on the reference element $\hat{T}$ for functions involving the standard basis functions:

$$\hat{\phi}_{e_0} = 1 - \hat{x} - \hat{y} - \hat{z}, \quad \hat{\phi}_{e_1} = \hat{x}, \quad \hat{\phi}_{e_2} = \hat{y}, \quad \hat{\phi}_{e_3} = \hat{z}.$$

The stiffness matrix $\mathbf{A}$ is sparse but in general has more nonzero entries in each row than those in the stiffness matrix associated with the 2D example (1.16). The finite element solution $u_h = \sum_{i=1}^{N} u_i \phi_i$ then is obtained by finding the coefficient vector $\boldsymbol{u} = \mathbf{A}^{-1}\boldsymbol{b}$.

In (1.13), (1.16), and (1.18), we have demonstrated the finite element method approximating model elliptic problems in different dimensions. These are Galerkin methods based on the variational form for functions in the finite element space built upon a triangulation. It is implied in these examples that one can achieve higher accuracy in the numerical solution on triangulations with smaller mesh size h. This, however, often gives rise to a larger system of linear equations that can be difficult to solve. In numerical analysis and scientific computing, it is important to invent new algorithms that produce better approximations with less computational complexity. The rest of this book is dedicated to developing finite element algorithms for elliptic boundary value problems with singularities.

Chapter 2
The Function Space

We introduce function spaces useful for the analysis of elliptic equations and for the error estimate of the finite element method. In particular, we review Sobolev spaces and the important properties of functions in these spaces, including the extension, embedding, and trace theorems. We also discuss the well-posedness and regularity estimates for elliptic boundary value problems in Sobolev spaces. In turn, we present key steps to derive the finite element error analysis. The trace and regularity results usually depend on the smoothness of the domain, and the nonsmooth points on the boundary can lead to singularities in the solution. For the conciseness of the presentation, some results are summarized without proofs. Readers will be referred to specific references for more details. This chapter is suitable for readers who need a review of basic results in Sobolev spaces and who are starting to work on finite element error analysis for elliptic equations.

Throughout this book, by $a \sim b$ (resp. $a \lesssim b$), we mean that there exists a constant $C > 0$ independent of a and b, such that $C^{-1}a \leq b \leq Ca$ (resp. $a \leq Cb$). The generic constant $C > 0$ in the estimates may be different at different occurrences. It may depend on the underlying domain, but not on the functions involved.

2.1 Vector Spaces

We review some basic concepts of vector spaces and useful results in functional analysis [91, 105, 118]. A *vector space* is a collection of objects called vectors, which is closed under finite vector addition and scalar multiplication. For a vector space V, the scalars are members of a field F, in which case V is called a vector space over F. In this section, let V and W be two vector spaces over the complex field $\mathbb{C}$ or over the real field $\mathbb{R}$.

H. Li, *Graded Finite Element Methods for Elliptic Problems in Nonsmooth Domains*
Surveys and Tutorials in the Applied Mathematical Sciences 10,
https://doi.org/10.1007/978-3-031-05821-9_2

Definition 2.1 (Normed Space) A vector space V is a *normed space* if for every $v \in V$, there is a real-valued function $\|\cdot\| : v \longrightarrow \mathbb{R}$, called the *norm* of v, such that for $v, w \in V$

1. $\|v + w\| \leq \|v\| + \|w\|$,
2. $\|\alpha v\| = |\alpha|\|v\|$ for any $\alpha \in \mathbb{R}$,
3. $\|v\| \geq 0$ and $\|v\| = 0$ if and only if (iff) $v = \mathbf{0}$, where $\mathbf{0}$ is the zero vector in V.

A *seminorm* on V is a function with the properties 1 and 2 above.

Definition 2.2 (Banach Space and Hilbert Space) Let V be a normed space. It is said to be *complete* if each Cauchy sequence in V converges to a vector in V. A *Banach space* is a complete and normed space. An *inner product* on a real vector space V is a bilinear mapping $(\cdot, \cdot) : V \times V \longrightarrow \mathbb{R}$, such that for $v, w \in V$,

1. $(v, v) \geq 0$ and $(v, v) = 0$ iff $v = \mathbf{0}$,
2. $(v, w) = (w, v)$,
3. the mapping $v \longrightarrow (v, w)$ is linear for every $w \in V$.

Given an inner product, the associated norm is defined as

$$\|v\| := (v, v)^{1/2} \qquad \forall\, v \in V.$$

A *Hilbert space* is a Banach space equipped with an inner product that induces the norm.

Let W be a closed linear subset of a Hilbert space V with inner product $(\cdot, \cdot)$. Here, W being *linear* means that for any $x, y \in W$ and $\alpha \in \mathbb{R}$ we have $x + \alpha y \in W$. Define its orthogonal complement $W^\perp$ by

$$W^\perp := \{y \in V : \ (x, y) = 0, \ \forall\, x \in W\}. \tag{2.1}$$

Then W and $W^\perp$ are subspaces of V satisfying $V = W \oplus W^\perp$ (i.e., $W \cup W^\perp = V$ and $W \cap W^\perp = \{\mathbf{0}\}$).

Theorem 2.1 (*Norm Equivalence) Let V be a finite-dimensional vector space. Assume $\|\cdot\|_1$ and $\|\cdot\|_2$ are two norms on V. Then these two norms are equivalent in the sense that there exist $c, C > 0$ such that for any $v \in V$,*

$$c\|v\|_1 \leq \|v\|_2 \leq C\|v\|_1.$$

A mapping $A : V \longrightarrow V$ is said to be a *linear* operator if

$$A(\alpha v + \beta w) = \alpha A(v) + \beta A(w), \qquad \forall\, v, w \in V \text{ and } \forall\, \alpha, \beta \in \mathbb{R}.$$

The *range* of A is $R(A) := \{w : \ w = Av$ for some $v \in V\}$. The *null space* of A is $N(A) := \{v \in V : \ Av = \mathbf{0}\}$. Denote by $L(V, W)$ (resp. $L(V)$) the collection of all the linear operators from V to W (resp. from V to itself). For a linear operator $A \in L(V, W)$, its norm can be defined by the norms on V and W.

Definition 2.3 (Operator Norm) Let $\|\cdot\|_V$ and $\|\cdot\|_W$ be the norms on Banach spaces V and W, respectively. Then

$$\|A\|_{V\to W} := \sup_{0\neq v\in V} \frac{\|Av\|_W}{\|v\|_V}, \qquad \text{for any linear operator } A: V\to W$$

defines a norm on $L(V, W)$.

An operator P on a linear space is a *projection* if $P^2 = P$, namely $Px = x$ for all x in the image of P. Let V be a Banach space. A bounded linear operator $v^*: V \to \mathbb{R}$ is called a *bounded linear functional* on V. Denote by V^* the *dual space* of V that consists of all bounded linear functionals on V. Let $\langle\cdot,\cdot\rangle$ be the pairing of V^* and V. Then for $v^* \in V^*$ and $w \in V$, $\langle v^*, w\rangle$ is the value of $v^*(w)$. The following result holds for Hilbert spaces, which ensures that any element in the dual space has a unique representation in the Hilbert space.

Theorem 2.2 (*Riesz Representation*) *Let V be a real Hilbert space with inner product $(\cdot,\cdot)$ and let V^* be its dual space. Then for any $v^* \in V^*$, there exists a unique $v \in V$ such that for any $w \in V$,*

$$(v, w) = \langle v^*, w\rangle.$$

In addition, this mapping $v^* \to v$ is an isomorphism from V^* onto V.

Definition 2.4 A bilinear form $a(\cdot,\cdot)$ on a normed linear space V is *bounded* (or *continuous*) if there exists $\alpha_1 > 0$ such that

$$|a(v, w)| \leq \alpha_1 \|v\|_V \|w\|_V \qquad \forall\, v, w \in V$$

and it is *coercive* on $W \subseteq V$ if there exists $\alpha_2 > 0$ such that

$$a(v, v) \geq \alpha_2 \|v\|_V^2 \qquad \forall\, v \in W.$$

Theorem 2.3 (*Lax–Milgram*) *Given a Hilbert space V, assume $a(\cdot,\cdot)$ is a continuous and coercive bilinear form on V, and $F \in V^*$ is a continuous linear functional. Then there exists a unique $u \in V$ such that*

$$a(u, v) = F(v) := \langle F, v\rangle \qquad \forall\, v \in V.$$

The Lax–Milgram Theorem is an extension of the Riesz Representation Theorem to continuous and coercive bilinear forms. Note that here $a(\cdot,\cdot)$ can be either symmetric or nonsymmetric.

A class of special bounded linear operators can be defined as follows.

Definition 2.5 (Compact Operator) Let V and W be Banach spaces. A continuous linear operator $A: V \to W$ is said to be *compact* if for each bounded sequence $\{v_\ell\}_{\ell=1}^\infty$ in V, there exists a subsequence $\{v_{\ell_i}\}_{i=1}^\infty$ such that the sequence $\{Av_{\ell_i}\}_{i=1}^\infty$ converges in W.

An interesting property related to compact operators is the Fredholm alternative.

Theorem 2.4 (*Fredholm Alternative*) *For a real Hilbert space V, let $A : V \to V$ be a compact operator and let $I : V \to V$ be the identity operator. Then the following relations hold regarding the range $R(\cdot)$ and the null space $N(\cdot)$:*

1. *$N(I - A)$ is finite dimensional,*
2. *$R(I - A)$ is closed,*
3. *$R(I - A) = N(I - A^*)^{\perp}$,*
4. *$N(I - A) = \{\mathbf{0}\}$ iff $R(I - A) = V$,*
5. *$dim\big(N(I - A)\big) = dim\big(N(I - A^*)\big)$.*

Here, $dim\big(N(I - A)\big)$ is the dimension of the space $N(I - A)$, and $A^* : V \to V$ is the adjoint operator of A such that for any $v, w \in V$, $(Av, w) = (v, A^*w)$. For a subspace $W \subseteq V$, recall that $W^{\perp}$ is the orthogonal complement as defined in (2.1).

A function is said to be Hölder continuous with exponent γ based on the following definition.

Definition 2.6 (Hölder Space) For $d = 1, 2, 3$, let Ω be an open subset of $\mathbb{R}^d$ and let $m \geq 0$ be an integer. Denote by $\alpha = (\alpha_1, \ldots, \alpha_d) \in \mathbb{Z}^d_{\geq 0}$ a *multi-index* with nonnegative integer components. We set $\partial^{\alpha} = \partial^{\alpha_1}_{x_1} \cdots \partial^{\alpha_d}_{x_d}$ and $|\alpha| = \sum_{i=1}^{d} \alpha_i$. Define

$$C^m(\Omega) = \{v(x) \in \mathbb{R} : \ x \in \Omega \text{ and } \partial^{\alpha} v \text{ is continuous for } |\alpha| \leq m\},$$
$$C^m(\bar{\Omega}) = \{v \in C^m(\Omega) : \ \partial^{\alpha} v \text{ is uniformly continuous on bounded subsets of } \Omega\}.$$

It is clear that for $v \in C^m(\bar{\Omega})$, $\partial^{\alpha} v$ continuously extends to $\bar{\Omega}$ for $|\alpha| \leq m$. Let $C(\Omega) = C^0(\Omega)$ and $C(\bar{\Omega}) = C^0(\bar{\Omega})$. Assume $v : \Omega \to \mathbb{R}$ is a function that is bounded and continuous. Define the norm

$$\|v\|_{C(\bar{\Omega})} := \sup_{x \in \Omega} |v(x)|,$$

and the seminorm

$$[v]_{C^{0,\gamma}(\bar{\Omega})} := \sup_{x,y \in \Omega, x \neq y} \left(\frac{|v(x) - v(y)|}{|x - y|^{\gamma}} \right), \qquad 0 < \gamma \leq 1.$$

Then the Hölder space $C^{m,\gamma}(\bar{\Omega})$ consists of functions such that the following norm is finite

$$\|v\|_{C^{m,\gamma}(\bar{\Omega})} := \sum_{|\alpha| \leq m} \|\partial^{\alpha} v\|_{C(\bar{\Omega})} + \sum_{|\alpha| = m} [\partial^{\alpha} v]_{C^{0,\gamma}(\bar{\Omega})}.$$

Functions in $C^{0,1}(\bar{\Omega})$ are *Lipschitz-continuous*.

2.2 Sobolev Spaces

For elliptic boundary value problems, the Sobolev space is an important function space that has been widely used to investigate the quantitative properties of the solution. There is a rich literature (see [1, 56, 91, 100] and the references therein) that contains comprehensive studies of various properties of the Sobolev space.

2.2.1 Domains and Sobolev Spaces

We consider the domain Ω to be an open bounded subset of $\mathbb{R}^d$, $1 \leq d \leq 3$, and denote its boundary by $\Gamma := \partial\Omega$. The elaboration of the properties of the Sobolev space usually requires precise assumptions on the regularity of the boundary Γ. We first recall the following definition from [59].

Definition 2.7 (Continuous and Lipschitz-Continuous Boundary) For a domain $\Omega \subset \mathbb{R}^d$, and for an integer $m \geq 1$, we say its boundary Γ is continuous (resp. Lipschitz-continuous, of class C^m, of class $C^{m,1}$) if for every $x \in \Gamma$, there is a neighborhood $\mathcal{O} \subset \mathbb{R}^d$ of x and orthogonal coordinates $x = (x', x_d)$ where $x' = (x_1, \ldots, x_{d-1})$, such that the following conditions hold.

(i) $\mathcal{N}$ is a hypercube in the coordinates:

$$\mathcal{N} = \{x : -a_i < x_i < a_i,\ 1 \leq i \leq d, \text{ for some } a_i > 0\}.$$

(ii) There is a continuous (resp. Lipschitz-continuous, C^m, $C^{m,1}$) function χ defined in

$$\mathcal{N}' := \{x' : -a_i < x_i < a_i,\ 1 \leq i \leq d-1, \text{ for some } a_i > 0\}$$

such that $|\chi(x')| \leq \frac{a_d}{2}$ for any $x' \in \mathcal{N}'$ and

$$\Omega \cap \mathcal{N} = \{x : x_d < \chi(x')\}, \qquad \Gamma \cap \mathcal{N} = \{x : x_d = \chi(x')\}.$$

We say Ω has a Lipschitz boundary (or Ω is a Lipschitz domain) if the boundary locally is the graph of a Lipschitz-continuous function and the domain is on one side of the boundary.

In this book, we are mainly interested in domains with flat faces: the *polygonal domain* in $\mathbb{R}^2$ and the *polyhedral domain* in $\mathbb{R}^3$. It is clear that they are also Lipschitz domains. Therefore, the properties of the Sobolev space in Lipschitz domains also hold in polygonal and polyhedral domains.

Denote by $\mathcal{D}(\Omega)$ the space of infinitely differentiable functions in Ω with compact support. Let $\mathcal{D}(\bar{\Omega})$ be the space of restrictions to Ω of functions in $\mathcal{D}(\mathbb{R}^d)$. We denote the Lebesgue integral of a function v by $\int_\Omega v(x)\,dx$. For $1 \leq p < \infty$, define the norm

$$\|v\|_{L^p(\Omega)} := \left(\int_\Omega |v(x)|^p\,dx\right)^{1/p},$$

and for $p = \infty$, let

$$\|v\|_{L^\infty(\Omega)} := \text{ess sup}\{|v(x)| : x \in \Omega\}.$$

Here ess sup$\{\cdot\}$ is the essential supremum of a function, which is the smallest value that is greater than or equal to the function values everywhere except on a set of measure zero. Then the $L^p(\Omega)$ space for $1 \leq p \leq \infty$ is defined by

$$L^p(\Omega) := \{v : \|v\|_{L^p(\Omega)} < \infty\}.$$

In addition, denote the set of locally integrable functions by

$$L^1_{loc}(\Omega) := \{v : v \in L^1(G), \forall \text{ compact } G \subset \Omega\}.$$

Definition 2.8 (Distributional and Weak Derivative) The *space of distributions* $\mathcal{D}'(\Omega)$ is the dual space of $\mathcal{D}(\Omega)$. Let $\langle \cdot, \cdot \rangle$ be the duality pairing between $\mathcal{D}'(\Omega)$ and $\mathcal{D}(\Omega)$. For $v \in \mathcal{D}'(\Omega)$, we define the *distributional derivative* $\partial^\alpha v$ by

$$\langle \partial^\alpha v, \phi \rangle := (-1)^{|\alpha|} \langle v, \partial^\alpha \phi \rangle \qquad \forall \phi \in \mathcal{D}(\Omega). \tag{2.2}$$

Note that $\langle \cdot, \cdot \rangle$ is an extension of the inner product of $L^2(\Omega)$. For example, a function $v \in L^1_{loc}(\Omega)$ can be identified with a distribution by

$$\langle v, \phi \rangle = \int_\Omega v(x)\phi(x)\, dx \qquad \forall \phi \in \mathcal{D}(\Omega).$$

Suppose $v, w \in L^1_{loc}(\Omega)$. Then, w is said to be the αth-*weak derivative* of v, written as $w = \partial^\alpha v$, provided

$$\int_\Omega v \partial^\alpha \phi\, dx = (-1)^{|\alpha|} \int_\Omega w\phi\, dx \qquad \forall \phi \in \mathcal{D}(\Omega). \tag{2.3}$$

We remark that the weak derivative is defined for functions in $L^1_{loc}(\Omega)$. Namely, if $v \in L^1_{loc}(\Omega)$ and if there exists a function $w \in L^1_{loc}(\Omega)$ that satisfies (2.3) for any function in $\mathcal{D}(\Omega)$, the derivative $w = \partial^\alpha v$ exists in the weak sense. The distributional derivative (2.2) is defined for any distribution in $\mathcal{D}'(\Omega)$ which is a larger class than $L^1_{loc}(\Omega)$. The distributional derivative coincides with the weak derivative if the weak derivative exists, and both of them coincide with the regular derivative if the function is sufficiently smooth.

Definition 2.9 (Sobolev Space) Given a domain $\Omega \subset \mathbb{R}^d$, let $m \geq 0$ be an integer. For any $v \in L^1_{loc}(\Omega)$, recall the weak derivative $\partial^\alpha v$ in Definition 2.8. Then define the Sobolev norm

$$\|v\|_{W^m_p(\Omega)} := \Big(\sum_{|\alpha| \leq m} \|\partial^\alpha v\|^p_{L^p(\Omega)} \Big)^{1/p}, \qquad 1 \leq p < \infty.$$

In the case $p = \infty$, define

$$\|v\|_{W^m_\infty(\Omega)} := \max_{|\alpha|\le m} \|\partial^\alpha v\|_{L^\infty(\Omega)}.$$

In either case, the Sobolev space is defined as

$$W^m_p(\Omega) := \{v : \|v\|_{W^m_p(\Omega)} < \infty\}.$$

Here, it is clear that $W^0_p(\Omega) = L^p(\Omega)$. In addition, the corresponding seminorms are

$$|v|_{W^m_p(\Omega)} := \Big(\sum_{|\alpha|=m} \|\partial^\alpha v\|^p_{L^p(\Omega)}\Big)^{1/p}, \qquad 1 \le p < \infty,$$

$$|v|_{W^m_\infty(\Omega)} := \max_{|\alpha|=m} \|\partial^\alpha v\|_{L^\infty(\Omega)}.$$

For $0 < t < 1$ and $1 \le p < \infty$, let $s = m + t$. Then the *fractional order* Sobolev space $W^s_p(\Omega)$ consists of distributions v in Ω such that

$$\|v\|_{W^s_p(\Omega)} := \left(\|v\|^p_{W^m_p(\Omega)} + \sum_{|\alpha|=m} \int_\Omega\int_\Omega \frac{|\partial^\alpha v(x) - \partial^\alpha v(y)|^p}{|x-y|^{d+tp}}\, dxdy\right)^{1/p} < \infty.$$

The fractional order Sobolev spaces can also be defined by interpolation [27].

Some related spaces are further defined as follows [61].

Definition 2.10 For any $s > 0$ and $1 \le p \le \infty$, we define $\mathring{W}^s_p(\Omega)$ to be the closure of $\mathcal{D}(\Omega)$ in $W^s_p(\Omega)$. For $1 < p < \infty$, define $W^{-s}_q(\Omega)$ to be the dual space of $\mathring{W}^s_p(\Omega)$, where $p^{-1} + q^{-1} = 1$. For a function v in a Lipschitz domain Ω, let $\tilde{v}$ be the extension of v by zero outside Ω. Then we denote by $\widetilde{W}^s_p(\Omega)$, $s \ge 0$ and $1 < p < \infty$, the space of all v such that $\tilde{v} \in W^s_p(\mathbb{R}^d)$ with the natural norm $\|v\|_{\widetilde{W}^s_p(\Omega)} = \|\tilde{v}\|_{W^s_p(\mathbb{R}^d)}$. It can be further shown that for $v \in \widetilde{W}^s_p(\Omega)$,

$$\|v\|_{\widetilde{W}^s_p(\Omega)} = \begin{cases} \|v\|_{W^s_p(\Omega)} & \text{if } s \in \mathbb{N}, \\ \left(\|v\|^p_{W^s_p(\Omega)} + \sum_{|\alpha|=m} \int_\Omega |\partial^\alpha v(x)|^p \rho(x)^{-tp}\, dx\right)^{1/p} & \text{if } s = m + t, \end{cases}$$

where $m \ge 0$ is an integer, $0 < t < 1$, and ρ is the distance to the boundary Γ. For $p = 2$, we use the notation $H^s_0(\Omega)$, $H^s(\Omega)$, $H^{-s}(\Omega)$, and $\tilde{H}^s(\Omega)$ instead of $\mathring{W}^s_p(\Omega)$, $W^s_p(\Omega)$, $W^{-s}_q(\Omega)$, and $\widetilde{W}^s_p(\Omega)$ to represent these Hilbert spaces.

2.2.2 Extension and Embedding Theorems

We summarize important properties of the Sobolev space from the classical results in [1, 61, 91]. We mainly focus on the properties regarding approximation by smooth functions, properties regarding embedding, extension, and trace mappings, and regularity estimates for elliptic boundary value problems. In this section, we assume $1 < p < \infty$ unless specified otherwise.

It is understood that the functions in Sobolev spaces can be approximated by smooth functions.

Theorem 2.5 (*Density of Smooth Functions*) *Let $\Omega \subset \mathbb{R}^d$ be a Lipschitz domain. Then for $s > 0$, $\mathcal{D}(\bar{\Omega})$ is dense in $W_p^s(\Omega)$ and $\mathcal{D}(\Omega)$ is dense in $\tilde{W}_p^s(\Omega)$. In addition, $\mathcal{D}(\Omega)$ is dense in $W_p^s(\Omega)$ for $0 < s \le \frac{1}{p}$.*

The next theorem [2, 100, 109] ensures that each function $v \in W_p^s(\Omega)$ is the restriction of a function in $W_p^s(\mathbb{R}^d)$.

Theorem 2.6 (*Extension*) *Let $\Omega \subset \mathbb{R}^d$ be a Lipschitz domain. Then for $s > 0$, there exists a bounded linear operator $E : W_p^s(\Omega) \to W_p^s(\mathbb{R}^d)$ such that for any $v \in W_p^s(\Omega)$,*

$$Ev = v \quad \text{in} \quad \Omega \qquad \text{and} \qquad \|Ev\|_{W_p^s(\mathbb{R}^d)} \le C\|v\|_{W_p^s(\Omega)}.$$

Such an extension is useful for applying results in $\mathbb{R}^d$ to a Lipschitz domain Ω. For example, for $0 \le s_1 < s_2$, the injection $W_p^{s_2}(\Omega)$ in $W_p^{s_1}(\Omega)$ is compact. Another related result is often referred to as the Poincaré inequality.

Theorem 2.7 (*Poincaré's Inequality*) *Let $\Omega \subset \mathbb{R}^d$ be a Lipschitz domain. There exists a constant $C > 0$ that depends on Ω, such that for any $v \in \mathring{W}_p^1(\Omega)$, $1 \le p \le \infty$,*

$$\|v\|_{L^p(\Omega)} \le C|v|_{W_p^1(\Omega)},$$

where the $W_p^1(\Omega)$ seminorm is $|v|_{W_p^1(\Omega)} := \left(\sum_{|\alpha|=1} \int_\Omega |\partial^\alpha v|^p \, dx\right)^{1/p}$ for $1 \le p < \infty$ or $|v|_{W_\infty^1(\Omega)} := \max_{|\alpha|=1} \|\partial^\alpha v\|_{L^\infty(\Omega)}$.

Recall the Hölder space $C^{m,\gamma}$ in Definition 2.6. The Sobolev embedding theorem summarizes the inclusions between certain Sobolev spaces [61].

Theorem 2.8 (*Embedding*) *Let $\Omega \subset \mathbb{R}^d$ be a Lipschitz domain. Then*

$$\begin{aligned} W_p^s(\Omega) &\subset W_q^t(\Omega), & & t \le s,\ q \ge p,\ s - d/p = t - d/q, \\ W_p^s(\Omega) &\subset C^{m,\gamma}(\bar{\Omega}), & & m < s - d/p < m + 1,\ \gamma = s - m - d/p, \end{aligned}$$

where $s > 0$, and $m \ge 0$ is an integer.

Using the Sobolev embedding results and the Hardy inequality, it is possible to derive more estimates for Sobolev spaces [61].

Theorem 2.9 *Let $\Omega \subset \mathbb{R}^d$ be a Lipschitz domain. Let $\rho(x)$ be the distance from $x \in \Omega$ to the boundary Γ. Then for $v \in \mathring{W}_p^s(\Omega)$ such that $s - 1/p$ is not an integer, we have for all $|\alpha| \le s$,*

$$\rho^{|\alpha|-s}\partial^\alpha v \in L^p(\Omega).$$

Consequently, for $0 < s < 1/p$, one has $\tilde{W}_p^s(\Omega) = \mathring{W}_p^s(\Omega) = W_p^s(\Omega)$. Let $m \ge 0$ be an integer and assume $s = m + t$ for $0 < t < 1$. Then if $s - 1/p$ is not an integer, $\tilde{W}_p^s(\Omega) = \mathring{W}_p^s(\Omega)$; and if $s - 1/p$ is an integer,

$$\tilde{W}_p^s(\Omega) = \left\{ v \in \mathring{W}_p^s(\Omega) : \frac{\partial^\alpha v}{\rho^t} \in L^p(\Omega), \ \forall \, |\alpha| = m \right\}.$$

It is clear that for $s \geq 0$, $H^{-s}(\Omega)$ is the dual space of $H_0^s(\Omega)$ and therefore, is also the dual space of $\tilde{H}^s(\Omega)$ if $s - 1/2$ is not an integer. When $s - 1/2$ is an integer, we shall denote the dual space of $\tilde{H}^s(\Omega)$ by $\tilde{H}^{-s}(\Omega)$.

2.2.3 Trace Theorems

Another consequence of Theorem 2.8 is that any function in $W_p^s(\Omega)$ for $sp > d$ is continuous up to the boundary. Let n be the unit outward normal on the boundary Γ. Note that for $d = 2, 3$, the vector field n is of class $C^{m-1,1}$ if Γ is of class $C^{m,1}$. Denote by γ the restriction operator to the boundary such that $\gamma v = v|_\Gamma$ when v is a continuous function. Then it is possible to give an accurate description for the boundary restriction of a function in the Sobolev space [61].

Theorem 2.10 (*Trace*) *Let $\Omega \subset \mathbb{R}^d$ be a bounded subset with a $C^{m,1}$ boundary Γ. Let $\ell \geq 0$ be an integer. Assume $s \leq m + 1$, where $s - 1/p$ is not an integer and $\ell < s - 1/p < \ell + 1$. Then for $v \in C^{m,1}(\bar{\Omega})$, the mapping*

$$v \to \{\gamma v, \gamma(\partial_n v), \ldots, \gamma(\partial_n^\ell v)\}$$

has a unique continuous extension as a trace operator from

$$W_p^s(\Omega) \quad \text{onto} \quad \prod_{i=0}^{\ell} W_p^{s-i-1/p}(\Gamma),$$

which also has a right continuous inverse independent of p. Here, $\partial_n v := \nabla v \cdot n$.

Recall from Theorem 2.9 that if $s - 1/p$ is not an integer, $\tilde{W}_p^s(\Omega) = \mathring{W}_p^s(\Omega)$. Then the kernel of the trace operator can be characterized as follows.

Corollary 2.1 *Given the conditions in Theorem 2.10, we have $v \in \tilde{W}_p^s(\Omega)$ ($v \in \mathring{W}_p^s(\Omega)$) iff*

$$\gamma v = \gamma(\partial_n v) = \cdots = \gamma(\partial_n^\ell v) = 0.$$

The trace estimates in Theorem 2.10 and in Corollary 2.1 require certain smoothness conditions on the boundary Γ. In what follows, we provide extended trace theorems [28, 61] for polytopal domains in different dimensions: interval (1D), polygon (2D), and polyhedron (3D).

Theorem 2.11 (*Trace for an Interval*) *Let $\Omega = (a_1, a_2)$ be a bounded interval. Let $\ell \geq 0$ be an integer. Assume $s - 1/p$ is not an integer and $\ell < s - 1/p < \ell + 1$. Then the trace mapping defined on $\mathcal{D}(\bar{\Omega})$*

$$v \to \{v(a_i), \partial_x v(a_i), \ldots, \partial_x^\ell v(a_i)\} \quad \text{for} \quad i = 1, 2$$

is continuous on $W_p^s(\Omega)$.

Note that for a Lipschitz domain Ω, the trace result in Theorem 2.10 is restricted to the case when $s = 1$ and $\ell = 0$. Therefore, only the Dirichlet boundary condition is well defined even for smooth functions.

Corollary 2.2 (Global Trace for a Lipschitz Domain) *Let* Ω *be a bounded Lipschitz domain in* $\mathbb{R}^d$, $d = 2, 3$. *For* $1 < p < \infty$ *and* $1/p < s < 1 + 1/p$, *the trace operator* $v \to \gamma v$ *is continuous from* $W_p^s(\Omega)$ *onto* $W_p^{s-1/p}(\Gamma)$, *and it has a continuous inverse. Moreover, we have*

$$\mathring{W}_p^s(\Omega) = \{v \in W_p^s(\Omega) : \gamma v = 0\}.$$

For a polygon or a polyhedron, the spaces $W_p^{s-1/p}(\Gamma)$ are defined by a partition of unity. The Sobolev space in a side or a face on the boundary are defined as in Definition 2.10. Near the nonsmooth points of the boundary, namely vertices in 2D or vertices and edges in 3D, a transformation can be designed to map the two sides of a sector to a straight line and map the two faces of a dihedron or all the faces sharing a common vertex into a plane. Such a transformation in turn leads to *compatibility conditions* of the traces at the nonsmooth boundary points.

Given a polygon (resp. polyhedron), we denote each side (resp. face) of Γ by $\overline{\Gamma_j}$ where Γ_j is open and j ranges from 1 to N (the number of sides (resp. faces) of Ω). Denote the normal and tangential derivatives on Γ_j by ∂_{n_j} and ∂_{τ_j}, respectively. Let m and k be two integers such that $m \geq 1$ and $k \geq 0$. Define the space of differential operators on Γ_j with constant coefficients that is homogeneous of degree k by

$$E_j^{k,m} := \{L_j = \sum_{\ell=0}^{\min\{k,m-1\}} a_j^\ell \partial_{n_j}^\ell \partial_{\tau_j}^{k-\ell}\}. \tag{2.4}$$

Note that the operators in $E_j^{k,m}$ are written in terms of ∂_{n_j} and ∂_{τ_j}, and m indicates the order of the normal derivative on Γ_j.

We first consider the case when Ω is a polygon. Let $c_j = \overline{\Gamma_j} \cap \overline{\Gamma_{j+1}}$ be the vertex shared by the two sides. Denote by (r_j, θ_j) the polar coordinates such that c_j is the origin and $\theta_j = 0$ (resp. $\theta_j = \omega_j$) corresponds to Γ_{j+1} (resp. Γ_j). See Fig. 2.1.

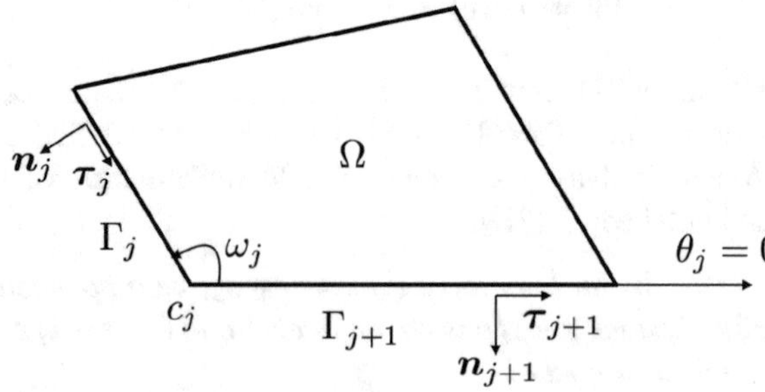

Fig. 2.1 A polygonal domain.

For a function v, let

$$\gamma_j^\ell(v) := (\partial_{n_j}^\ell v)|_{\Gamma_j} \tag{2.5}$$

be the restriction of the ℓth-order normal derivative of v to the side Γ_j. For $s \geq 0$ and $1 < p < \infty$, define

$$M = s - \frac{1}{p} - 1, \quad \text{if } s - \frac{1}{p} \text{ is an integer;} \qquad M = [s - \frac{1}{p}] \quad \text{otherwise,}$$

where $[x]$ is the largest integer $\leq x$. Although the global trace operator in Theorem 2.10 does not extend to high-order Sobolev spaces due to lack of smoothness of the boundary, it can be shown that the trace operator $(\gamma_j^0, \ldots, \gamma_j^M)_{1\leq j\leq N}$ is continuous from $W_p^s(\Omega)$ into $\prod_{j=1}^N \prod_{\ell=0}^M W_p^{s-\ell-1/p}(\Gamma_j)$. The following theorem characterizes the range of this trace mapping [28, 61]. Recall from (2.4) that any $L_j \in E_j^{k,m}$ can be written as $L_j = \sum_\ell P_j^\ell \partial_{n_j}^\ell$, where $P_j^\ell = a_j^\ell \partial_{\tau_j}^{k-\ell}$ involves $(k-\ell)$th-order tangential derivatives on Γ_j.

Theorem 2.12 (*Trace for a Polygon*) *Let $\Omega \subset \mathbb{R}^2$ be a bounded polygon. Let $1 < p < \infty$, $s \geq 1/p$, and m be an integer. Then for $0 \leq m - 1 < s - 1/p$, the trace mapping $(\gamma_j^0, \ldots, \gamma_j^{m-1})_{1\leq j\leq N}$ is continuous from $W_p^s(\Omega)$ onto a subspace of*

$$W_p^{s,(m)}(\Gamma) := \prod_{j=1}^{N} \prod_{\ell=0}^{m-1} W_p^{s-\ell-1/p}(\Gamma_j). \tag{2.6}$$

Denote this subspace by $\mathbb{W}_p^{s,(m)}(\Gamma) \subset W_p^{s,(m)}(\Gamma)$. Let $L_j = \sum_{\ell=0}^{\min\{k,m-1\}} P_j^\ell \partial_{n_j}^\ell \in E_j^{k,m}$ and $L_{j+1} = \sum_{\ell=0}^{\min\{k,m-1\}} P_{j+1}^\ell \partial_{n_{j+1}}^\ell \in E_{j+1}^{k,m}$ be any two operators on Γ_j and Γ_{j+1}, respectively. Then the functions in $\mathbb{W}_p^{s,(m)}(\Gamma)$ satisfy two compatibility conditions at the vertices.

(i) Suppose $k < s - 2/p$ and $L_j = L_{j+1}$. Then

$$\sum_\ell (P_j^\ell \gamma_j^\ell(v))(c_j) = \sum_\ell (P_{j+1}^\ell \gamma_{j+1}^\ell(v))(c_j) \qquad \forall\, v \in W_p^s(\Omega). \tag{2.7}$$

(ii) Suppose $s - 2/p$ is an integer and $k = s - 2/p$. If $L_j = L_{j+1}$, for any $v \in W_p^s(\Omega)$, we have

$$\int_0^\epsilon |\sum_\ell (P_j^\ell \gamma_j^\ell(v))(c_j - t\tau_j) - \sum_\ell (P_{j+1}^\ell \gamma_{j+1}^\ell(v))(c_j + t\tau_{j+1})|^p \frac{dt}{t} < \infty, \tag{2.8}$$

where $0 < \epsilon < \min_j\{|\Gamma_j|\}$. In both (2.7) and (2.8), we require $j+1 = 1$ if $j = N$. If $s - 2/p \geq 2m - 1$, condition (ii) is not necessary.

In a nutshell, a differential operator L with constant coefficients that is homogeneous of degree k can be written in terms of ∂_{n_j} and ∂_{τ_j}, and also in terms of $\partial_{n_{j+1}}$ and $\partial_{\tau_{j+1}}$. Nonetheless, these different expressions shall have the same value at the vertex coming from both sides if $L(v)$ is a continuous function. Condition (ii) in Theorem 2.12 represents the limit case for the Sobolev embedding result.

Let D_j be the dimension of $E_j^{k,m} \cap E_{j+1}^{k,m}$. Let $\{\mathcal{L}_{j,i}\}$, $1 \leq i \leq D_j$, be the basis of $E_j^{k,m} \cap E_{j+1}^{k,m}$. Suppose $\mathcal{L}_{j,i} = L_{j,i} = L_{j+1,i}$, where $L_{j,i} = \sum_\ell P_{j,i}^\ell \partial_{n_j}^\ell \in E_j^{k,m}$ and $L_{j+1,i} = \sum_\ell P_{j+1,i}^\ell \partial_{n_{j+1}}^\ell \in E_{j+1}^{k,m}$. Based on the definition, the space $W_p^{s,(m)}(\Gamma)$ in (2.6) is provided with the natural norm $\|\cdot\|_{W_p^{s,(m)}(\Gamma)}$. With a suitable norm for the space $\mathbb{W}_p^{s,(m)}(\Gamma)$, it is possible to show that the trace mapping has a continuous inverse [28].

Corollary 2.3 *Equipped with the same norm* $\|\cdot\|_{W_p^{s,(m)}(\Gamma)}$, *the space* $\mathbb{W}_p^{s,(m)}(\Gamma)$ *satisfying the conditions in (2.7) and (2.8) is closed iff* $s - 2/p \notin \{m-1, \ldots, 2m-2\}$. *When* $s - 2/p \in \{m-1, \ldots, 2m-2\}$, *let* $\mathbb{W}_p^{s,(m)}(\Gamma)$ *be the space equipped with the norm of* $W_p^{s,(m)}(\Gamma)$ *augmented with the left hand side in (2.8) for* $1 \leq j \leq N$ *and for the basis* $L_{j,i} = L_{j+1,i}$ *of* $E_j^{k,m} \cap E_{j+1}^{k,m}$, $1 \leq i \leq D_j$. *Then the trace operator* $(\gamma_j^0, \ldots, \gamma_j^{m-1})_{1 \leq j \leq N}$ *is continuous from* $W_p^s(\Omega)$ *onto* $\mathbb{W}_p^{s,(m)}(\Gamma)$ *and it admits a continuous inverse from* $\mathbb{W}_p^{s,(m)}(\Gamma)$ *into* $W_p^s(\Omega)$.

Remark 2.1 For $v \in W_p^s(\Omega)$ and $1 \leq \ell \leq m-1$, recall that $\gamma_j^\ell(v) = (\partial_{n_j}^\ell v)|_{\Gamma_j}$ is the restriction of the ℓth-order normal derivative of v to the side Γ_j. In Table 2.1, we display the compatibility conditions at $c_j = \overline{\Gamma_j} \cap \overline{\Gamma_{j+1}}$ for $m = 1, 2, 3$ assuming $k < s - 2/p$ (see also [28]). Here, $c := \cos\omega_j$ and $s := \sin\omega_j$, where ω_j is the interior angle at c_j. In the case where $\omega_j = \pi/2$, the compatibility conditions are much simpler. See Table 2.2 for the case $m = 3$.

We proceed to consider the case where $\Omega \subset \mathbb{R}^3$ is a bounded polyhedral domain. The trace estimate for Ω is more involved, largely due to the complexity in the 3D geometry. For completeness, we closely follow the presentation in [28]. See also [60]. Recall that we denote by Γ_j, $1 \leq j \leq N$, the jth face of the boundary Γ. As in the case of polygonal domains (Theorem 2.12), we use the same notation for the general trace space

$$W_p^{s,(m)}(\Gamma) := \prod_{j=1}^{N} \prod_{\ell=0}^{m-1} W_p^{s-\ell-1/p}(\Gamma_j). \tag{2.9}$$

Denote the trace operator by

$$(\gamma_j^0, \ldots, \gamma_j^{m-1})_{1 \leq j \leq N}, \tag{2.10}$$

where $\gamma_j^\ell(v) := (\partial_{n_j}^\ell v)|_{\Gamma_j}$ is the restriction of the ℓth-order normal derivative of v to the face Γ_j. The compatibility conditions involve the condition for each edge and for each vertex of Ω.

Consider a polyhedral cone Λ with vertex c. Denote the edges of $\partial\Lambda$ by e_l, $1 \leq l \leq L$, and denote the faces of $\partial\Lambda$ by Γ_j, $1 \leq j \leq J$. Let σ_l be the unit tangential vector to e_l pointing to c. Denote by Γ_{l_-} and Γ_{l_+} the two faces intersecting at e_l. In addition, let $n_{l_\pm}$ stand for the unit outward normal vector to $\Gamma_{l_\pm}$ and $\tau_{l_\pm}$ for a unit vector tangent to $\Gamma_{l_\pm}$ and orthogonal to σ_l. See Fig. 2.2.

$m = 1,\ k = 0$	$\gamma^0_{j+1}(v)(c_j) = \gamma^0_j(v)(c_j)$
$m = 2, k = 0$	$\gamma^0_{j+1}(v)(c_j) = \gamma^0_j(v)(c_j)$
$m = 2, k = 1$	$\partial_{\tau_{j+1}}\gamma^0_{j+1}(v)(c_j) = -[c\partial_{\tau_j}\gamma^0_j(v) + s\gamma^1_j(v)](c_j)$ $\gamma^1_{j+1}(v)(c_j) = [s\partial_{\tau_j}\gamma^0_j(v) - c\gamma^1_j(v)](c_j)$
$m = 2, k = 2$	$[c\partial^2_{\tau_{j+1}}\gamma^0_{j+1}(v) - s\partial_{\tau_{j+1}}\gamma^1_{j+1}(v)](c_j) = [c\partial^2_{\tau_j}\gamma^0_j(v) + s\partial_{\tau_i}\gamma^1_j(v)](c_j)$
$m = 3, k = 0$	$\gamma^0_{j+1}(v)(c_j) = \gamma^0_j(v)(c_j)$
$m = 3, k = 1$	$\partial_{\tau_{j+1}}\gamma^0_{j+1}(v)(c_j) = -[c\partial_{\tau_j}\gamma^0_j(v) + s\gamma^1_j(v)](c_j)$ $\gamma^1_{j+1}(v)(c_j) = [s\partial_{\tau_j}\gamma^0_j(v) - c\gamma^1_j(v)](c_j)$
$m = 3, k = 2$	$\partial^2_{\tau_{j+1}}\gamma^0_{j+1}(v)(c_j) = [c^2\partial^2_{\tau_j}\gamma^0_j(v) + 2sc\partial_{\tau_j}\gamma^1_j(v) + s^2\gamma^2_j(v)](c_j)$ $\partial_{\tau_{j+1}}\gamma^1_{j+1}(v)(c_j) = [-sc\partial^2_{\tau_j}\gamma^0_j(v) + (c^2 - s^2)\partial_{\tau_j}\gamma^1_j(v) + sc\gamma^2_j(v)](c_j)$ $\gamma^2_{j+1}(v)(c_j) = [s^2\partial^2_{\tau_j}\gamma^0_j(v) - 2sc\partial_{\tau_j}\gamma^1_j(v) + c^2\gamma^2_j(v)](c_j)$
$m = 3, k = 3$	$[c\partial^3_{\tau_{j+1}}\gamma^0_{j+1}(v) - s\partial^2_{\tau_{j+1}}\gamma^1_{j+1}(v)](c_j) =$ $-[c^2\partial^3_{\tau_j}\gamma^0_j(v) + 2sc\partial^2_{\tau_j}\gamma^1_j(v) + s^2\partial_{\tau_j}\gamma^2_j(v)](c_j)$ $[c^2\partial^3_{\tau_{j+1}}\gamma^0_{j+1}(v) - 2sc\partial^2_{\tau_{j+1}}\gamma^1_{j+1}(v) + s^2\partial_{\tau_{j+1}}\gamma^2_{j+1}](c_j) =$ $-[c\partial^3_{\tau_j}\gamma^0_j(v) + s\partial^2_{\tau_j}\gamma^1_j(v)](c_j)$
$m = 3, k = 4$	$[c^2\partial^4_{\tau_{j+1}}\gamma^0_{j+1}(v) - 2sc\partial^3_{\tau_{j+1}}\gamma^1_{j+1}(v) + s^2\partial^2_{\tau_{j+1}}\gamma^2_{j+1}(v)](c_j) =$ $[c^2\partial^4_{\tau_j}\gamma^0_j(v) + 2sc\partial^3_{\tau_j}\gamma^1_j(v) + s^2\partial^2_{\tau_j}\gamma^2_j(v)](c_j)$

Table 2.1 Compatibility conditions for a polygonal domain, $j + 1 = 1$ when $j = N$.

$m = 3, k = 0$	$\gamma^0_{j+1}(v)(c_j) = \gamma^0_j(v)(c_j)$
$m = 3, k = 1$	$\partial_{\tau_{j+1}}\gamma^0_{j+1}(v)(c_j) = -\gamma^1_j(v)(c_j)$ $\gamma^1_{j+1}(v)(c_j) = \partial_{\tau_j}\gamma^0_j(v)(c_j)$
$m = 3, k = 2$	$\partial^2_{\tau_{j+1}}\gamma^0_{j+1}(v)(c_j) = \gamma^2_j(v)(c_j)$ $\partial_{\tau_{j+1}}\gamma^1_{j+1}(v)(c_j) = -\partial_{\tau_j}\gamma^1_j(v)(c_j)$ $\gamma^2_{j+1}(v)(c_j) = \partial^2_{\tau_j}\gamma^0_j(v)(c_j)$
$m = 3, k = 3$	$\partial^2_{\tau_{j+1}}\gamma^1_{j+1}(v)(c_j) = \partial_{\tau_j}\gamma^2_j(v)(c_j)$ $\partial_{\tau_{j+1}}\gamma^2_{j+1}(c_j) = -\partial^2_{\tau_j}\gamma^1_j(v)(c_j)$
$m = 3, k = 4$	$\partial^2_{\tau_{j+1}}\gamma^2_{j+1}(v)(c_j) = \partial^2_{\tau_j}\gamma^2_j(v)(c_j)$

Table 2.2 Compatibility conditions for $\omega_j = \pi/2$, $j + 1 = 1$ when $j = N$.

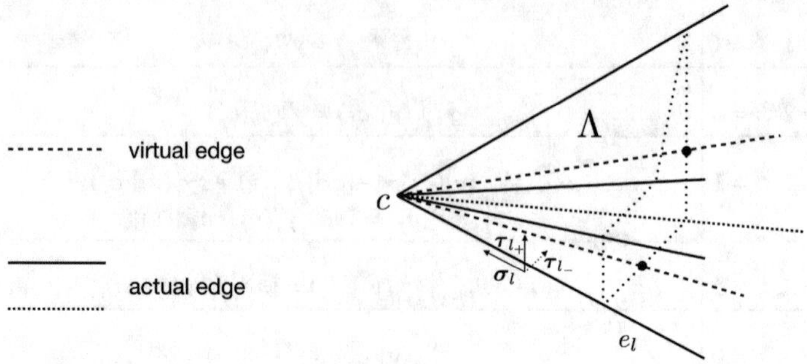

Fig. 2.2 A polyhedral cone Λ.

Associated with each face Γ_j of Λ, $1 \le j \le J$, define the spaces $E_j^{k,m}$ of homogeneous differential operators on $\mathbb{R}^3$ of degree k with constant coefficients:

$$E_j^{k,m} := \left\{ L = \sum_{\ell=0}^{\min\{k,m-1\}} a_j^\ell \partial_{n_j}^\ell \partial_{\tau_j}^{k-\ell} \right\}, \tag{2.11}$$

where ∂_{n_j} and ∂_{τ_j} are the directional derivatives in the directions of n_j and τ_j, respectively. Thus, for an edge e_l of $\partial\Lambda$, the two spaces (2.11) associated with Γ_{l_+} and Γ_{l_-} are denoted by $E_{l_+}^{k,m}$ and $E_{l_-}^{k,m}$, respectively. Write each $L_{l_+} \in E_{l_+}^{k,m}$ (resp. $L_{l_-} \in E_{l_-}^{k,m}$) as $L_{l_+} = \sum_\ell P_{l_+}^\ell \partial_{n_{l_+}}^\ell$ (resp. $L_{l_-} = \sum_\ell P_{l_-}^\ell \partial_{n_{l_-}}^\ell$), where $P_{l_+}^\ell = a_{l_+}^\ell \partial_{\tau_{l_+}}^{k-\ell}$ (resp. $P_{l_-}^\ell = a_{l_-}^\ell \partial_{\tau_{l_-}}^{k-\ell}$) involves $(k-\ell)$th-order derivatives in the τ_{l_+} (resp. τ_{l_-}) direction.

In the neighborhood of the edge e_l, the domain Ω coincides with a dihedral angle which is a tensor product of e_l and a plane angle with sides parallel to τ_{ℓ_+} and τ_{ℓ_-}. Thus, the compatibility condition for an edge can be derived using the results in Theorem 2.12.

Lemma 2.1 (*Edge Compatibility*) *Let $\Lambda \subset \mathbb{R}^3$ be a polyhedral cone. Let $1 < p < \infty$, and let $s > 1/p$, where $s - 3/p$ is not an integer. Suppose m is an integer and $0 \le m-1 < s - 1/p$. Then the image of the trace mapping $(\gamma_j^0, \ldots, \gamma_j^{m-1})_{j=l_+,l_-}$ for any function in $W_p^s(\Lambda)$ belongs to*

$$\prod_{\ell=0}^{m-1} W_p^{s-\ell-1/p}(\Gamma_{l_+}) \times \prod_{\ell=0}^{m-1} W_p^{s-\ell-1/p}(\Gamma_{l_-})$$

and satisfies the following compatibility conditions.

(i) For all $0 \le k < s - 2/p$, and any pair of operators $L_{l_+} \in E_{l_+}^{k,m}$ and $L_{l_-} \in E_{l_-}^{k,m}$ such that $L_{l_+} = L_{l_-}$, it holds that for a.e. $x \in e_l$ and $0 \le \ell \le \min\{k, m-1\}$,

$$\sum_\ell P_{l_+}^\ell \gamma_{l_+}^\ell(v)(x) = \sum_\ell P_{l_-}^\ell \gamma_{l_-}^\ell(v)(x), \quad \forall\, v \in W_p^s(\Lambda). \tag{2.12}$$

(ii) When $s-2/p$ is an integer and $k = s-2/p$, for any pair of operators $L_{l_+} \in E_{l_+}^{k,m}$ and $L_{l_-} \in E_{l_-}^{k,m}$ such that $L_{l_+} = L_{l_-}$, it holds that for a.e. $x \in e_l$, $v \in W_p^s(\Lambda)$, and $0 \le \ell \le \min\{k, m-1\}$,

$$\int_0^{\epsilon(x)} \Big| \sum_\ell P_{l_+}^\ell \gamma_{l_+}^\ell(v)(x + t\tau_{l_+}) - \sum_\ell P_{l_-}^\ell \gamma_{l_-}^\ell(v)(x + t\tau_{l_-}) \Big|^p \frac{dt}{t} < \infty, \quad (2.13)$$

for a small $\epsilon(x)$ depending on x. The condition in (2.12) is satisfied everywhere when $k < s-3/p$.

We next describe the compatibility conditions for a vertex. Consider the polyhedral cone Λ with the vertex c. Let Γ_j, $1 \le j \le J$, be the faces of $\partial\Lambda$. For two faces Γ_j and $\Gamma_{j'}$, $1 \le j < j' \le J$, let $D_{jj'}$ be the intersection line of the two planes $\mathcal{P}_j$ and $\mathcal{P}_{j'}$ containing Γ_j and $\Gamma_{j'}$, respectively. Thus, when Γ_j and $\Gamma_{j'}$ are adjacent, namely sharing an edge e_l, it is clear that $e_l \subset D_{jj'}$; otherwise, we call $D_{jj'}$ a *virtual edge* common to Γ_j and $\Gamma_{j'}$. See Fig. 2.2. Note that the functions in Γ_j and in $\Gamma_{j'}$ have extensions in $\mathcal{P}_j$ and $\mathcal{P}_{j'}$ and are, therefore, connected via their traces on $D_{jj'}$. Let $\sigma_{jj'}$ be the unit vector in $D_{jj'}$ pointing to c. Define the set

$$\Sigma_c := \{\sigma_{jj'} : 1 \le j \le j' \le J\}.$$

For any $\sigma \in \Sigma_c$, let $J(\sigma)$ be the set of indices j, $1 \le j \le J$, such that σ is tangential to Γ_j. For $j \in J(\sigma)$, τ_j stands for the unit vector tangential to $\mathcal{P}_j$ and orthogonal to σ. For example, if Λ has three faces, Σ_c has three elements and there is no virtual edge; if the cone has four faces, Σ_c has six elements and $J(\sigma)$ has two elements for any $\sigma \in \Sigma_c$. Define the trace space for the cone Λ

$$W_p^{s,(m)}(\partial\Lambda) := \prod_{j=1}^{J} \prod_{\ell=0}^{m-1} W_p^{s-\ell-1/p}(\Gamma_j)$$

with the natural norm. Use the trace operator (2.10) for Λ where $1 \le j \le J$. Denote by $\mathcal{W}$ a plane sector with vertex at the origin. Assume the angle and radius of $\mathcal{W}$ are sufficiently small and for any $1 \le j \le J$, there exists a linear transformation F_j that involves a translation and a rotation, such that the sector $F_j(\mathcal{W})$ is contained in Γ_j and its vertex coincides with c.

Lemma 2.2 (*Vertex Compatibility*) *Consider a polyhedral cone $\Lambda \subset \mathbb{R}^3$. Let $1 < p < \infty$ and $s > 1/p$. Suppose m is an integer, $0 \le m-1 < s-1/p$, and $0 \le \ell \le \min\{k, m-1\}$. For any $L_j \in E_j^{k,m}$, write $L_j = \sum_\ell P_j^\ell \partial_{n_j}^\ell$ as in Lemma 2.1. Then the image of the trace mapping $(\gamma_j^0, \ldots, \gamma_j^{m-1})_{1\le j\le J}$ for any function in $W_p^s(\Lambda)$ belongs to $W_p^{s,(m)}(\partial\Lambda)$ and satisfies the following compatibility conditions at the vertex c.*

(i) Suppose $0 \le k < s-3/p$ and $0 \le \nu \le s-k-3/p$. Then for any $\sigma \in \Sigma_c$ and any $(L_j)_{j\in J(\sigma)} \in \prod_{j\in J(\sigma)} E_j^{k,m}$ such that $\sum_{j\in J(\sigma)} L_j = 0$, we have

$$\partial_\sigma^\nu \sum_{j \in J(\sigma)} \sum_\ell P_j^\ell \gamma_j^\ell(v)(c) = 0, \quad \forall\, v \in W_p^s(\Lambda). \tag{2.14}$$

(ii) Suppose $s - 3/p$ is an integer, $k = s - 3/p$, and $\nu = s - k - 3/p$. For all $\sigma \in \Sigma_c$ and any $(L_j)_{j \in J(\sigma)} \in \prod_{j \in J(\sigma)} E_j^{k,m}$ such that $\sum_{j \in J(\sigma)} L_j = 0$, we have

$$\int_{\mathcal{W}} \left| \partial_\sigma^\nu \sum_{j \in J(\sigma)} \sum_\ell P_j^\ell \gamma_j^\ell(v)(F_j(\boldsymbol{t})) \right|^p \frac{dt}{|\boldsymbol{t}|^2} < \infty, \quad \forall\, v \in W_p^s(\Lambda), \tag{2.15}$$

where $\boldsymbol{t} = (t_1, t_2) \in \mathcal{W}$.

With the results in Lemmas 2.1 and 2.2, we proceed with the compatibility conditions for a polyhedral domain.

Theorem 2.13 (*Trace for a Polyhedron*) *Let $\Omega \subset \mathbb{R}^3$ be a bounded polyhedron. Let $1 < p < \infty$ and $s > 1/p$, and m be an integer. Recall the general trace space $W_p^{s,(m)}(\Gamma)$ in (2.9). Then for $0 \le m - 1 < s - 1/p$, the trace mapping $(\gamma_j^0, \ldots, \gamma_j^{m-1})_{1 \le j \le N}$ in (2.10) is continuous from $W_p^s(\Omega)$ onto a subspace of $W_p^{s,(m)}(\Gamma)$. Denote this subspace by $\mathbb{W}_p^{s,(m)}(\Gamma) \subset W_p^{s,(m)}(\Gamma)$. Then the functions in $\mathbb{W}_p^{s,(m)}(\Gamma)$ satisfy the following compatibility conditions.*

(i) Condition (2.12) holds at all edges of $\partial\Omega$ for $0 \le k < s - 2/p$. In addition, when $s - 2/p$ is an integer, condition (2.13) holds at all edges for $k = s - 2/p$.

(ii) Condition (2.14) holds for each vertex of $\partial\Omega$ and for $0 \le k < s - 3/p$. In addition, when $s - 3/p$ is an integer, condition (2.15) holds for each vertex and for $k = s - 3/p$.

The space $W_p^{s,(m)}(\Gamma)$ in (2.9) is provided with the natural norm. With a suitable norm for the space $\mathbb{W}_p^{s,(m)}(\Gamma)$, the trace mapping also has a continuous inverse.

Corollary 2.4 *With the assumptions in Theorem 2.13, the space $\mathbb{W}_p^{s,(m)}(\Gamma)$ defined in the same theorem is closed in $W_p^{s,(m)}(\Gamma)$ when neither $s - 2/p$ nor $s - 3/p$ is an integer. When $\mathbb{W}_p^{s,(m)}(\Gamma)$ is equipped with the norm $\|\cdot\|_{W_p^{s,(m)}(\Gamma)}$ augmented with appropriate terms from (2.13) and (2.15), the trace mapping $(\gamma_j^0, \ldots, \gamma_j^{m-1})_{1 \le j \le N}$ is continuous from $W_p^s(\Omega)$ onto $\mathbb{W}_p^{s,(m)}(\Gamma)$ and admits a continuous inverse from $\mathbb{W}_p^{s,(m)}(\Gamma)$ into $W_p^s(\Omega)$.*

Example 2.1 The trace theorem for a general polyhedron is rather technical. We summarize the results in the case of the cube Ω. Each edge e_l, $1 \le l \le 12$, is the intersection of two faces Γ_{j_+} and Γ_{j_-} and is parallel to the unit vector σ_l. We choose the vectors $n_{l_\pm}$ and $\tau_{l_\pm}$ such that $n_{l_-} = \tau_{l_+}$ and $n_{l_+} = \tau_{l_-}$. Recall for a function v, $\gamma_j^\ell(v) = (\partial_{n_j}^\ell v)|_{\Gamma_j}$. Let $1 < p < \infty$, $s > 1/p$, and m be an integer such that $0 \le m - 1 < s - 1/p$. Then for any $v \in W_p^s(\Omega)$, the trace $\left(\gamma_j^0(v), \ldots, \gamma_j^{m-1}(v)\right)_{1 \le j \le 6}$ belongs to $\mathbb{W}_p^{s,(m)}(\Gamma)$ and satisfies the following conditions.

(i) For $1 \le l \le 12$ and for $0 \le k \le 2m-2$ with $k < s - 2/p$, it holds for a.e. x on e_l that

$$\partial_{\tau_{l_-}}^{k-\ell} \gamma_{l_-}^{\ell}(v) = \partial_{\tau_{l_+}}^{\ell} \gamma_{l_+}^{k-\ell}(v), \qquad 0 \le \ell, k-\ell \le m-1.$$

(ii) Suppose $s - 2/p \in \{m-1, \ldots, 2m-2\}$. Then for $1 \le l \le 12$, $k = s - 2/p$, and $0 \le \ell, k-\ell \le m-1$, it holds for a.e. x on e_l that

$$\int_0^{\epsilon(x)} \left| \partial_{\tau_{l_-}}^{k-\ell} \gamma_{l_-}^{\ell}(v)(x - t\tau_{l_-}) - \partial_{\tau_{l_+}}^{\ell} \gamma_{l_+}^{k-\ell}(v)(x - t\tau_{l_+}) \right|^p \frac{dt}{t} < \infty,$$

for a proper small $\epsilon(x)$ depending on x.

2.3 Regularity Theorems

We investigate the solvability of second-order elliptic boundary value problems in $\Omega \subset \mathbb{R}^d$. We are particularly interested in the smoothness of the solution pertaining to the regularity of the given data. Let L be a second-order differential operator written in the *divergence form*

$$Lu = -\sum_{i,k=1}^{d} \partial_{x_i}\left(a_{i,k}(x)\partial_{x_k} u\right) + \sum_{i=1}^{d} a_i(x)\partial_{x_i} u + a_0(x)u, \tag{2.16}$$

where the coefficients are measurable functions and $a_{i,k} = a_{k,i}$. When $a_{i,k}$ are differentiable functions, L can also be written in the *nondivergence form*

$$Lu = -\sum_{i,k=1}^{d} a_{i,k}(x)\partial_{x_i}\partial_{x_k} u + \sum_{i=1}^{d} a_i'(x)\partial_{x_i} u + a_0(x)u. \tag{2.17}$$

We say L is *elliptic* if the coefficient matrix $(a_{i,k})$ is positive definite in Ω, and L is *uniformly (strictly) elliptic* if there exists $C > 0$ such that for any $(\xi_1, \ldots, \xi_d) \in \mathbb{R}^d$,

$$\sum_{i,k=1}^{d} a_{i,k}\xi_i\xi_k \ge C \sum_{i=1}^{d} \xi_i^2.$$

In this section, we assume L is uniformly elliptic in Ω. We first recall Green's formula in a Lipschitz domain [61, 100].

Theorem 2.14 (*Green's Formula*) *Let $\Omega \subset \mathbb{R}^d$ be a bounded domain with a Lipschitz boundary Γ. Then for any $u \in W_p^1(\Omega)$ and $v \in W_q^1(\Omega)$ with $1/p + 1/q = 1$,*

$$\int_\Omega \partial_{x_i} uv\, dx + \int_\Omega u\partial_{x_i} v\, dx = \int_\Gamma uvn_i\, ds,$$

where n_i denotes the ith component of the unit outward normal vector n.

Define the space associated with the Laplace operator

$$D\big(\Delta, L^2(\Omega)\big) = \{v \in L^2(\Omega) : \Delta v \in L^2(\Omega)\}$$

with norm $\|v\|_D^2 := \|v\|_{L^2(\Omega)}^2 + \|\Delta v\|_{L^2(\Omega)}^2$. As a consequence of Green's formula, we can extend the trace theorem in the following two corollaries (see [61]).

Corollary 2.5 *Let $\Omega \subset \mathbb{R}^2$ be a bounded polygonal domain. Recall the trace operator (2.5). Then the trace mapping $v \longrightarrow \{\gamma_j^0 v, \gamma_j^1 v\}$ that is defined for $v \in H^2(\Omega)$ has a unique continuous extension from $D\big(\Delta, L^2(\Omega)\big)$ into $\tilde{H}^{-1/2}(\Gamma_j) \times \tilde{H}^{-3/2}(\Gamma_j)$. In addition, for $v \in D\big(\Delta, L^2(\Omega)\big)$ and $u \in H^2(\Omega)$ such that $\gamma_0 u \in \tilde{H}^{3/2}(\Gamma_j)$ and $\gamma_j^1 u \in \tilde{H}^{1/2}(\Gamma_j)$, we have*

$$\int_\Omega u\Delta v\, dx - \int_\Omega v\Delta u\, dx = \sum_j \Big(\langle \gamma_j^0 u, \gamma_j^1 v\rangle - \langle \gamma_j^0 v, \gamma_j^1 u\rangle\Big),$$

where $\tilde{H}^{-s}(\Omega)$ is the dual space of $\tilde{H}^s(\Omega)$ when $s - 1/2$ is an integer.

Similarly, define

$$E\big(\Delta, L^p(\Omega)\big) = \{v \in H^1(\Omega) : \Delta v \in L^p(\Omega)\} \qquad p > 1.$$

Corollary 2.6 *Let $\Omega \subset \mathbb{R}^2$ be a bounded polygonal domain. Then $H^2(\Omega)$ is dense in $E(\Delta, L^p(\Omega))$ and the trace mapping $v \longrightarrow \gamma_j^1 v$ has a unique continuous extension from $E\big(\Delta, L^p(\Omega)\big)$ into $\tilde{H}^{-1/2}(\Gamma_j)$. In addition, for $v \in E\big(\Delta, L^p(\Omega)\big)$ and $u \in H^1$ such that $\gamma_j^0 u \in \tilde{H}^{1/2}(\Gamma_j)$, we have*

$$\int_\Omega u\Delta v\, dx = -\int_\Omega \nabla u \cdot \nabla v\, dx + \sum_j \langle \gamma_j^0 u, \gamma_j^1 v\rangle.$$

We now survey classical regularity results for the elliptic equation with the Dirichlet boundary condition. Equations with other boundary conditions will be discussed in later chapters. Let $\Omega \subset \mathbb{R}^d$ be a bounded domain with the boundary $\Gamma = \partial\Omega$. Let L be the elliptic operator in the divergence form (2.16). Consider the elliptic boundary value problem

$$Lu = f \quad \text{in} \quad \Omega \qquad \text{and} \qquad u = 0 \quad \text{on} \quad \Gamma. \tag{2.18}$$

In the case of nonhomogeneous boundary data, namely $u|_\Gamma = g \neq 0$, one can construct a function u_g in Ω based on the trace theorems such that $u_g|_\Gamma = g$. It is clear that $w := u - u_g$ has the zero boundary condition and satisfies $Lw = f - Lu_g$. Using Green's formula, one derives the bilinear form for (2.18)

$$a(u, v) := \int_\Omega \left(\sum_{i,k=1}^d a_{i,k}\partial_{x_i} u \partial_{x_k} v + \sum_{i=1}^d a_i \partial_{x_i} uv + a_0 uv\right) dx, \tag{2.19}$$

for $u, v \in H_0^1(\Omega)$. We say that $u \in H_0^1(\Omega)$ is a *weak solution* of (2.18) if

$$a(u, v) = \langle f, v\rangle \qquad \forall\, v \in H_0^1(\Omega), \tag{2.20}$$

where $\langle \cdot, \cdot\rangle$ is the pairing of $H^{-1}(\Omega)$ and $H_0^1(\Omega)$ and $\langle f, v\rangle = \int_\Omega f v\, dx$ if $f \in L^1(\Omega)$.

The existence and uniqueness of the weak solution (2.20) depends on the differential operator L. For example, consider the Poisson equation ($L = -\Delta$) in (2.18). Applying Hölder's inequality and the Poincaré inequality (Theorem 2.7), one sees that there exist constants $C, c > 0$ such that for any $u, v \in H_0^1(\Omega)$

$$|a(u, v)| \le C\|u\|_{H^1(\Omega)}\|v\|_{H^1(\Omega)}, \qquad a(u, u) \ge c\|u\|^2_{H^1(\Omega)}. \tag{2.21}$$

Therefore, the Lax–Milgram Theorem ensures that for each $f \in H^{-1}(\Omega)$, there exists a unique weak solution $u \in H_0^1(\Omega)$. For a general second-order elliptic operator L, the existence and uniqueness of the weak solution is not guaranteed. One can, however, refer to the Fredholm alternative (Theorem 2.4) and maximum principles [56, 91].

The regularity estimate for the solution of (2.18) depends on the smoothness of the coefficients in L and the domain geometry. We survey the regularity results for the interior of the domain and the entire domain in Sobolev spaces [56, 58, 91, 112]. We start with the interior estimate.

Theorem 2.15 (*Interior Regularity*) *Let $u \in H^1(\Omega)$ be a weak solution of the equation $Lu = f$ in Ω. Suppose $f \in H^m(\Omega)$ for $m \ge 0$. Suppose the coefficients $a_{i,k} \in C^{0,1}(\bar{\Omega})$ and $a_i, a_0 \in L^\infty(\Omega)$ for $m = 0$, and $a_{i,k} \in C^{m,1}(\bar{\Omega})$, $a_i, a_0 \in C^{m-1,1}(\bar{\Omega})$ for $m \ge 1$. Then for any subdomain away from the boundary $\Omega' \subset\subset \Omega$, we have*

$$\|u\|_{H^{m+2}(\Omega')} \le C\big(\|u\|_{L^2(\Omega)} + \|f\|_{H^m(\Omega)}\big).$$

It is worth noting that the solution in the interior of the domain is smoother than the given data f provided that the coefficients are sufficiently smooth. The estimate in an interior region does not require the information from the boundary. However, it is not the case when we extend the regularity estimates to the entire domain.

Theorem 2.16 (*Global Regularity*) *Assume, in addition to the conditions in Theorem 2.15, that in (2.18) the boundary Γ is of class C^{m+2} for $m \ge 0$. Then we have*

$$\|u\|_{H^{m+2}(\Omega)} \le C\big(\|u\|_{L^2(\Omega)} + \|f\|_{H^m(\Omega)}\big).$$

Thus, extra smoothness conditions on the boundary are needed for the global regularity estimates in high-order Sobolev spaces. It is often the case in practice that the boundary of the domain is less regular. For example, a polygonal domain merely has a $C^{0,1}$ boundary. This in turn can result in *singularities* in the solution although the given data is smooth. For the $L^p(\Omega)$ regularity estimates of the elliptic problem, we consider the Dirichlet problem associated with the operator in the nondivergence form (2.17). We first have the existence and uniqueness result [58].

Theorem 2.17 *Let $\Omega \subset \mathbb{R}^d$ be a bounded domain with a $C^{1,1}$ boundary. Assume the elliptic operator L (2.17) has coefficients $a_{i,k} \in C^0(\bar{\Omega})$, a_i', $a_0 \in L^\infty(\Omega)$, and $a_0 \geq 0$. Then for $1 < p < \infty$, given $f \in L^p(\Omega)$ and $u_g \in W_p^2(\Omega)$, the Dirichlet problem $Lu = f$ in Ω, $u - u_g \in \mathring{W}_p^1(\Omega)$ has a unique solution $u \in W_p^2(\Omega)$.*

There are two special cases of Theorem 2.17 that are particularly interesting. In the case $p = 2$, it gives a well-posedness result in the Hilbert space $H^2(\Omega)$. In the case $u_g = 0$, one obtains the well-posedness of the homogeneous Dirichlet problem in $W_p^2(\Omega)$. The next theorem is the L^p counterpart of the high-order regularity results in Theorems 2.15 and 2.16. Recall the elliptic operator L in (2.17).

Theorem 2.18 (*L^p Regularity*) *Let $W^s_{p,loc}(\Omega)$ be the space consisting of functions in $W_p^s(\Omega')$ for any $\Omega' \subset\subset \Omega$. Suppose $1 < p, q < \infty$, $m \geq 1$, and $0 < \gamma < 1$. Let $u \in W^2_{p,loc}(\Omega)$ be a solution of the elliptic equation $Lu = f$ in Ω. Suppose the coefficients of L belong to $C^{m-1,1}(\Omega)$ and $f \in W^m_{q,loc}(\Omega)$. Then, $u \in W^{m+2}_{q,loc}(\Omega)$. In addition, if Ω has a $C^{m+1,1}$ ($C^{m+1,\gamma}$) boundary, the coefficients of L belong to $C^{m-1,1}(\bar{\Omega})$ ($C^{m-1,\gamma}(\bar{\Omega})$), and $f \in W_q^m(\Omega)$ ($C^{m-1,\gamma}(\bar{\Omega})$), then $u \in W_q^{m+2}(\Omega)$ ($C^{m+1,\gamma}(\bar{\Omega})$).*

It is clear from Theorems 2.15–2.18 that the regularity of the solution depends on the elliptic operator and the smoothness of the boundary. It in fact also depends on the boundary conditions. We have focused on the regularity results for the Dirichlet boundary condition to present the essence of the problem and avoid unnecessary technical details. Equations associated with other boundary conditions will be discussed in later sections.

2.4 Basic Finite Element Error Estimates

In this section, we revisit the finite element algorithm described in Chap. 1 for second-order elliptic boundary value problems. We are particularly interested in the error analysis for the numerical approximation.

The finite element method is defined by a variational formulation on a finite-dimensional space built upon a triangulation. We first introduce the general formulation for the Galerkin method. Assume the following conditions:

$$\begin{cases} H \text{ is a Hilbert space with the inner product } (\cdot,\cdot). \\ V \subset H \text{ is a closed subspace.} \\ a(\cdot,\cdot) \text{ is a bilinear form that is continuous and coercive on } V. \end{cases} \tag{2.22}$$

Then consider the following variational problem.

$$\text{Given } F \in V^*, \text{ find } u \in V \text{ such that } a(u, v) = F(v), \quad \forall\, v \in V, \tag{2.23}$$

where V^* is the dual space of V. The Galerkin (finite element) approximation to (2.23) is the following. Given a finite-dimensional subspace $V_h \subset V$, find $u_h \in V_h$ such that

$$a(u_h, v) = F(v), \quad \forall\, v \in V_h. \tag{2.24}$$

We now apply this abstract formulation to second-order elliptic equations. Note that $a(\cdot,\cdot)$ is not necessarily symmetric. Let $\Omega \subset \mathbb{R}^d$, $d = 1, 2, 3$, be a bounded *polytope*. In particular, a 1D polytope is an interval, a 2D polytope is a polygon, and a 3D polytope is a polyhedron. Consider the elliptic problem with the mixed boundary condition

$$Lu = f \quad \text{in} \quad \Omega, \qquad u = 0 \quad \text{on} \quad \Gamma_D \quad \text{and} \quad \partial_n u = 0 \quad \text{on} \quad \Gamma_N. \tag{2.25}$$

Here L is in the divergence form (2.16), Γ_D and Γ_N are open subsets of the boundary $\Gamma := \partial\Omega$ such that $\overline{\Gamma_D} \cup \overline{\Gamma_N} = \Gamma$. For simplicity, we suppose that $\Gamma_D \neq \emptyset$, and suppose each side ($d = 2$) or face ($d = 3$) of Γ is included either in $\overline{\Gamma_D}$ or in $\overline{\Gamma_N}$. Define the space

$$H^1_D(\Omega) := \{v \in H^1(\Omega) : v|_{\Gamma_D} = 0\}. \tag{2.26}$$

Then for $f \in L^2(\Omega)$, by Green's formula, the weak solution $u \in H^1_D(\Omega)$ for (2.25) is defined by

$$a(u, v) = (f, v) := \int_\Omega f v \, dx \qquad \forall\, v \in H^1_D(\Omega), \tag{2.27}$$

where the bilinear form $a(\cdot,\cdot)$ is the same as in (2.19) but defined here for functions in $H^1_D(\Omega)$. Recall that L in general possesses low-order differential operators and may not be symmetric. Nonetheless, we assume the associated bilinear form $a(\cdot,\cdot)$ in (2.27) is continuous and coercive on $H^1_D(\Omega)$. Therefore, by the Lax–Milgram Theorem, there exists a unique solution $u \in H^1_D(\Omega)$ for (2.25). Let $\mathcal{T}$ be a triangulation of Ω defined as in Definition 1.1 and let S_L be the associated Lagrange finite element space (Definition 1.2). Then the Lagrange finite element space for (2.25) is

$$S_h := S_L \cap \{v|_{\Gamma_D} = 0\} = \{v \in C(\bar{\Omega}) : v|_{\Gamma_D} = 0,\ v|_T = P_m(T),\ \forall\, T \in \mathcal{T}\}. \tag{2.28}$$

Then the finite element approximation $u_h \in S_h$ of (2.25) satisfies

$$a(u_h, v) = (f, v) \qquad \forall\, v \in S_h, \tag{2.29}$$

where $a(\cdot,\cdot)$ and $(\cdot,\cdot)$ are the same bilinear forms as in (2.27). Since $S_h \subset H^1_D(\Omega)$ is a Hilbert space itself, by the Lax–Milgram Theorem, the continuity and coercivity of $a(\cdot,\cdot)$ ensure a unique finite element solution defined by (2.29). Hence, the finite element approximation is obtained by restricting the variational formulation to the finite element space.

Example 2.2 In the particular case when $L = -\Delta$ is the Laplace operator and $\Gamma_N = \emptyset$, (2.25) is the Poisson equation with the Dirichlet data. As mentioned in Chap. 1, the corresponding variational form is

$$a(u, v) = \int_\Omega \nabla u \cdot \nabla v \, dx = \int_\Omega f v \, dx = (f, v) \qquad \forall\, v \in H_0^1(\Omega)$$

and the finite element solution $u_h \in S_h \subset H_0^1(\Omega)$ is defined by

$$\int_\Omega \nabla u_h \cdot \nabla v \, dx = \int_\Omega f v \, dx \qquad \forall\, v \in S_h.$$

Both equations are well posed due to the Poincaré inequality and Lax–Milgram Theorem. See Chap. 1 for more detailed discussions on the numerical algorithm.

For the error analysis, we are mainly interested in the error $\|u - u_h\|_{H^1(\Omega)}$ in the energy norm, and we will also mention the error estimates in some other norms.

Theorem 2.19 (*Céa*) *Recall the conditions (2.22). Let u and u_h be the solutions of the variational equation (2.23) and the finite element equation (2.24), respectively. Then*

$$\|u - u_h\|_V \le C \min_{v \in V_h} \|u - v\|_V.$$

Proof Since $a(u, v) = F(v)$ for any $v \in V$, $a(u_h, v) = F(v)$ for any $v \in V_h$, and $V_h \subset V$, we have

$$a(u - u_h, v) = 0 \qquad \forall\, v \in V_h. \tag{2.30}$$

Let $\alpha_1, \alpha_2 > 0$ be the continuity and coercivity constants for $a(\cdot, \cdot)$. Then by Definition 2.4 and the condition (2.30), we have for any $v \in V_h$,

$$\begin{aligned} \|u - u_h\|_V^2 &\le \alpha_2^{-1} a(u - u_h, u - u_h) = \alpha_2^{-1} a(u - u_h, u - v) \\ &\le \frac{\alpha_1}{\alpha_2} \|u - u_h\|_V \|u - v\|_V. \end{aligned}$$

Therefore, for $C = \alpha_1/\alpha_2$, we have $\|u - u_h\|_V \le C \min_{v \in V_h} \|u - v\|_V$. □

Remark 2.2 In the Lagrange finite element approximation of (2.25), $V = H_D^1(\Omega)$ and $V_h = S_h$. The Céa Theorem implies that the finite element solution to the elliptic problem (2.25) is comparable with the best approximation in the H^1 norm by functions in the finite element space. (2.30) is also called the *orthogonality* condition. When $a(\cdot, \cdot)$ is symmetric and induces the norm on V, the finite element solution u_h is the projection of the solution u on the finite element space and the estimate in the Céa Theorem becomes $\|u - u_h\|_V = \min_{v \in V_h} \|u - v\|_V$.

We follow [35, 43] to give a definition of the finite element, which generalizes the finite element described in Chap. 1.

Definition 2.11 (Finite Element) A finite element $(T, \mathcal{P}, \mathcal{N})$ has three components.

(i) $T \subset \mathbb{R}^d$ is a bounded closed set with non-empty interior and piecewise smooth boundary (the element domain).

(ii) $\mathcal{P}$ is a finite-dimensional space of functions on T (the space of shape functions).
(iii) $\mathcal{N} = \{N_1, \ldots, N_k\}$ is the basis for the dual space $\mathcal{P}^*$ of $\mathcal{P}$ (the set of nodal variables).

The basis $\{\phi_1, \ldots, \phi_k\}$ of $\mathcal{P}$ that is dual to $\mathcal{N}$ (i.e., $N_i(\phi_j) = \delta_{ij}$) is called the *nodal basis* of $\mathcal{P}$. In addition, one can show that condition (iii) is equivalent to the condition that $\mathcal{N}$ determines $\mathcal{P}$. Namely, given $v \in \mathcal{P}$ with $N_i(v) = 0$ for $1 \leq i \leq k$ then $v = 0$.

Example 2.3 (2D Linear Lagrange Finite Element) Let T be the triangle with vertices (0, 0), (1, 0), and (0,1), $\mathcal{P}$ be the set of linear polynomials, and $\mathcal{N} = \{N_1, N_2, N_3\}$ be such that for any $v \in \mathcal{P}$, $N_1(v) = v(0,0)$, $N_2(v) = v(1,0)$, and $N_3(v) = v(0,1)$. Then $(T, \mathcal{P}, \mathcal{N})$ is a 2D Lagrange finite element and the nodal basis consists of $\phi_1 = 1 - x - y$, $\phi_2 = x$, and $\phi_3 = y$.

The Lagrange finite element method utilizes continuous piecewise polynomials on a simplicial mesh. It is worth noting that this is not the only option available. Different selections of the finite element $(T, \mathcal{P}, \mathcal{N})$ have led to various finite element methods. For example, the piecewise polynomial space can consist of C^1 or discontinuous functions, and the mesh can consist of non-simplicial elements such as quadrilaterals, prisms, and other polytopal subsets. To present the main ideas and meanwhile simplify the exposition, we restrict our attention to the continuous Lagrange finite element method on simplicial meshes. We refer readers to [35, 37, 43] and the references therein for a variety of finite elements. Hence, choosing $V = H^1_D(\Omega)$ and choosing $V_h = S_h$ to be the Lagrange finite element space in (2.28), we shall derive the error analysis $\|u - u_h\|_{H^1(\Omega)}$ for the elliptic problem (2.25).

We say a bounded domain $G \subset \mathbb{R}^d$ is *star-shaped* with respect to a ball B if for any $x \in G$, the closed convex hall of $\{x\} \cup B$ is a subset of G. Suppose that $\rho_{\max}$ is the supremum of the diameters of such balls. The ratio $\gamma = d_G/\rho_{\max}$ is called the *chunkiness* of G, where d_G is the diameter of G. Then a fundamental polynomial approximation result in Sobolev spaces is summarized as follows [35, 43, 54, 55].

Lemma 2.3 (*Deny-Lions) Suppose $G \subset \mathbb{R}^d$ is star-shaped with respect to a ball with $\rho_{\max}$ and γ specified above. Then for any $v \in W^m_p(G)$ with $p \geq 1$, there exists a polynomial $p \in P_{m-1}$ such that for all $0 \leq k \leq m$,*

$$|v - p|_{W^k_p(G)} \leq C\big(\text{diam}(G)\big)^{m-k} |v|_{W^m_p(G)},$$

where the constant C depends on m and on the domain G through the dimension d and the chunkiness γ.

Note that another version of this approximation result is also well known.

Lemma 2.4 (*Bramble–Hilbert) Assume the conditions in Lemma 2.3. Then for all $w \in \big(W^m_p(G)\big)^*$ vanishing on P_{m-1},*

$$|w(v)| \leq C\|w\|_{(W^m_p(G))^*} |v|_{W^m_p(G)}, \quad \forall\, v \in W^m_p(G).$$

Using Lemma 2.3 and a scaling argument, one can obtain an upper bound for the best local approximation error by finite element functions on each $T \in \mathcal{T}$. Note that the finite element solution $u_h \in S_h \subset H_D^1(\Omega)$ is continuous in Ω. Although u_h is comparable to the best approximation of u over Ω (Theorem 2.19), it may not be the best local approximation of u on each T. Therefore, we need to piece the local information together to derive an error estimate $\|u-u_h\|_{H^1(\Omega)}$ over the entire domain. We begin by defining the interpolation of a function.

Definition 2.12 (Interpolant) Given a finite element $(T, \mathcal{P}, \mathcal{N})$, let $\{\phi_1, \ldots, \phi_k\}$ be the basis of $\mathcal{P}$ dual to $\mathcal{N}$. If v is a function for which all $N_i \in \mathcal{N}$, $1 \le i \le k$, are defined, then the local interpolant on T can be defined by

$$\mathcal{I}_T v := \sum_{i=1}^{k} N_i(v)\phi_i.$$

Consider the linear Lagrange finite element in Example 2.3. The nodal interpolation for a function is a linear polynomial on T such that $v(x_i) = \mathcal{I}_T v(x_i)$ where x_i, $1 \le i \le k$, are the nodes. In particular, for $v(x, y) = x^2y^2$, $\mathcal{I}_T v = \sum_{i=1}^{3} N_i(v)\phi_i = 0(1-x-y) + 0(x) + 0(y) = 0$. A different construction of the degrees of freedom $\mathcal{N}$ can result in a different finite element and a different interpolant [37, 43, 59]. Using the local interpolant, we can define the global interpolant.

Definition 2.13 (Global Interpolation) Let $\mathcal{T}$ be a triangulation of the domain Ω. Assume each element domain $T \in \mathcal{T}$ is equipped with some shape functions, $\mathcal{P}$, and nodal variables, $\mathcal{N}$, such that $(T, \mathcal{P}, \mathcal{N})$ forms a finite element. Let m be the order of the highest partial derivatives involved in the nodal variables. For $v \in C^m(\bar{\Omega})$, the global interpolant is

$$\mathcal{I}_{\mathcal{T}} v|_{T_i} = \mathcal{I}_{T_i} v \quad \forall\, T_i \in \mathcal{T}.$$

For the Lagrange finite element space (2.28), only the function value ($m = 0$) is involved in the nodal variables. Therefore, for $v \in C^0(\bar{\Omega}) \cap H_D^1(\Omega)$, its global interpolation $\mathcal{I}_{\mathcal{T}} v \in S_h$ is a continuous piecewise polynomial such that $v(x_i) = \mathcal{I}_{\mathcal{T}} v(x_i)$ at all the nodes x_i in the triangulation $\mathcal{T}$.

Remark 2.3 We now proceed to derive the finite element error analysis $\|u-u_h\|_{H^1(\Omega)}$ for the elliptic problem (2.25). To simplify the presentation, we make the following assumptions. Consider a bounded domain $\Omega \subset \mathbb{R}^d$, $1 \le d \le 3$, and let $\mathcal{T}$ be a triangulation of Ω with shape regular d-simplexes. Let S_h (2.28) be the Lagrange finite element space and let $\mathcal{I}_{\mathcal{T}}$ be the nodal interpolant. Since $u_h, \mathcal{I}_{\mathcal{T}} u \in S_h \subset H^1(\Omega)$, based on the Céa Theorem, we have

$$\|u - u_h\|_{H^1(\Omega)} \le C\|u - \mathcal{I}_{\mathcal{T}} u\|_{H^1(\Omega)} \le C\left(\sum_{T\in\mathcal{T}} \|u - \mathcal{I}_{\mathcal{T}} u\|_{H^1(T)}^2\right)^{1/2}. \tag{2.31}$$

Therefore, it is sufficient to analyze the interpolation error $\|u - \mathcal{I}_{\mathcal{T}} u\|_{H^1(T)}$ on each simplex T.

Lemma 2.5 *Let T be a shape regular d-simplex and let $P_m(T)$ be the space of all polynomials of degree $\leq m$ on T. Recall the reference element $\hat{T}$ (Fig. 1.4) and the affine mapping $F : \hat{T} \to T$ (1.12). For $p \geq 1$, suppose $m + 1 > d/p$ when $p > 1$ and $m + 1 \geq d$ when $p = 1$. Then, for $0 \leq i \leq m + 1$ and $v \in W_p^{m+1}(T)$, we have*

$$\|v - \mathcal{I}_T v\|_{W_p^i(T)} \leq C h_T^{m+1-i} |v|_{W_p^{m+1}(T)},$$

where $h_T = \mathrm{diam}(T)$ and C depends on d, m, and the chunkiness γ.

Proof With the affine mapping $F : \hat{T} \to T$, for any $\hat{x} \in \hat{T}$, we have

$$x = F(\hat{x}) = \mathbf{B}\hat{x} + \boldsymbol{c},$$

where $\mathbf{B}$ is a $d \times d$ matrix and $\boldsymbol{c}$ is a vector. See (1.17) and (1.19) for the 2D and 3D cases, respectively. For a function $v \in W_p^m(T)$, we further define the function on $\hat{T}$: $\hat{v}(\hat{x}) := v(x)$. That is, $\hat{v} = v \circ F$. By a straightforward calculation, we obtain $\hat{v} \in W_p^m(\hat{T})$ and the following estimates:

$$|\hat{v}|_{W_p^m(\hat{T})} \leq C\|\mathbf{B}\|^m |\det(\mathbf{B})|^{-1/p} |v|_{W_p^m(T)}, \tag{2.32}$$

$$|v|_{W_p^m(T)} \leq C\|\mathbf{B}^{-1}\|^m |\det(\mathbf{B})|^{1/p} |\hat{v}|_{W_p^m(\hat{T})}, \tag{2.33}$$

where $\|\cdot\|$ and $\det(\cdot)$ are the l^2 norm and the determinant of the matrix, respectively. In addition, since T is shape regular, by the scaling argument, we have

$$\|\mathbf{B}\| \leq Ch_T, \quad \|\mathbf{B}^{-1}\| \leq Ch_T^{-1}, \quad \text{and} \quad |\det(\mathbf{B})| = \frac{|T|}{|\hat{T}|} \leq Ch_T^d,$$

where $|T|$ (resp. $|\hat{T}|$) is the measure of T (resp. $\hat{T}$).

Let $\hat{\mathcal{I}} : C^0(\hat{T}) \to P_m(\hat{T})$ be the nodal interpolation operator on the reference element $\hat{T}$. Thus, $\widehat{\mathcal{I}_T v}|_{\hat{T}} = \hat{\mathcal{I}}\hat{v}|_{\hat{T}}$. Let σ be the operator norm of $\hat{\mathcal{I}} : C^0(\hat{T}) \to W_p^{m+1}(\hat{T})$. Then on the reference element $\hat{T}$, there is a polynomial $p \in P_m(\hat{T})$ such that

$$\begin{aligned}
\|\hat{v} - \hat{\mathcal{I}}\hat{v}\|_{W_p^{m+1}(\hat{T})} &\leq \|\hat{v} - p\|_{W_p^{m+1}(\hat{T})} + \|p - \hat{\mathcal{I}}\hat{v}\|_{W_p^{m+1}(\hat{T})} \\
&= \|\hat{v} - p\|_{W_p^{m+1}(\hat{T})} + \|\hat{\mathcal{I}}(p - \hat{v})\|_{W_p^{m+1}(\hat{T})} \\
&\leq \|\hat{v} - p\|_{W_p^{m+1}(\hat{T})} + \sigma\|p - \hat{v}\|_{C^0(\hat{T})} \\
&\leq (1 + C\sigma)\|\hat{v} - p\|_{W_p^{m+1}(\hat{T})} \leq C|\hat{v}|_{W_p^{m+1}(\hat{T})}.
\end{aligned} \tag{2.34}$$

Here we used the triangle inequality, the fact $p - \hat{\mathcal{I}}\hat{v} = \hat{\mathcal{I}}(p - \hat{v})$ because $p \in P_m(\hat{T})$, the Sobolev embedding (Theorem 2.8), and Lemma 2.3.

Then using the scaling argument based on (2.32), (2.33), and (2.34), we have

$$\begin{aligned}
\|v - \mathcal{I}_T v\|_{W_p^i(T)} &\leq Ch_T^{d/p-i}\|\hat{v} - \hat{\mathcal{I}}\hat{v}\|_{W_p^i(\hat{T})} \leq Ch_T^{d/p-i}|\hat{v}|_{W_p^{m+1}(\hat{T})} \\
&\leq Ch_T^{d/p-i} h_T^{m+1-d/p}|v|_{W_p^{m+1}(T)} = Ch_T^{m+1-i}|v|_{W_p^{m+1}(T)},
\end{aligned}$$

where $0 \leq i \leq m+1$. This completes the proof. □

As a direct consequence of (2.31) and Lemma 2.5, we have obtained the global H^1 finite element error analysis for the elliptic equation.

Theorem 2.20 *Recall the assumptions in Remark 2.3. Suppose $u \in H^{m+1}(\Omega)$. Then the finite element solution u_h satisfies*

$$\|u - u_h\|_{H^1(\Omega)} \leq Ch^m \|u\|_{H^{m+1}(\Omega)},$$

where $h = \max_{T \in \mathcal{T}}\{\mathrm{diam}(T)\}$ is the mesh size.

Remark 2.4 Theorem 2.20 contains the error analysis for the finite element solution in the H^1-norm. This is the optimal convergence rate of the finite element method when the solution has the desired regularity $u \in H^{m+1}(\Omega)$. However, as mentioned in the regularity theorems, besides the given data f in the elliptic equation, the regularity of the solution depends on several other factors, including the smoothness of the boundary, the boundary conditions, and the regularity of the coefficients in the differential operator. In the case where the solution u only has limited regularity, say $u \in H^{s+1}(\Omega)$ for $0 < s < m$, we can use operator interpolation theory [27, 35] to obtain a reduced convergence rate for the numerical approximation u_h:

$$\|u - u_h\|_{H^1(\Omega)} \leq Ch^s \|u\|_{H^{s+1}(\Omega)}. \tag{2.35}$$

Starting in the next chapter, we shall shift our attention to analyzing singular solutions of elliptic boundary value problems and developing special finite element algorithms to approximate different singular solutions in 2D and 3D domains.

In addition to the analysis in the H^1-norm, it is possible to evaluate the error $u - u_h$ in other norms. We conclude this chapter by deriving the negative-norm error estimate using a duality argument. Recall that for $s \geq 0$, the H^{-s}-norm is defined by

$$\|v\|_{H^{-s}(\Omega)} = \sup_{0 \neq \phi \in \mathcal{D}(\Omega)} \frac{(v, \phi)}{\|\phi\|_{H^s(\Omega)}}. \tag{2.36}$$

Corollary 2.7 *Given the conditions in Theorem 2.20, recall that $u \in H^1_D(\Omega)$ is the solution of $a(u, v) = (f, v)$ for any $v \in H^1_D(\Omega)$. Let $w \in H^1_D(\Omega)$ be the solution of the adjoint problem $a(v, w) = (v, \phi)$ for any $v \in H^1_D(\Omega)$. For $s \geq 0$, suppose the regularity estimate*

$$\|w\|_{H^{s+2}(\Omega)} \leq C\|\phi\|_{H^s(\Omega)},$$

and the approximation property

$$\|w - w_h\|_{H^1(\Omega)} \leq Ch^{s+1}\|w\|_{H^{s+2}(\Omega)}$$

hold, where $w_h \in S_h$ is the finite element solution of $a(v, w_h) = (v, \phi)$ for any $v \in S_h$ and $h = \max_{T \in \mathcal{T}}\{\mathrm{diam}(T)\}$ is the mesh size. Then we have

$$\|u - u_h\|_{H^{-s}(\Omega)} \leq Ch^{s+1}\|u - u_h\|_{H^1(\Omega)}.$$

Proof For any $\phi \in \mathcal{D}(\Omega)$, by the definition of w and by the orthogonality condition (2.30), we have

$$\begin{aligned}(u - u_h, \phi) &= a(u - u_h, w) = a(u - u_h, w - w_h)\\ &\leq C\|u - u_h\|_{H^1(\Omega)}\|w - w_h\|_{H^1(\Omega)}\\ &\leq Ch^{s+1}\|u - u_h\|_{H^1(\Omega)}\|w\|_{H^{s+2}(\Omega)},\end{aligned} \tag{2.37}$$

where the last two inequalities are due to the continuity of $a(\cdot,\cdot)$ and to the approximation property of w_h. Then by the definition of the negative norm (2.36), the estimate (2.37), and the regularity estimate of w, we obtain

$$\begin{aligned}\|u - u_h\|_{H^{-s}(\Omega)} &= \sup_{0 \neq \phi \in \mathcal{D}(\Omega)} \frac{(u - u_h, \phi)}{\|\phi\|_{H^s(\Omega)}}\\ &\leq Ch^{s+1}\|u - u_h\|_{H^1(\Omega)}\frac{\|w\|_{H^{s+2}(\Omega)}}{\|\phi\|_{H^s(\Omega)}} \leq Ch^{s+1}\|u - u_h\|_{H^1(\Omega)}. \qquad \square\end{aligned}$$

Chapter 3
Singularities and Graded Mesh Algorithms

Consider the Poisson equation with mixed boundary conditions in a polygonal or a polyhedral domain as the model problem. We first describe the singular solution of elliptic equations that are due to the nonsmoothness of the domain and to the change of boundary conditions. Then we present principles that lead to graded mesh algorithms for effective finite element methods approximating these singular solutions. These graded mesh algorithms are simple, explicit, and applicable to general polytopal domains in $\mathbb{R}^d$, $d = 1, 2, 3$. The resulting meshes are conforming but can also be unconventional in the sense that they may be anisotropic and do not satisfy the maximum angle condition. This chapter is suitable for readers who look for an overview of corner singularities (2D) and vertex and edge singularities (3D), and who are interested in their effect on numerical approximations and ideas to improve the accuracy of the numerical solution.

3.1 A Numerical Example

We present a numerical example to demonstrate that the performance of the finite element method can be affected by the domain geometry and by the boundary conditions. Let $\Omega \subset \mathbb{R}^2$ be a polygonal domain. Let Γ_D and Γ_N be open subsets of the boundary $\Gamma = \partial\Omega$. Consider the elliptic equation

$$-\Delta u = 1 \quad \text{in} \quad \Omega \tag{3.1}$$

in the following three situations:

Case 1 $\Omega = (0, 1)^2$ with the Dirichlet boundary condition $u|_\Gamma = 0$;
Case 2 $\Omega = (0, 1)^2 \setminus (0, 0.5]^2$ with the Dirichlet boundary condition $u|_\Gamma = 0$;
Case 3 $\Omega = (0, 1)^2$ with the mixed boundary condition $u|_{\Gamma_D} = 0, \partial_n u|_{\Gamma_N} = 0$.

We use the linear Lagrange finite element method to solve these equations. See Fig. 3.1 for illustrations of the domain, the boundary condition, and the triangulation.

H. Li, *Graded Finite Element Methods for Elliptic Problems in Nonsmooth Domains*
Surveys and Tutorials in the Applied Mathematical Sciences 10,
https://doi.org/10.1007/978-3-031-05821-9_3

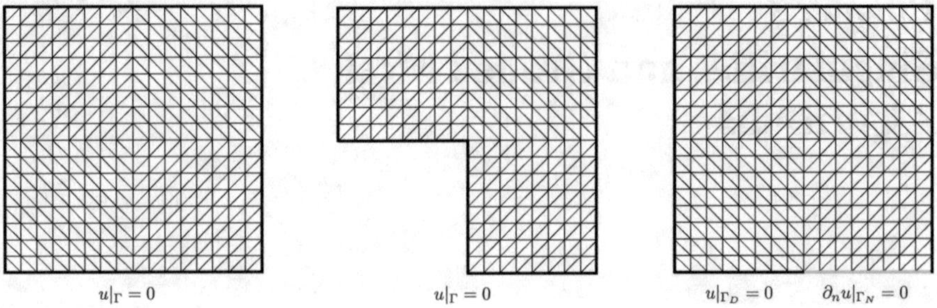

Fig. 3.1 Domains for (3.1). Dirichlet boundary (black); Neumann boundary (green).

Given a triangulation of the domain $\mathcal{T}_j$, we obtain a finer mesh $\mathcal{T}_{j+1}$ by the uniform midpoint refinement (see Fig. 1.6). Let $h_j = \max_{T \in \mathcal{T}_j}\{\mathrm{diam}(T)\}$ be the mesh size of $\mathcal{T}_j$. Starting with an initial triangulation $\mathcal{T}_0$, we obtain a sequence of nested triangulations $\{\mathcal{T}_j\}_{j\geq 0}$ by recursive refinements. Since the exact solution is generally not known, we compute the numerical convergence rate between successive mesh refinements by

$$\mathcal{R} = \log_2\left(\frac{|u_j - u_{j-1}|_{H^1(\Omega)}}{|u_{j+1} - u_j|_{H^1(\Omega)}}\right), \tag{3.2}$$

where u_j is the finite element solution associated with the mesh $\mathcal{T}_j$. As j increases, since $h_j = 2h_{j+1}$, the asymptotic rate in (3.2) is a reasonable indicator of s in the actual convergence $\|u - u_h\|_{H^1(\Omega)} \leq Ch^s\|u\|_{H^{1+s}(\Omega)}$ for the finite element method (see (2.35)). For linear elements ($m = 1$), the optimal convergence rate is $\mathcal{R} = 1$ (Theorem 2.20). We display the numerical convergence rates (3.2) for different cases

$j\backslash\mathcal{R}$	Case 1	Case 2	Case 3
6	1.00	0.80	0.60
7	1.00	0.77	0.56
8	1.00	0.74	0.53
9	1.00	0.72	0.52
10	1.00	0.70	0.51

Table 3.1 Convergence history of the finite element method for (3.1) in different cases.

of (3.1) in Table 3.1. One observes that the finite element method is optimal ($\mathcal{R} = 1$) solving Case 1, but far less than optimal ($\mathcal{R} < 1$) for Case 2 and Case 3. In fact, one has $u \in H^{1+s}(\Omega)$ for $s < \frac{2}{3}$ when Ω is the L-shaped domain, and $u \in H^{1+s}(\Omega)$ for $s < \frac{1}{2}$ when the mixed boundary condition is imposed. The numerical convergence rates in these two cases become closer and closer to the theoretical rates as the mesh size $h_j \to 0$. Nevertheless, these results show that the convergence rate of the finite element solution can be greatly reduced by the domain geometry and by the change of boundary conditions in elliptic problems. See Fig. 3.2 for finite element solutions of (3.1) in different cases. In problems associated with more complex differential operators [45, 47, 95, 99, 116], the finite element method may even cause false approximations.

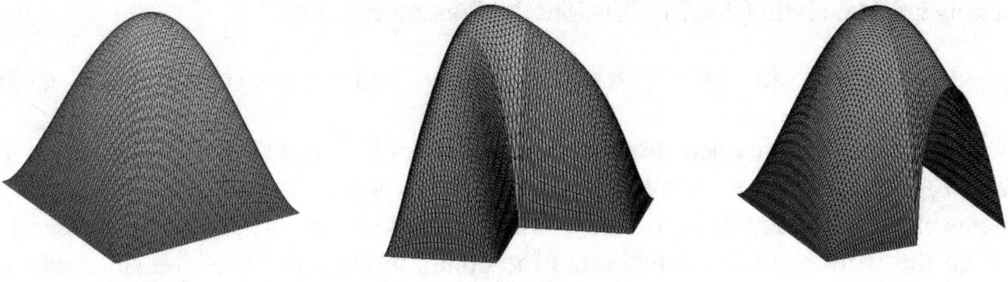

Fig. 3.2 Numerical solutions for (3.1): Case 1–Case 3.

We now take a pause to explain why the numerical convergence rate in (3.2) is a reasonable indicator for the actual convergence rate solving (3.1). In any of the three cases of (3.1) above, recall the space $H^1_D(\Omega)$ in (2.26). For $u, v \in H^1_D(\Omega)$, let

$$a(u, v) := \int_\Omega \nabla u \cdot \nabla v \, dx.$$

Assume $|u - u_j|_{H^1(\Omega)} = O(h_j^s) = C2^{-sj}$, where $0 < s \leq 1$ is the convergence rate and $C > 0$ is independent of j. Recall the Galerkin orthogonality

$$a(u, u_j) = a(u_j, u_j) \quad \text{and} \quad a(u - u_{j-1}, u_{j-1}) = a(u_j - u_{j-1}, u_{j-1}) = 0.$$

Then

$$\begin{aligned} |u - u_j|^2_{H^1(\Omega)} &= a(u - u_j, u - u_j) = a(u - u_j, u) \\ &= a(u, u) - a(u_j, u_j) = |u|^2_{H^1(\Omega)} - |u_j|^2_{H^1(\Omega)}, \end{aligned}$$

and

$$\begin{aligned} |u_j - u_{j-1}|^2_{H^1(\Omega)} &= a(u_j - u_{j-1}, u_j - u_{j-1}) = a(u_j - u_{j-1}, u_j) \\ &= a(u_j, u_j) - a(u_{j-1}, u_{j-1}) = |u_j|^2_{H^1(\Omega)} - |u_{j-1}|^2_{H^1(\Omega)}. \end{aligned}$$

Therefore, we have

$$\begin{aligned} |u_j - u_{j-1}|^2_{H^1(\Omega)} &= |u_j|^2_{H^1(\Omega)} - |u_{j-1}|^2_{H^1(\Omega)} \\ &= |u|^2_{H^1(\Omega)} - |u_{j-1}|^2_{H^1(\Omega)} - \left(|u|^2_{H^1(\Omega)} - |u_j|^2_{H^1(\Omega)}\right) \\ &= |u - u_{j-1}|^2_{H^1(\Omega)} - |u - u_j|^2_{H^1(\Omega)} \\ &= C^2 2^{2s(1-j)} - C^2 2^{-2sj} = \frac{3C^2}{4} 2^{2s(1-j)}. \end{aligned}$$

This leads to $|u_j - u_{j-1}|_{H^1(\Omega)} = \frac{\sqrt{3}C}{2} 2^{s(1-j)}$. Hence,

$$\mathcal{R} = \log_2 \left(\frac{|u_j - u_{j-1}|_{H^1(\Omega)}}{|u_{j+1} - u_j|_{H^1(\Omega)}} \right) \to s \quad \text{as } j \text{ increases.}$$

In the rest of this chapter, we let $\Omega \subset \mathbb{R}^d$ be either a polygonal domain ($d = 2$) or a polyhedral domain ($d = 3$). Consider the Poisson equation

$$-\Delta u = f \quad \text{in} \quad \Omega, \qquad u = 0 \quad \text{on} \quad \Gamma_D \quad \text{and} \quad \partial_n u = 0 \quad \text{on} \quad \Gamma_N, \tag{3.3}$$

where Γ_D and Γ_N are open subsets of the boundary $\Gamma := \partial\Omega$ such that $\overline{\Gamma_D} \cup \overline{\Gamma_N} = \Gamma$ and Γ_D has positive measure. For $d = 2$, we suppose on each side of Ω, every connected component in Γ_D or in Γ_N is a line segment; and the vertex set C consists of all the vertices of the domain and the points in the interior of the sides where the boundary condition changes. For $d = 3$, we suppose on each face of Ω, every connected component in Γ_D or in Γ_N is a polygon; the edge set $\mathcal{E}$ consists of all the open edges of the domain and the open sides of the polygons in Γ_D where the boundary condition changes; the vertex set C consists of all the points where the edges in $\mathcal{E}$ meet, namely the vertices of the domain and the vertices of the polygons in Γ_D. For example, when Ω is a polygon, the boundary condition can change type at an interior point of a boundary side. This interior point is considered as a vertex in C with an opening angle π. For the pure Dirichlet problem $\overline{\Gamma_D} = \Gamma$, C consists of all the vertices of the domain ($d = 2, 3$) and $\mathcal{E}$ consists of all the edges of the domain ($d = 3$). In addition, define the set $\mathcal{S}$ of "nonsmooth" points on Γ, such that $\mathcal{S} = C$ when $d = 2$ and $\mathcal{S} = C \cup \mathcal{E}$ when $d = 3$. Then the weak solution $u \in H^1_D(\Omega)$ of (3.3) satisfies

$$a(u, v) = \int_\Omega \nabla u \cdot \nabla v \, dx = \int_\Omega f v \, dx = (f, v) \qquad \forall\, v \in H^1_D(\Omega).$$

For any $f \in L^2(\Omega)$, $u \in H^1_D(\Omega)$ is uniquely determined. However, the points in $\mathcal{S}$ can give rise to singularities in the solution even when the function f is smooth. Therefore, u may not have the global high-order regularity as indicated in Theorem 2.16. For instance, the solution of (3.1) belongs to $H^2(\Omega)$ in Case 1 but has singularities associated with the reentrant corner and with the change of boundary conditions in Case 2 and Case 3, respectively. Consequently, the finite element convergence (Table 3.1) deteriorates due to the presence of the singular solutions in the last two cases.

Regularity is a local property. Using a smooth cut-off function, one can show that the possible singularity affects only the neighborhood of $\mathcal{S}$ [61].

Theorem 3.1 *Let $u \in H^1_D(\Omega)$ be the weak solution of (3.3) with $f \in L^2(\Omega)$. Then $\chi u \in H^2(\Omega)$ for any $\chi \in \mathcal{D}(\bar{\Omega})$ whose support intersects Γ only at points that do not belong to $\mathcal{S}$.*

The study of singular solutions of elliptic boundary value problems has received significant attention in recent decades. See the works [4, 6, 19, 24, 31, 33, 39, 49, 61, 71, 73, 101] and references therein. Starting from the next section, we shall summarize the key results that are particularly important for finite element approximations.

3.2 Singularities in Polygonal Domains

We focus on the 2D case: the Poisson equation (3.3) in a polygonal domain. We shall give a detailed description of the singular solutions near the vertex set $\mathcal{S} = C$ and study their properties. Many results in this section can also be found in the works [49, 61, 71]. We first recall a useful identity for functions in H^2 [65].

Theorem 3.2 *Let Ω be a bounded polygonal domain. For any $v \in H^2(\Omega)$ such that $v|_{\Gamma_D} = 0$ and $\partial_n v|_{\Gamma_N} = 0$, it holds that*

$$\|\Delta v\|^2_{L^2(\Omega)} = \|\partial_x^2 v\|^2_{L^2(\Omega)} + \|\partial_y^2 v\|^2_{L^2(\Omega)} + 2\|\partial_x \partial_y v\|^2_{L^2(\Omega)}.$$

Suppose that v satisfies the conditions in Theorem 3.2 and in addition $\Gamma_D \neq \emptyset$. Note that by integration by parts, one has

$$\int_\Omega \nabla v \cdot \nabla v \, dx = -\int_\Omega v \Delta v \, dx \leq \|v\|_{L^2(\Omega)} \|\Delta v\|_{L^2(\Omega)}.$$

Then by the Poincaré inequality,

$$\|v\|_{H^1(\Omega)} \leq C\|\nabla v\|_{L^2(\Omega)} \leq C\|\Delta v\|_{L^2(\Omega)}.$$

Using the identity in Theorem 3.2, one can further show a Poincaré type inequality for v:

$$\|v\|_{H^2(\Omega)} \leq C\|\Delta v\|_{L^2(\Omega)}.$$

Remark 3.1 According to Theorem 2.16, the Laplace operator

$$\Delta : H^2(\Omega) \cap H^1_0(\Omega) \to L^2(\Omega)$$

defines a bijective mapping provided that the boundary is sufficiently smooth. However, given a polygonal domain and the mixed boundary condition, the mapping

$$\Delta : M := H^2(\Omega) \cap \{v : v|_{\Gamma_D} = 0, \ \partial_n v|_{\Gamma_N} = 0\} \to L^2(\Omega) \tag{3.4}$$

in general is only injective and has a closed range in $L^2(\Omega)$. Denote by $\mathfrak{N}$ this range and by $\mathfrak{N}^\perp$ its orthogonal in $L^2(\Omega)$ ($L^2(\Omega) = \mathfrak{N} \oplus \mathfrak{N}^\perp$). Write $f = f_1 + f_2$ such that $f_1 \in \mathfrak{N}$ and $f_2 \in \mathfrak{N}^\perp$. Then in (3.3), the solution $u \in H^2(\Omega)$ only when $f \in \mathfrak{N}$; and u will have non-H^2 components whenever $f_2 \neq 0$. These non-H^2 components are the *singularities* in the solution. Let $\mathbb{S}$ be the collection of these singularities. Note that the singularities in $\mathbb{S}$ are still in $H^1(\Omega)$ since $u \in H^1(\Omega)$. Consequently, the Laplace operator defines a bijection in the following sense:

$$\Delta : M \cup \mathbb{S} \to L^2(\Omega). \tag{3.5}$$

Hence, it is critical to characterize the space $\mathfrak{N}^\perp$ and the space $\mathbb{S}$. Fortunately, as one shall see, these two spaces have the same dimension and are intrinsically connected.

For any $v \in \mathfrak{N}^\perp$, based on the definition, one has

$$\int_\Omega v\Delta w \, dx = 0 \qquad \forall\, w \in M.$$

Since this equation holds for all $w \in \mathcal{D}(\Omega)$, we conclude that v is harmonic. In fact, using Green's formula (Corollary 2.5) and the boundary condition of w, one sees that v is the solution of the following equation

$$-\Delta v = 0 \quad \text{in} \quad \Omega, \qquad v = 0 \quad \text{on} \quad \Gamma_D \quad \text{and} \quad \partial_n v = 0 \quad \text{on} \quad \Gamma_N. \tag{3.6}$$

Note that if $v \in H^1(\Omega)$, (3.6) leads to $v = 0$. However, since we are looking for L^2 solutions of (3.6), v is not necessarily zero and may not be uniquely determined. Nevertheless, one can show that v is smooth in $\bar{\Omega} \backslash N_S$, where N_S is any neighborhood of the set $\mathcal{S} = C$. The possible singular component of v concentrates at the vertices of the domain, on which we shall elaborate below.

Use the same notation Γ_j and Γ_{j+1} as in Fig. 2.1 to denote the adjacent "sides" that meet at $c_j \in \mathcal{S}$ and use ω_j to denote the angle they form. Recall that c_j can be a vertex of the domain or an interior point on the side of the domain. In the latter case, $\omega_j = \pi$. Let (r_j, θ_j) be the polar coordinates centered at c_j such that $\theta_j = 0$ on Γ_{j+1} and $\theta_j = \omega_j$ on Γ_j. For a fixed $\delta > 0$, denote by

$$N_{\delta,j} = \{(r_j, \theta_j) : \; 0 < r_j < \delta, \; 0 < \theta_j < \omega_j\} \tag{3.7}$$

the small neighborhood of c_j such that $N_{\delta,j} \cap N_{\delta,k} = \emptyset$ if $j \neq k$. We will omit the index j in r_j, θ_j, and $N_{\delta,j}$ when the discussion is clearly in $N_{\delta,j}$. To see the impact of c_j on v, we change the variables $(x, y) \to (r, \theta)$ in N_δ. Then in N_δ, (3.6) becomes

$$(\partial_r^2 + r^{-1}\partial_r + r^{-2}\partial_\theta^2)v = 0 \tag{3.8}$$

with boundary conditions on Γ_j and Γ_{j+1} inherited from (3.6):

$$\begin{cases} \text{at } \theta = 0, \; v = 0 \text{ if } \Gamma_{j+1} \subset \Gamma_D \text{ and } \partial_\theta v = 0 \text{ if } \Gamma_{j+1} \subset \Gamma_N; \\ \text{at } \theta = \omega_j, \; v = 0 \text{ if } \Gamma_j \subset \Gamma_D \text{ and } \partial_\theta v = 0 \text{ if } \Gamma_j \subset \Gamma_N. \end{cases} \tag{3.9}$$

We study (3.8) based on separation of variables and the spectrum property of ∂_θ^2. It is worth noting that by setting $t = \ln r$, (3.8) can be written as $(\partial_t^2 + \partial_\theta^2)v = 0$ for $t < \ln \delta$. The aforementioned series of transformations $(x, y) \to (r, \theta) \to (t, \theta)$ shares the same spirit as the Mellin transform [49, 71–73, 93] in analyzing corner singularities.

First, for $k \geq 1$, the eigenvalues $\lambda_{j,k}^2$ and normalized eigenfunctions $\phi_{j,k}(\theta)$ of $-\partial_\theta^2$ in $(0, \omega_j)$ are defined by

$$-\partial_\theta^2 \phi_{j,k} = \lambda_{j,k}^2 \phi_{j,k} \tag{3.10}$$

with appropriate boundary conditions in (3.9). Assume $\lambda_{j,k} \geq 0$ (and thus $-\lambda_{j,k} \leq 0$) and assume the eigenvalues are in increasing order with respect to k (i.e., $\lambda_{j,k+1}^2 \geq \lambda_{j,k}^2$). Below we drop the index j in $\lambda_{j,k}$ and $\phi_{j,k}$ to simplify the exposition. In particular, λ_k and ϕ_k satisfy

$$\begin{cases} \lambda_k = \frac{k\pi}{\omega_j} \text{ and } \phi_k = c\sin(\lambda_k\theta) \text{ if } \Gamma_j, \Gamma_{j+1} \subset \Gamma_D, \\ \lambda_k = \frac{(k-1/2)\pi}{\omega_j} \text{ and } \phi_k = c\sin(\lambda_k\theta) \text{ if } \Gamma_j \subset \Gamma_N \text{ and } \Gamma_{j+1} \subset \Gamma_D, \\ \lambda_k = \frac{(k-1/2)\pi}{\omega_j} \text{ and } \phi_k = c\sin\left(\lambda_k(\omega_j - \theta)\right) \text{ if } \Gamma_j \subset \Gamma_D \text{ and } \Gamma_{j+1} \subset \Gamma_N, \\ \lambda_k = \frac{(k-1)\pi}{\omega_j},\ \phi_1 = \frac{c}{\sqrt{2}}, \text{ and } \phi_k = c\cos(\lambda_k\theta)\ (k \geq 2) \text{ if } \Gamma_j, \Gamma_{j+1} \subset \Gamma_N, \end{cases} \tag{3.11}$$

where $c = \sqrt{2/\omega_j}$. Recall v in (3.6) and $v \in L^2(N_\delta)$. Then by separation of variables using θ and the auxiliary variable t, one derives [65] that in N_δ,

$$v = \sum_{k\geq 1} \alpha_k r^{\lambda_k} \phi_k + \sum_{0<\lambda_k<1} \beta_k r^{-\lambda_k} \phi_k, \tag{3.12}$$

where α_k, β_k are constants depending on v and N_δ. We observe that each function in the form $r^{\pm\lambda_k}\phi_k$ in (3.12) is harmonic and satisfies the corresponding boundary conditions. This implies that near c_j, the function $v \in \mathfrak{N}^\perp$ is the sum of an H^1 function and an L^2 function that involves $r^{-\lambda_k}\phi_k$ for $0 < \lambda_k < 1$. In addition, the local behavior (3.12) of v can be used to estimate the dimension of the space $\mathfrak{N}^\perp$, for which the term $r^{-\lambda_k}\phi_k$ plays an important role [65].

Theorem 3.3 *For each $c_j \in \mathcal{S}$, let d_j be the number of the parameters $\lambda_{j,k}$ defined in (3.11) such that $0 < \lambda_{j,k} < 1$. Then the dimension of $\mathfrak{N}^\perp$ satisfies*

$$d_{\mathfrak{N}^\perp} = \dim(\mathfrak{N}^\perp) = \sum_j d_j.$$

Namely, the dimension of $\mathfrak{N}^\perp$ is the sum of the contributions from all the corners as follows. (I) The contribution of a "Dirichlet vertex" c_j ($\Gamma_j, \Gamma_{j+1} \subset \Gamma_D$) is 0 if $\omega_j \leq \pi$ and is 1 if $\omega_j > \pi$. (II) The contribution of a "Neumann vertex" c_j ($\Gamma_j, \Gamma_{j+1} \subset \Gamma_N$) is 0 if $\omega_j \leq \pi$ and is 1 if $\omega_j > \pi$. (III) The contribution of a "mixed vertex" c_j ($\Gamma_j \subset \Gamma_D$, $\Gamma_{j+1} \subset \Gamma_N$ or vice versa) is 0 if $\omega_j \leq \pi/2$, is 1 if $\pi/2 < \omega_j \leq 3\pi/2$, and is 2 if $\omega_j > 3\pi/2$.

Remark 3.2 Let $\chi_j \in \mathcal{D}(\bar{\Omega})$ be a cut-off function such that $\chi_j = 1$ in the neighborhood of c_j and vanishes near all the other sides except for Γ_j and Γ_{j+1}. In addition, we require that the supports of χ_i and χ_j do not intersect if $i \neq j$. Then one can show that [65] for each c_j and $0 < \lambda_{j,k} < 1$, there exists $v_{j,k} \in \mathfrak{N}^\perp$ such that

$$\sigma_{j,k} := v_{j,k} - \chi_j r_j^{-\lambda_{j,k}} \phi_{j,k} \in H^1(\Omega).$$

In fact, let $\sigma_{j,k} \in H^1(\Omega)$ be the function that satisfies the following equation with the boundary condition in (3.6):

$$-\Delta\sigma_{j,k} = \Delta(\chi_j r_j^{-\lambda_{j,k}} \phi_{j,k}) \quad \text{in} \quad \Omega.$$

Note that $\Delta(\chi_j r_j^{-\lambda_{j,k}} \phi_{j,k}) \in L^2(\Omega)$ and therefore, $\sigma_{j,k} \in H^1(\Omega)$ is uniquely defined. Then $v_{j,k} = \sigma_{j,k} + \chi_j r_j^{-\lambda_{j,k}} \phi_{j,k}$ satisfies (3.6) and they are linearly independent. Thus, based on Theorem 3.3, $\{v_{j,k}\}_{1\leq j\leq N}$ with $0 < \lambda_{j,k} < 1$ is a basis of $\mathfrak{N}^\perp$.

It is worth pointing out that the support of the basis functions $v_{j,k}$ can be the entire domain. Therefore, it involves restrictive global conditions to determine whether a function f belongs to $\mathfrak{N}$. When $\mathfrak{N}^\perp \neq \emptyset$ for (3.3), it turns out f often has non-zero components in $\mathfrak{N}^\perp$, which implies $u \notin H^2(\Omega)$. We now identify the space $\mathbb{S}$ of solution singularities with which the isomorphism (3.5) holds.

Theorem 3.4 *Using the notation in Remark 3.2, we have*

$$\mathbb{S} = \{\chi_j r_j^{\lambda_{j,k}} \phi_{j,k} : 1 \leq j \leq N,\ 0 < \lambda_{j,k} < 1\}.$$

Namely, for $f \in L^2(\Omega)$, the solution of (3.3) can be written as $u = u_R + u_S$ such that the regular component $u_R \in H^2(\Omega)$ and the singular component $u_S \in \mathbb{S}$.

Proof When $\mathbb{S} = \emptyset$, we have $\mathfrak{N}^\perp = \emptyset$ and the theorem is a direct consequence of the discussions above.

We now assume $\mathbb{S} \neq \emptyset$. For each vertex c_j, let

$$S_{j,k} := \chi_j r_j^{\lambda_{j,k}} \phi_{j,k}, \quad 0 < \lambda_{j,k} < 1.$$

Then $S_{j,k} \in H^1(\Omega)$ and satisfies the boundary condition in (3.3), but $S_{j,k} \notin H^2(\Omega)$. Write $S_{j,k} = S^0_{j,k} + S^1_{j,k}$ such that $\Delta S^0_{j,k} \in \mathfrak{N}$ and $\Delta S^1_{j,k} \in \mathfrak{N}^\perp$. Then $\Delta S^1_{j,k} \neq 0$ because otherwise we have $\Delta S_{j,k} = \Delta S^0_{j,k} \in \mathfrak{N}$ and therefore, obtain the contradiction $S_{j,k} \in H^2(\Omega)$. In addition, the functions $\Delta S^1_{j,k}$ with $1 \leq j \leq N$ and $0 < \lambda_{j,k} < 1$ are linearly independent. This can be seen by assuming linear dependence: there are constants $a_{j,k}$, not all zero, such that $\sum_{j,k} a_{j,k}\Delta S^1_{j,k} = 0$. Thus,

$$\sum_{j,k} a_{j,k}\Delta S_{j,k} = \sum_{j,k} a_{j,k}\Delta S^0_{j,k} \in \mathfrak{N}.$$

This means $\sum_{j,k} a_{j,k} S_{j,k} \in H^2(\Omega)$. On the other hand, the functions $S_{j,k}$ are locally supported and linearly independent. Therefore, $\sum_{j,k} a_{j,k} S_{j,k} \notin H^2(\Omega)$ if $a_{j,k}$ are not all zero. Hence $\Delta S^1_{j,k}$ are linearly independent by contradiction. Since the spaces $\mathbb{S}$ and $\mathfrak{N}^\perp$ have the same dimension, the number of linearly independent functions $\Delta S^1_{j,k}$ is the same as $\dim(\mathfrak{N}^\perp)$. Therefore, the functions $\Delta S^1_{j,k}$ can be a set of basis functions in $\mathfrak{N}^\perp$, and $L^2(\Omega) = \mathfrak{N} \cup \text{span}\{\Delta S_{j,k}\}$. Thus, we conclude that the mapping (3.5) is a bijection. □

Example 3.1 Recall the different types of vertices (Dirichlet, Neumann, and mixed) in Theorem 3.3. Then for (3.3), based on (3.11) and Theorem 3.4, we have the following explicit expressions of the solution singularities near each vertex c_j.

$$u = u_{R,j} + \alpha_{j,1} r_j^{\pi/\omega_j} \sin(\frac{\pi}{\omega_j}\theta_j) \qquad \text{if } c_j \text{ is a Dirichlet vertex and } \omega_j > \pi,$$

$$u = u_{R,j} + \alpha_{j,1} r_j^{\pi/\omega_j} \cos(\frac{\pi}{\omega_j}\theta_j) \qquad \text{if } c_j \text{ is a Neumann vertex and } \omega_j > \pi.$$

For a mixed vertex c_j, we have

$$u = u_{R,j} + \alpha_{j,1} r_j^{\pi/(2\omega_j)} \phi_{j,1} \qquad \text{if } \frac{\pi}{2} < \omega_j \le \frac{3\pi}{2},$$

$$u = u_{R,j} + \alpha_{j,1} r_j^{\pi/(2\omega_j)} \phi_{j,1} + \alpha_{j,2} r_j^{3\pi/(2\omega_j)} \phi_{j,2} \qquad \text{if } \omega_j > \frac{3\pi}{2}.$$

Here $u_{R,j}$ is a function in H^2, $\alpha_{j,1}, \alpha_{j,2} \in \mathbb{R}$, and $\phi_{j,k} \in C^\infty[0, \omega_j]$ is the eigenfunction in (3.11). Recall the small neighborhood $N_{\delta,j}$ of c_j. Then the solution $u \in H^{1+s}(N_{\delta,j})$ for any $s \le 1$ and $s < \min_k\{\lambda_{j,k} > 0\}$. For example, assume the Dirichlet boundary condition $\overline{\Gamma_D} = \Gamma$ and suppose the maximum opening angle $\omega_{max} = \max_j\{\omega_j\} > \pi$. Then we have $u \in H^{1+s}(\Omega)$ for $s < \pi/\omega_{max}$.

Remark 3.3 According to Theorem 3.4, given $f \in L^2(\Omega)$ in (3.3), there exist constants $\alpha_{j,k}$ such that

$$u - \sum_{1\le j\le N} \sum_{0<\lambda_{j,k}<1} \alpha_{j,k} \chi_j r_j^{\lambda_{j,k}} \phi_{j,k} \in H^2(\Omega). \tag{3.13}$$

Similar formulas hold for the solution of 2D elasticity problems [62, 64, 65, 78, 111], where $\alpha_{j,k}$'s are known as *stress intensity factors*. It is also possible to obtain similar results in high-order Sobolev spaces by adding additional singular terms in the expansion [61]. In particular, assume the Dirichlet boundary condition $\overline{\Gamma_D} = \Gamma$ in (3.3). Define

$$S_{j,k} := \chi_j r_j^{k\pi/\omega_j} \sin(\frac{k\pi\theta_j}{\omega_j}) \qquad \text{if } \lambda_{j,k} = \frac{k\pi}{\omega_j} \notin \mathbb{N}, \tag{3.14}$$

$$S_{j,k} := \chi_j r_j^{k\pi/\omega_j} \left(\ln r_j \sin(\frac{k\pi\theta_j}{\omega_j}) + \theta_j \cos(\frac{k\pi\theta_j}{\omega_j}) \right) \text{ if } \lambda_{j,k} = \frac{k\pi}{\omega_j} \in \mathbb{N}. \tag{3.15}$$

Then for $f \in H^s(\Omega)$, $s \ge -1$, there exist constants $\alpha_{j,k}$ such that

$$u - \sum_j \sum_{0<\lambda_{j,k}<s+1} \alpha_{j,k} S_{j,k} \in H^{s+2}(\Omega),$$

provided that $\lambda_{j,k} \ne s+1$ for any j and k. In addition, if $f \in W_p^m(\Omega)$ for an integer $m \ge 0$, there exist constants $\alpha_{j,k}$ such that

$$u - \sum_j \sum_{0<\lambda_{j,k}<m+2-2/p} \alpha_{j,k} S_{j,k} \in W_p^{m+2}(\Omega),$$

provided that $\lambda_{j,k} \neq m + 2 - 2/p$ for any j and k. The singular function (3.14) satisfies the Dirichlet boundary condition. In the neighborhood of the vertex, the function (3.15) is 0 on $\theta_j = 0$ and is $(-1)^k \omega_j r^{\lambda_{j,k}}$ on $\theta_j = \omega_j$. Nevertheless, both functions in (3.14) and (3.15) are harmonic and coincide with a smooth function on the boundary. For example, suppose $\theta_j = 0$ corresponds to part of $ay = x$ and $\theta_j = \omega_j$ corresponds to $\{y = 0, x > 0\}$. Then on the boundary near the vertex, the function (3.15) coincides with the polynomial $(-1)^k \omega_j x^{\lambda_{j,k}-1}(x - ay)$. We refer the readers to [49, 61, 71, 73] for more detailed studies on 2D corner singularities.

We conclude this section by a formula for the coefficient of the singular functions.

Proposition 3.1 *Assume the Dirichlet boundary condition* $\overline{\Gamma_D} = \Gamma$ *in (3.3) and assume that the vertex* c_j *has an angle* $\omega_j > \pi$. *Let* $S_{j,-} := \chi_j r_j^{-\pi/\omega_j} \sin(\frac{\pi\theta_j}{\omega_j})$. *Then the coefficient* $\alpha_{j,1}$ *in the singular expansion (3.13) is determined by*

$$\alpha_{j,1} = \frac{1}{\pi} \int_\Omega f S_{j,-} + u\Delta S_{j,-} \, dx.$$

Proof Given the pure Dirichlet boundary condition, in the neighborhood of c_j, the singular expansion (3.13) of the solution becomes $u = u_{R,j} + \alpha_{j,1} S_{j,1}$, where $u_{R,j}$ is in H^2 and $S_{j,1}$ is given in (3.14). Let $B(c_j, \delta)$ be the ball centered at the vertex c_j with radius δ. We choose δ such that $\chi_j = 1$ in the neighborhood $N_\delta := \Omega \cap B(c_j, \delta)$. Thus by a direct calculation, we obtain $\Delta S_{j,1} = \Delta S_{j,-} = 0$ in N_δ. Then we have

$$\begin{aligned}
\int_\Omega f S_{j,-} \, dx &= -\int_{N_\delta} \Delta u S_{j,-} \, dx - \int_{\Omega \setminus N_\delta} \Delta u S_{j,-} \, dx \\
&= -\int_{N_\delta} \Delta u_{R,j} S_{j,-} \, dx - \int_{\Omega \setminus N_\delta} \Delta u S_{j,-} \, dx.
\end{aligned}$$

Using integration by parts and Corollary 2.5, we have

$$\int_\Omega f S_{j,-} + u\Delta S_{j,-} \, dx = \int_0^{\omega_j} \Big((u_{R,j} - u)(\partial_r S_{j,-}) + S_{j,-} \partial_r (u - u_{R,j}) \Big) r \, d\theta|_{r=\delta},$$

where we used the polar coordinates (r, θ) centered at c_j. Therefore,

$$\begin{aligned}
\int_\Omega f S_{j,-} + u\Delta S_{j,-} \, dx &= \alpha_{j,1} \int_0^{\omega_j} \Big(- S_{j,1}(\partial_r S_{j,-}) + S_{j,-} \partial_r S_{j,1} \Big) r \, d\theta|_{r=\delta} \\
&= \alpha_{j,1} \pi.
\end{aligned}$$

The proof is hence complete. □

3.3 Singularities in Polyhedral Domains

In this section, we study the Poisson equation (3.3) where $\Omega \subset \mathbb{R}^3$ is a polyhedral domain and $f \in L^2(\Omega)$. Recall the set of "nonsmooth" points $\mathcal{S} = \mathcal{C} \cup \mathcal{E}$. In the presence of the mixed boundary condition, the vertex set $\mathcal{C}$ and the edge set $\mathcal{E}$ are expanded to also include points where the boundary condition changes.

As discussed in the last section, the 2D corner singularities are isotropic, concentrating in the neighborhood of the vertex of the domain. It involves singular terms $r_j^{\lambda_{j,k}} \phi_{j,k}$ near the vertex c_j, where $\phi_{j,k}$ is a smooth function of θ_j. In a 3D polyhedral domain Ω, it turns out that the solution can have different types of singularities: the vertex singularity and the edge singularity. In particular, the edge singularity is *anisotropic*—singular in directions orthogonal to the edge and smoother in the edge direction.

3.3.1 The 3D Edge Singularity

Let Ω_Q be a bounded 2D polygonal domain and let $Q := \Omega_Q \times \mathbb{R}$ be a 3D domain with only edges. For $f \in L^2(Q)$, consider the Poisson equation

$$-\Delta u = f \quad \text{in} \quad Q. \tag{3.16}$$

We assume on each face of the boundary ∂Q, either the Dirichlet boundary condition $u = 0$ or the Neumann boundary condition $\partial_n u = 0$ is imposed. Denote by $\partial_D Q$ and $\partial_N Q$ the set of Dirichlet faces and the set of Neumann faces, respectively. Suppose $\partial_D Q \neq \emptyset$. Although the domain Q is unbounded, using a localization argument, the solution in (3.16) can give useful insights on the local behavior of the solution in (3.3) along the edge. Note that each edge e_j of Q, $1 \leq j \leq N$, corresponds to a vertex c_j of the polygonal domain Ω_Q.

We use the Cartesian coordinates (x, y, z) for (3.16) such that the edges of Q are parallel to the z-axis. Let

$$H_D^1(Q) = \{v \in H^1(Q) : v|_{\partial_D Q} = 0\}.$$

Then the weak solution $u \in H_D^1(Q)$ of (3.16) is defined by

$$\int_Q \nabla u \cdot \nabla v \, dx = \int_Q f v \, dx \qquad \forall\, v \in H_D^1(Q). \tag{3.17}$$

Recall that the partial Fourier transform with respect to z for u is

$$\hat{u}(x, y, \zeta) = \frac{1}{\sqrt{2\pi}} \int_{-\infty}^{\infty} e^{-i\zeta z} u(x, y, z)\, dz, \tag{3.18}$$

where $\mathrm{i} = \sqrt{-1}$. Applying it to (3.16), we obtain an elliptic problem with the parameter ζ in a polygonal domain:

$$-\Delta'\hat{u} + \zeta^2\hat{u} = \hat{f} \quad \text{in} \quad \Omega_Q, \tag{3.19}$$

where $\zeta \in \mathbb{R}$ and $\Delta' = \partial_x^2 + \partial_y^2$ is the Laplace operator in x and y. Let Γ_D and Γ_N be parts of the boundary $\partial\Omega_Q$ such that $\partial_D Q = \overline{\Gamma_D} \times \mathbb{R}$ and $\partial_N Q = \overline{\Gamma_N} \times \mathbb{R}$. Assign the following boundary condition to $\hat{u}$:

$$\hat{u} = 0 \quad \text{on} \quad \Gamma_D \quad \text{and} \quad \partial_n\hat{u} = 0 \quad \text{on} \quad \Gamma_N. \tag{3.20}$$

Recall the space $H^1_D(\Omega_Q)$ in (2.26). Then the 2D problem (3.19) and the 3D problem (3.16) are connected as follows.

Proposition 3.2 *Let $u \in H^1_D(Q)$ be the weak solution of (3.16). Then for almost every $\zeta \in \mathbb{R}$, $\hat{u} \in H^1_D(\Omega_Q)$ and is the solution of*

$$\int_{\Omega_Q} \nabla'\hat{u} \cdot \nabla' v \, dx' + \zeta^2 \int_{\Omega_Q} \hat{u} v \, dx' = \int_{\Omega_Q} \hat{f} v \, dx' \qquad \forall\, v \in H^1_D(\Omega_Q),$$

where $x' = (x, y)$ and ∇' is the gradient in x'.

Proof For $f \in L^2(Q)$, due to the Poincaré inequality and the Lax–Milgram Theorem, the variational formulation (3.17) defines a unique weak solution $u \in H^1_D(Q)$ of (3.16). In addition, according to Theorem 3.1, $u \in H^2$ in any subdomain of Q such that its closure does not include the edge of Q.

Defined in the partial Fourier transform (3.18), the function $\hat{u}$ has the following properties for almost every ζ: $\hat{u} \in H^1(\Omega_Q)$, and $\hat{u} \in H^2(G)$ for any $G \subset \Omega_Q$ such that $\bar{G}$ does not include any vertex of Ω_Q. Meanwhile, for almost every ζ, $\hat{u}$ satisfies the boundary value problem (3.19)–(3.20). Let $\overline{\Gamma_j}$ be a boundary side of Ω_Q. Then if $\Gamma_j \subset \Gamma_D$, $\hat{u}|_{\Gamma_j} = 0$. Note that $\hat{u} \in E(\Delta', L^2(\Omega_Q))$. Based on Corollary 2.6, if $\Gamma_j \subset \Gamma_N$, $\partial_n\hat{u}|_{\Gamma_j} = 0$ in the sense of $\tilde{H}^{-1/2}(\Gamma_j)$. Therefore, for any $v \in H^1_D(\Omega_Q)$, $\hat{u}$ satisfies

$$\int_{\Omega_Q} \nabla'\hat{u} \cdot \nabla' v \, dx' + \zeta^2 \int_{\Omega_Q} \hat{u} v \, dx' = \int_{\Omega_Q} \hat{f} v \, dx'. \qquad \square$$

It is clear from Proposition 3.2 that the regularity study in Sect. 3.2 for the 2D equation shall help obtain regularity estimates for the edge singularity in the 3D problem (3.16). For the 2D boundary value problem (3.19)–(3.20), the solution regularity in the neighborhood of a vertex c_j is determined by the Laplace operator Δ' (the principle component of the elliptic operator) and the associated boundary conditions. According to the singular expansion (3.13), we have

$$\hat{u}_R := \hat{u} - \sum_{1 \le j \le N} \sum_{0 < \lambda_{j,k} < 1} \alpha_{j,k}(\zeta) e^{-|\zeta| r_j} \chi_j r_j^{\lambda_{j,k}} \phi_{j,k} \in H^2(\Omega_Q).$$

This is equivalent to (3.13) since $(1 - e^{-|\zeta|r_j})\chi_j r_j^{\lambda_{j,k}} \phi_{j,k} \in H^2(\Omega_Q)$. In addition, the following estimates on $\hat{u}_R$ and $\alpha_{j,k}(\zeta)$ can be found in [65].

Theorem 3.5 *There exists $C > 0$ such that*

$$\|\hat{u}_R\|_{H^2(\Omega_Q)} + |\zeta| \|\hat{u}_R\|_{H^1(\Omega_Q)} + \zeta^2 \|\hat{u}_R\|_{L^2(\Omega_Q)} \leq \|\hat{f}\|_{L^2(\Omega_Q)}.$$

Meanwhile, the coefficient $\alpha_{j,k}(\zeta)$ satisfies

$$|\alpha_{j,k}(\zeta)| \leq C(1 + |\zeta|)^{\lambda_{j,k}-1} \|\hat{f}\|_{L^2(\Omega_Q)}.$$

It can been seen that $\alpha_{j,k}(\zeta)e^{-|\zeta|r_j}$ is the partial Fourier transform in z of the convolution of two functions $K_j * \psi_{j,k}$, where $\psi_{j,k} \in H^{1-\lambda_{j,k}}(\mathbb{R})$ is such that $\hat{\psi}_{j,k} = \alpha_{j,k}(\zeta)$ and $K_j = r_j / \left(\pi(r_j^2 + z^2)\right)$ satisfies $\hat{K}_j = e^{-|\zeta|r_j}$. In particular, the convolution is

$$K_j * \psi_{j,k} = \int_{-\infty}^{\infty} \frac{\psi_{j,k}(z-t)}{\pi r_j (1 + t^2 r_j^{-2})} dt, \tag{3.21}$$

and $\widehat{K_j * \psi_{j,k}} = \alpha_{j,k}(\zeta)e^{-|\zeta|r_j}$. Consequently, we obtain the singular expansion for the edge behavior of the solution.

Theorem 3.6 *For $f \in L^2(Q)$, the variational formulation (3.17) defines a unique solution $u \in H^1_D(Q)$ of (3.16). Moreover, there exist functions $\psi_{j,k}(z) \in H^{1-\lambda_{j,k}}(\mathbb{R})$ and $K_j = r_j / \left(\pi(r_j^2 + z^2)\right)$ such that*

$$u_R := u - \sum_{1 \leq j \leq N} \sum_{0 < \lambda_{j,k} < 1} \chi_j (K_j * \psi_{j,k}) r_j^{\lambda_{j,k}} \phi_{j,k} \in H^2(Q),$$

where the functions χ_j and $\phi_{j,k}$ are the same as in (3.13).

The regularity results in Theorems 3.5 and 3.6 are for an Eq. (3.16) in a domain Q with edges. Using a localization argument, one obtains the following regularity estimates along an edge for the solution of (3.3) in a bounded polyhedral domain Ω.

Corollary 3.1 *In (3.3), let $e_j \in \mathcal{E}$ be an edge. Suppose e_j is on the z-axis. Note that the region in Ω that is formed by e_j and its two adjacent faces is part of a dihedral angle D with angle measure ω_j. Let G be the region in the xy-plane that is in D with the vertex at $(0, 0)$. Let B_δ be a ball in the xy-plane centered at the origin with radius δ and let $N_{\delta,j} = G \cap B_\delta$. Let $l \subset e_j$ be an open line segment on e_j. Suppose that $\delta > 0$ is sufficiently small such that the subset $\Omega_l := N_{\delta,j} \times l \subset \Omega$ in the neighborhood of l does not include any other vertices or edges of Ω. Suppose $f \in L^2(\Omega)$. Then the solution u belongs to $H^{1+s}(\Omega_l)$ for any $s \leq 1$ and $s < \min_k\{\lambda_{j,k} > 0\}$, where $\lambda_{j,k}$ is defined in (3.11) based on the boundary conditions on the two adjacent faces of e_j. In addition, if f is infinitely differentiable in z, u is also infinitely differentiable in z with values in $H^{1+s}(N_{\delta,j})$.*

Proof Denote by $(0,0,a)$ and $(0,0,b)$ the two endpoints of l with $a < b$. Choose $\epsilon > 0$ and let l_ϵ be a larger open line segment with endpoints $(0,0,a-\epsilon)$ and $(0,0,b+\epsilon)$ such that $\Omega_{l_\epsilon} = N_{2\delta,j} \times l_\epsilon \subset \Omega$ is a neighborhood of l_ϵ and it does not include any other vertices or edges of Ω. It is clear that $\Omega_l \subset \Omega_{l_\epsilon}$. Let (r,θ,z) be the cylindrical coordinates with e_j on the z-axis. Let χ be a smooth function independent of θ that is equal to 1 in Ω_l and equal to 0 outside Ω_{l_ϵ}. Then χu satisfies (3.16) with a properly chosen domain, boundary conditions, and an L^2 right hand side function. Based on the singular expansion in Theorem 3.6, we obtain that $u \in H^{1+s}(\Omega_l)$ for any $s \le 1$ and $s < \min_k\{\lambda_{j,k} > 0\}$.

In addition, when f is infinitely differentiable in z, $\chi_j(K_j * \psi_{j,k})r_j^{\lambda_{j,k}}\phi_{j,k}$ is infinitely differentiable in z with values in $H^{1+s}(N_{\delta,j})$. Meanwhile, Theorems 3.5 and 3.6 imply that u_R is infinitely differentiable in z with values in $H^2(N_{\delta,j})$. The proof is hence complete. □

Remark 3.4 In the case where the Dirichlet boundary condition is imposed on both adjacent faces of e_j, by (3.11), we have $\lambda_{j,k} = k\pi/\omega_j$. Thus, when $\omega_j > \pi$, the solution $u \notin H^2$ near e_j ($u \in H^{1+s}$ for $s < \pi/\omega_j$). Moreover, Theorem 3.6 and Corollary 3.1 imply that the edge singularity is anisotropic. It occurs in the direction perpendicular to the edge, while the solution is relatively smooth in the edge direction.

3.3.2 The 3D Vertex Singularity

In Sect. 3.3.1, we discussed the Poisson equation in a domain with only edges to analyze the edge singularity. We adopt a similar strategy here to study the solution behavior in the neighborhood of a vertex in (3.3) by considering an elliptic problem in an unbounded domain with only one vertex.

Recall the domain Ω in (3.3). For a vertex $c \in C$ of Ω, let Λ_c be an infinite cone that coincides with Ω in the neighborhood of c. In addition, we assume c coincides with the origin o of the coordinate system. Denote by F_ℓ, $1 \le \ell \le L$, the faces of Λ_c that meet at c and denote by $e_{\ell,m}$ the edge of Λ_c where the faces F_ℓ and F_m intersect. Note that near o, each F_ℓ coincides with a face Γ_j of Ω. Assign the same boundary condition on F_ℓ as that on Γ_j. Then for $f \in L^2(\Lambda_c)$, we consider the problem in the infinite cone

$$-\Delta u = f \quad \text{in} \quad \Lambda_c. \tag{3.22}$$

Let $\omega_{\ell,m}$ be the interior angle of the edge $e_{\ell,m}$ and let $\lambda_{\ell,m,k} \ge 0$, $k \ge 1$, be such that $\lambda^2_{\ell,m,k}$ are the eigenvalues of the elliptic operator associated with the corresponding boundary conditions (see (3.10) and (3.11)). In particular, one has

$$\lambda_{\ell,m,k} = \frac{k\pi}{\omega_{\ell,m}} \text{ if } F_\ell \text{ and } F_m \text{ are Dirichlet faces;}$$

$$\lambda_{\ell,m,k} = \frac{(k-1)\pi}{\omega_{\ell,m}} \text{ if } F_\ell \text{ and } F_m \text{ are Neumann faces;}$$

$$\lambda_{\ell,m,k} = \frac{(2k-1)\pi}{2\omega_{\ell,m}} \text{ if one of } F_\ell,\ F_m \text{ is a Dirichlet face and one is a Neumann face.}$$

Then Theorem 3.6 and Corollary 3.1 give rise to the following properties of the edge singularity in (3.22).

Proposition 3.3 *For a function $\chi \in \mathcal{D}(\bar{\Lambda}_c)$ whose support is away from the vertex c, the solution of (3.22) satisfies $\chi u \in H^{1+s}(\Lambda_c)$ for $s \le 1$ and $s < \min_{\ell,m,k}\{\lambda_{\ell,m,k} > 0\}$.*

We now proceed to study the solution singularity associated with the vertex c. We adopt the following notation. Denote by G the intersection between the cone Λ_c and the unit sphere S^2 centered at c. Thus, G is a "polygon" on S^2 and its boundary is the union of the arcs of great circles, denoted by e'_ℓ, $1 \le \ell \le L$. Namely, $e'_\ell = F_\ell \cap S^2$. Let (ρ, θ, φ) be the *spherical coordinates* with the origin $o = c$. For a point p with the coordinates (ρ, θ, φ), $\rho = |op|$ is the distance to o, θ (azimuthal angle) is the angle between op' and the positive x-axis, where p' is the projection of p in the xy-plane, and φ (polar angle) is the angle between op and the positive z-axis. Thus one has

$$x = \rho\cos\theta\sin\varphi, \quad y = \rho\sin\theta\sin\varphi, \quad z = \rho\cos\varphi.$$

Note that the Lebesgue measure on S^2 is $\sin\varphi d\theta d\varphi$. Then in terms of the spherical coordinates, we rewrite (3.22) as

$$-\left(\partial_\rho^2 + \frac{2}{\rho}\partial_\rho + \frac{1}{\rho^2}\Delta'\right)u = f, \tag{3.23}$$

where

$$\Delta' := \frac{1}{\sin\varphi}\partial_\varphi(\sin\varphi\,\partial_\varphi) + \frac{1}{(\sin\varphi)^2}\partial_\theta^2 \tag{3.24}$$

is the Laplace–Beltrami operator on S^2. Choose f such that (3.23) coincides with (3.3) in the neighborhood of o. In addition, we require that there is $R > 0$ such that $f = 0$ when $\rho > R$. Note that the boundary condition for u on e'_ℓ is $u = 0$ if F_ℓ is a Dirichlet face and is $\partial_n u = 0$ if F_ℓ is a Neumann face. To simplify the exposition, we assume that the Dirichlet boundary condition is imposed on at least one side of the boundary ∂G.

Let $H^1_D(G) = \{v \in H^1(G) : v|_{e'_\ell} = 0 \text{ if } e'_\ell \text{ is on a Dirichlet face } F_\ell\}$. Note that the gradient operator in spherical coordinates is

$$\nabla = \left(\partial_\rho, \frac{1}{\rho\sin\varphi}\partial_\theta, \frac{1}{\rho}\partial_\varphi\right).$$

Applying this to functions independent of ρ and setting $\rho = 1$, we obtain the bilinear form associated with the Laplace–Beltrami operator

$$b(v, w) = \int_G \left(\sin\varphi\, \partial_\varphi v \partial_\varphi w + \frac{1}{\sin\varphi} \partial_\theta v \partial_\theta w \right) d\varphi d\theta \tag{3.25}$$

for $w, v \in H^1_D(G)$. Define the space $E_{\Delta'} = \{v \in H^1_D(G) : \Delta' v \in L^2(G)$ and $b(v, w) = \int_G -w\Delta' v \sin\varphi\, d\varphi d\theta, \forall\, w \in H^1_D(G)\}$. It is clear that $b(\cdot, \cdot)$ is continuous and coercive on $H^1_D(G)$ which makes $-\Delta'$ a self-adjoint operator in $L^2(\Omega)$ with compact resolvent. Therefore, its spectrum consists of an infinite sequence $\lambda_i, i \geq 1$, where $0 \leq \lambda_i \leq \lambda_k$ for $i \leq k$. Denote by ϕ_i the orthonormalized sequence of eigenfunctions associated with the eigenvalue λ_i. Then $\phi_i \in E_{\Delta'}$ and

$$-\Delta' \phi_i = \lambda_i \phi_i \quad \text{in } G. \tag{3.26}$$

In contrast to the smooth eigenfunctions (3.11) of the Laplace operator in a one-dimensional domain, the eigenfunctions (3.26) of the Laplace–Beltrami operator can possess corner singularities at the vertex of G. We in fact have a rather general result for functions in $E_{\Delta'}$.

Lemma 3.1 *For any $v \in E_{\Delta'}$, we have $v \in H^{1+s}(G)$ where s satisfies the same conditions as in Proposition 3.3.*

Proof Let c_1 and c_2 be two constants such that $0 < c_1 < c_2$. Define a 3D domain $G_1 = \{(\rho, \theta, \varphi) : c_1 < \rho < c_2, (\theta, \varphi) \in G\}$. For $v \in E_{\Delta'}$, we extend it to a function q in G_1, such that $q(x, y, z) = q(\rho, \theta, \varphi) = v(\theta, \varphi)$. Namely, q is independent of ρ. Then for any $w \in H^1(G_1)$, we have

$$\begin{aligned}\int_{G_1} \nabla q \cdot \nabla w\, dx &= \int_{c_1}^{c_2} \int_G (\partial_\rho q \partial_\rho w)\rho^2 \sin\varphi\, d\rho d\varphi d\theta + \int_{c_1}^{c_2} b(q, w)\, d\rho \\ &= \int_{c_1}^{c_2} -\rho^{-2} w(\Delta' q)\rho^2 \sin\varphi\, d\rho d\varphi\, d\theta = -\int_{G_1} \rho^{-2} w \Delta' q\, dx.\end{aligned}$$

Thus $\int_{G_1} \nabla q \cdot \nabla w\, dx = \int_{G_1} f w\, dx$ for some $f \in L^2(G_1)$. According to Proposition 3.3, one has $q \in H^{1+s}(G_1)$, which implies $v \in H^{1+s}(G)$. □

Therefore, because the eigenfunctions ϕ_i in (3.26) belongs to $E_{\Delta'}$, one has $\phi_i \in H^{1+s}(G)$.

It is interesting to notice the following relation between the Laplace–Beltrami operator and the Laplace operator in a 2D domain. Introduce a new variable $r' = \sin\varphi$. Then the Laplace–Beltrami operator (3.24) can be written as

$$\partial^2_{r'} + \frac{1}{r'}\partial_{r'} + \frac{1}{r'^2}\partial^2_\theta - r'^2\partial^2_{r'} - 2r'\partial_{r'}. \tag{3.27}$$

Suppose that p is a vertex of G and that op is on the z-axis. Thus, in the small neighborhood of p, the corner singularity in ϕ_i is determined by the operator

$$\partial_{r'}^2 + r'^{-1}\partial_{r'} + r'^{-2}\partial_\theta^2 \tag{3.28}$$

with the associated boundary conditions. This is the case because the other terms in (3.27) are negligible as $r' \to 0$ in the small neighborhood of p. The operator (3.28) is the same as the Laplace operator written in polar coordinates (see (3.8)). Therefore, depending on the boundary condition near p, the corner singularity of ϕ_i at p has a structure like those in Example 3.1. For example, when p has the Dirichlet boundary condition on both adjacent sides of G, ϕ_i has a corner singularity involving the function

$$(r')^{\pi/\omega} \sin(\pi\theta/\omega) = (\sin\varphi)^{\pi/\omega} \sin(\pi\theta/\omega), \tag{3.29}$$

where ω is the interior angle corresponding to the edge op of the cone Λ_c.

To study the singularity of the solution associated with the vertex in (3.22), we further change the variable by letting $t = \ln\rho$. Then (3.23) can be written as

$$-(\partial_t^2 + \partial_t + \Delta')u = e^{2t} f \quad \text{in } \mathbb{R} \times G. \tag{3.30}$$

Denote by $N_{o,\delta} \subset \Lambda_c$ a small neighborhood of the vertex $o = c$. By a direct calculation involving change of variables, one can show the following relation between derivatives when different variables are used.

Proposition 3.4 *For $s \geq -1$, assume $v(t, \theta, \varphi) \in H^{1+s}(\mathbb{R}\times G)$. Then $\rho^{s-1/2}v(x, y, z) \in H^{1+s}(N_{o,\delta})$.*

Suppose that u_0 is a solution of (3.22). We introduce the function $v_0 = e^{(1/2-s)t}u_0$ for which (3.30) becomes

$$-\big(\partial_t^2 + 2s\partial_t + \Delta' + (s + 1/2)(s - 1/2)\big)v_0 = g \quad \text{in } \mathbb{R} \times G, \tag{3.31}$$

where $g = e^{(5/2-s)t} f$. According to Proposition 3.4, the regularity estimate for v_0 shall lead to a regularity estimate for u_0 in $N_{o,\delta}$ since $u_0 = \rho^{s-1/2}v_0$. Suppose that $(s + 1/2)(s - 1/2)$ is not an eigenvalue of $-\Delta'$ defined in (3.26). In (3.31), using the partial Fourier transform in t, we obtain

$$\hat{v}_0 = P\hat{g} := -\big(\Delta' - \zeta^2 + 2is\zeta + (s + 1/2)(s - 1/2)\big)^{-1}\hat{g}.$$

Note that the operator P is well defined because $\big(\Delta' - \zeta^2 + 2is\zeta + (s+1/2)(s-1/2)\big)$ is invertible for any $\zeta \in \mathbb{R}$ provided that Δ' is a self-adjoint and negative operator and $(s + 1/2)(s - 1/2)$ is not an eigenvalue of $-\Delta'$. For large ζ, the operator P has the same behavior as $-(\Delta' - \zeta^2)^{-1}$. Therefore, we have

$$\|\zeta^2 P\|_{L^2(G)\to L^2(G)} + \|\zeta P\|_{L^2(G)\to H^1_D(G)} + \|\Delta' P\|_{L^2(G)\to L^2(G)} \leq C,$$

where C is bounded and independent of ζ. Hence, we have

$$\partial_t^2 v_0 \in L^2(\mathbb{R} \times G), \quad \partial_t v_0 \in L^2(\mathbb{R}, H^1(G)), \quad \text{and} \quad v_0 \in L^2(\mathbb{R}, E_{\Delta'}).$$

According to the estimates in Lemma 3.1, one has $v_0 \in H^{1+s}(\mathbb{R} \times G)$. These observations can be further summarized as follows.

Proposition 3.5 *Assume that $(s + 1/2)(s - 1/2)$ is not an eigenvalue of $-\Delta'$. Then there exists a solution v_0 of (3.31) such that $v_0 \in H^{1+s}(\mathbb{R} \times G)$ and $v_0 \in E_{\Delta'}$ for every $t \in \mathbb{R}$, where s satisfies the same conditions as in Proposition 3.3.*

Consequently, we have found a solution $u_0 = \rho^{s-1/2} v_0$ of (3.30). It is clear that u_0 satisfies (3.3) in the neighborhood $N_{o,\delta}$ of the vertex. Then we have the main theorem regarding the solution singularity of (3.3) in the neighborhood of the vertex.

Theorem 3.7 *Let $u \in H^1_D(\Omega)$ be the solution of (3.3) with $f \in L^2(\Omega)$, where Ω is a polyhedral domain. Let $c \in C$ be a vertex of Ω and denote by $N_{o,\delta}$ its small neighborhood. Then there are unique numbers α_i such that*

$$u - \sum_i \alpha_i \rho^{-1/2+\sqrt{\lambda_i+1/4}} \phi_i \in H^{1+s}(N_{o,\delta}),$$

where (λ_i, ϕ_i) is the eigenpair of the Laplace–Beltrami operator in (3.26) and s satisfies the same conditions $s \leq 1$ and $s < \min_{\ell,m,k}\{\lambda_{\ell,m,k} > 0\}$ as in Proposition 3.3. The sum is over i such that $\lambda_i \leq s^2 - 1/4$.

Proof According to Propositions 3.4 and 3.5, in the neighborhood $N_{o,\delta}$ of a vertex, the function $u_0 = \rho^{s-1/2} v_0$ satisfies $-\Delta u_0 = f$ and possesses the same boundary conditions as those in (3.3). In addition, $u_0 \in H^{1+s}(N_{o,\delta})$. Therefore, $w = u - u_0$ satisfies the same boundary condition near the vertex and

$$-\left(\partial_\rho^2 + \frac{2}{\rho}\partial_\rho + \frac{1}{\rho^2}\Delta'\right) w = 0 \quad \text{in } N_{o,\delta}.$$

Expanding w using the eigenfunctions ϕ_i of $-\Delta'$ in (3.26), one obtains

$$w = \sum_{i \geq 1} \left(a_i \rho^{-1/2+\sqrt{\lambda_i+1/4}} + b_i \rho^{-1/2-\sqrt{\lambda_i+1/4}} \right) \phi_i.$$

Recall $w \in H^1(N_{o,\delta})$. Thus, we have $b_i = 0$ for all $i \geq 1$ and

$$w = \sum_{i \geq 1} a_i \rho^{-1/2+\sqrt{\lambda_i+1/4}} \phi_i.$$

Note that $u = w + u_0$. Then the proof follows from the fact that when $\lambda_i > s^2 - 1/4$, the term $a_i \rho^{-1/2+\sqrt{\lambda_i+1/4}} \phi_i$ belongs to $H^{1+s}(N_{o,\delta})$. □

Remark 3.5 We summarize the regularity results derived for the Poisson equation (3.3) in 2D polygonal and 3D polyhedral domains. Generally speaking, the solution can possess singularities due to the nonsmoothness of the domain and the change of boundary conditions. Although the specifics of the singularity depend on the local geometry of the domain and the associated boundary conditions, they all involve the distance function to the set of nonsmooth points.

- The 2D corner singularity consists of functions in the form of $r_j^{\lambda_{j,k}} \phi_{j,k}$ (Theorem 3.4), where r_j is the distance to the vertex and $(\lambda_{j,k}^2, \phi_{j,k})$ with $0 < \lambda_{j,k} < 1$ is the eigenpair of the Laplace operator in a 1D domain with associated boundary conditions (3.10).
- The 3D edge singularity, derived using partial Fourier transform and the 2D equation with parameters, involves singular terms like $(K_j * \psi_{j,k}) r_j^{\lambda_{j,k}} \phi_{j,k}$ (Theorem 3.6), where r_j is the distance to the edge, $(\lambda_{j,k}^2, \phi_{j,k})$ with $0 < \lambda_{j,k} < 1$ is the eigenpair of the Laplace operator in a 1D domain with associated boundary conditions, and $K_j * \psi_{j,k}$ is the convolution defined in (3.21). By requiring $0 < \lambda_{j,k} < 1$, the 2D corner singularity and the 3D edge singularity are components in the solution that do not belong to H^2. If one looks for the component that is not in H^m $(m > 2)$, more singular functions with $\lambda_{j,k} \geq 1$ may be included in the expansion. The edge singularity occurs in the direction perpendicular to the edge, while the solution is smoother along the edge.
- The 3D vertex singularity has a more complex structure. First, in the neighborhood of a 3D vertex, the edge singularity can still exist if the vertex is connected to edges of the domain. Therefore, even if we subtract the vertex singularity from the solution, it is possible that the remaining function is still not in H^2. In fact, it is in H^{1+s} where s is determined by the edge singularity. In particular, the 3D vertex singularity consists of functions in the form of $\rho^{-1/2+\sqrt{\lambda_i+1/4}} \phi_i$ (Theorem 3.7), where ρ is the distance to the vertex and (λ_i, ϕ_i) with $\lambda_i \leq s^2 - 1/4$ is the eigenpair of the Laplace–Beltrami operator in (3.26). Another distinguishable feature of the vertex singularity is that the eigenfunction $\phi_i(\theta, \varphi)$ itself can possess a corner singularity near the vertex of the "polygon" G on S^2. See the discussion about (3.29). Note that $-1/2 + \sqrt{\lambda_i + 1/4} \geq 0$. Therefore, there exists $s_0 > 3/2$ such that for the pure Dirichlet or pure Neumann boundary condition, the solution of the Poisson equation in a polyhedral domain belongs to $H^{1+s}(\Omega)$ for all $s < s_0$.

3.4 The Graded Mesh Algorithm

The standard finite element method solves an elliptic boundary value problem based on a quasi-uniform mesh. The convergence of the numerical solution closely depends on the regularity of the actual solution (see Theorem 2.20). As illustrated in Sect. 3.1, the convergence rate deteriorates in the presence of solution singularities largely because the finite element function associated with a quasi-uniform mesh cannot effectively capture the high-frequency singular components in the solution. These singularities are local. They often occur near the nonsmooth points of the boundary even when the given data is smooth, while the solution in other regions of the domain is relatively smooth. Consequently, one approach to improve the effectiveness of the numerical approximation is to increase the local mesh density in regions close to the singular points, such that the fine-tuned finite element space can better trace the singular solution. In this section, we describe the construction of a class of graded meshes for which the element size gradually decreases toward the singular points.

For simplicity, we consider the Poisson equation with the pure Dirichlet boundary condition

$$-\Delta u = f \quad \text{in} \quad \Omega, \qquad u = 0 \quad \text{on} \quad \Gamma = \partial\Omega, \tag{3.32}$$

where $\Omega \subset \mathbb{R}^d$ is either a polygonal domain ($d = 2$) or a polyhedral domain ($d = 3$). For the domain Ω, let C be the set of vertices. Let $\mathcal{E}$ be the set of boundary edges when $d = 3$ and let $\mathcal{E}$ be the set of straight sides of the boundary when $d = 2$. Denote by $\mathcal{S}$ the set of nonsmooth (singular) points of the domain. Then $\mathcal{S} = C$ for $d = 2$ and $\mathcal{S} = C \cup \mathcal{E}$ for $d = 3$. For each vertex $c \in C$ and each edge (or side) $e \in \mathcal{E}$, we assign the associated grading parameters $\kappa_c \in (0, 0.5]$ and $\kappa_e \in (0, 0.5]$, respectively. Let $\mathcal{T}$ be a triangulation of Ω with triangles ($d = 2$) or tetrahedra ($d = 3$). Let T be a triangle (resp. tetrahedron) with vertices a, b, c (resp. a, b, c, d). Then we denote T by its vertices: $\triangle^3 abc$ for the triangle and $\triangle^4 abcd$ for the tetrahedron, where the sup-index implies the number of vertices for T.

Definition 3.1 (Singular Vertex and Edge) The singular vertices and singular edges are the special vertices and edges of the triangles or tetrahedra in $\mathcal{T}$ defined as follows. Let pq be a closed edge of an element $T \in \mathcal{T}$ with p and q as the endpoints. We define different singular sets based on the location of the edge pq. We say pq is a *singular edge* if $pq \subset \bar{e}$, where $e \in \mathcal{E}$ and $\kappa_e < 0.5$. Namely, a singular edge in $\mathcal{T}$ lies on an edge (or side) e of the domain boundary for which the assigned parameter $\kappa_e < 0.5$. We further describe two types of *singular vertices*. We call p a *v-singular vertex* of pq if $p = c \in C$ and $\kappa_c < 0.5$. In this case, p is a singular vertex for all the element edges connecting to p. We call p an *e-singular vertex* of pq if the following three conditions are satisfied: 1. p is not a v-singular vertex; 2. p lies on a singular edge that is part of $\bar{e}$ for $e \in \mathcal{E}$ with $\kappa_e < 0.5$; 3. $pq \not\subset \bar{e}$. In this case, p is a singular vertex for all the element edges intersecting $\bar{e}$ at p but is not a singular vertex for the singular edges on $\bar{e}$. It is possible that an e-singular vertex p is an endpoint of e, which means $p = c \in C$ is a vertex of the domain but p is not a v-singular vertex (i.e., $\kappa_c = 0.5$). See Fig. 3.3 for examples of the singular vertices and edges.

We require that each element edge in the triangulation has at most one singular vertex as an endpoint. This condition can be enforced by, for example, requiring each element $T \in \mathcal{T}$ contains at most one singular edge and at most one v-singular vertex; if it contains both, the v-singular vertex is an endpoint of the singular edge; and if an edge in T has an e-singular vertex as an endpoint, the other endpoint must not be a singular vertex. Suppose p is a singular vertex of an edge pq in the triangulation. Then we assign p a grading parameter κ_p as follows:

$$\kappa_p := \begin{cases} \kappa_e, & p \in e \in \mathcal{E} \text{ is an } e\text{-singular vertex of } pq, \\ \min_{e_r \in \mathcal{E}_c \setminus \{e\}} \{\kappa_{e_r}\}, & p = c \in C \text{ but is an } e\text{-singular vertex of } pq, \\ \min_{e_r \in \mathcal{E}_c \setminus \{e\}} \{\kappa_c, \kappa_{e_r}\}, & p = c \in C \text{ is an } v\text{-singular vertex of } pq, \end{cases} \tag{3.33}$$

where in the last two cases, $\mathcal{E}_c \subset \mathcal{E}$ is the set of edges (or sides) that touch the vertex c and $e \in \mathcal{E}_c$ is the edge (or side) such that $pq \subset \bar{e}$. In the last two cases of (3.33), if pq is not on any edge (or side) in $\mathcal{E}_c$, we set $e = \emptyset$.

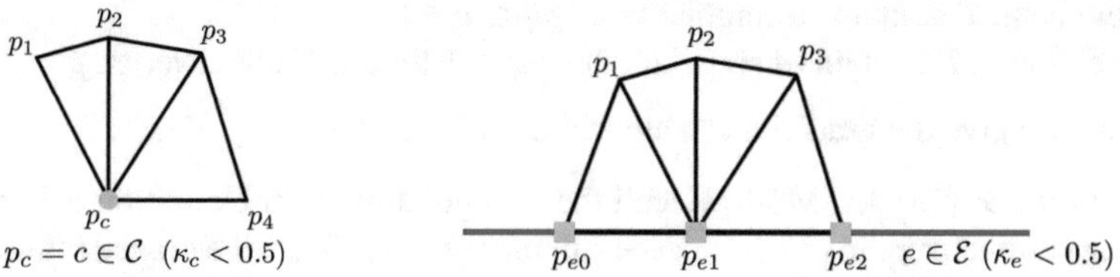

Fig. 3.3 Singular vertices and edges: a v-singular vertex $p_c = c \in \mathcal{C}$ with $\kappa_c < 0.5$ (left); e-singular vertices p_{e0}, p_{e1}, and p_{e2} on $e \in \mathcal{E}$ (red line segment = e) with $\kappa_e < 0.5$ (right). p_c is a singular vertex for all edges $p_c p_i$, $1 \le i \le 4$. p_{e0} is a singular vertex for $p_{e0}p_1$, p_{e1} is a singular vertex for $p_{e1}p_i$, $1 \le i \le 3$, and p_{e2} is a singular vertex for $p_{e2}p_3$. $p_{e0}p_{e1}$ and $p_{e1}p_{e2}$ are singular edges.

According to Definition 3.1, the singular vertices and singular edges are predetermined by the assigned grading parameters. Each singular vertex p of an element edge is given a parameter $\kappa_p < 0.5$. Then we describe the algorithm to produce new nodes in the triangulation.

Algorithm 3.8 (New Node) Let pq be an edge in the triangulation $\mathcal{T}$ with p and q as the endpoints. Then in a graded refinement, a new node r on pq is produced as follows:

1. (Neither p nor q is a singular vertex of pq.) We choose r as the midpoint ($|pr| = |qr|$).
2. (p is a singular vertex of pq.) We choose r such that $|pr| = \kappa_p |pq|$, where κ_p is defined in (3.33).

Here the singular vertex can be either a v-singular vertex or an e-singular vertex of pq. See Fig. 3.4 for an illustration. □

Fig. 3.4 Refinement of an edge pq (left–right): no singular vertices (midpoint); p is a singular vertex ($|pr| = \kappa_p |pq|$, $\kappa_p < 0.5$).

Based on the number and type of singular vertices and the number of singular edges in an element, we can classify the elements in $\mathcal{T}$ as follows.

Definition 3.2 (Element Type) Given the conditions on the singular vertices and singular edges in Definition 3.1, each element $T \in \mathcal{T}$ falls into one of the five categories.

1. o-element: T contains no singular vertex or singular edge.
2. v-element: T contains a vertex $c \in C$ that is either a v-singular vertex or an e-singular vertex ($\kappa_c = 0.5, \min_{e \in \mathcal{E}_c}\{\kappa_e\} < 0.5$) but no singular edge.
3. v_e-element: T contains an e-singular vertex that is in the interior of an edge (or side) $e \in \mathcal{E}$ but no singular edge.
4. e-element: T contains a singular edge $pq \subset e \in \mathcal{E}$.
5. ev-element: T contains a singular edge $pq \subset \bar{e}$ for $e \in \mathcal{E}$ and a vertex $p = c \in C$.

Now, we give the graded mesh algorithm in 2D and 3D.

Algorithm 3.9 (Graded Mesh) Recall the triangulation $\mathcal{T}$ in Definition 3.1 and the grading parameter κ_p in (3.33) for each singular vertex p. Then the graded refinement, denoted by $\boldsymbol{\kappa}(\mathcal{T})$, proceeds as follows.

- Triangular Elements ($d = 2$). For each triangle $T = \triangle^3 x_0 x_1 x_2 \in \mathcal{T}$, a new node is generated on each edge of T based on Algorithm 3.8. Then T is decomposed into four small triangles by connecting these new nodes (Fig. 3.5).
- Tetrahedral Elements ($d = 3$). For each tetrahedron $T = \triangle^4 x_0 x_1 x_2 x_3 \in \mathcal{T}$, a new node x_{kl} is generated on each edge $x_k x_l$, $0 \le k < l \le 3$, based on Algorithm 3.8. Connecting these new nodes x_{kl} on all the faces of T, we obtain four small tetrahedra and one octahedron. The octahedron then is cut into four tetrahedra using x_{13} as the common vertex. Therefore, after one refinement, we obtain eight sub-tetrahedra for each $T \in \mathcal{T}$ denoted by their vertices (Fig. 3.6):

$$\triangle^4 x_0 x_{01} x_{02} x_{03},\ \triangle^4 x_{01} x_1 x_{12} x_{13},\ \triangle^4 x_{02} x_{12} x_2 x_{23},\ \triangle^4 x_{03} x_{13} x_{23} x_3,$$
$$\triangle^4 x_{01} x_{02} x_{03} x_{13},\ \triangle^4 x_{01} x_{02} x_{12} x_{13},\ \triangle^4 x_{02} x_{03} x_{13} x_{23},\ \triangle^4 x_{02} x_{12} x_{13} x_{23}.$$

Given an initial mesh $\mathcal{T}_0$ satisfying the conditions in Definition 3.1, the associated family of graded meshes $\{\mathcal{T}_n : n \ge 0\}$ is defined recursively $\mathcal{T}_n = \boldsymbol{\kappa}(\mathcal{T}_{n-1})$. See Figs. 3.7–3.11 for example. □

Remark 3.6 Recall that the 2D corner singularity is isotropic, concentrating at the vertex of the domain; and the 3D edge singularity is anisotropic, singular in the direction perpendicular to the edge and smoother in the edge direction. Thus, Algorithm 3.9 is based on a simple and intuitive principle for mesh refinement: producing the new nodes closer to the singular point and consequently forming new elements that are small in the direction of the singularity. The successful application of Algorithm 3.9 relies on choosing the proper grading parameters for the elements in C and in $\mathcal{E}$. For example, with specific selections of κ_c and κ_e, Algorithm 3.9 leads to a variety of graded meshes in the literature, including the meshes in approximating 2D corner singularities [6, 25, 34, 84], approximating anisotropic degenerate elliptic operators [79], and solving 3D elliptic problems with vertex and edges singularities [68, 82, 85, 86]. In subsequent chapters, we shall discuss in detail how these parameters can be selected in Algorithm 3.9 for 2D and 3D elliptic problems.

Remark 3.7 The grading parameter κ_p can be regarded as an indicator of the severity of the singularity at p. A smaller value of κ_p leads to a higher mesh density near p,

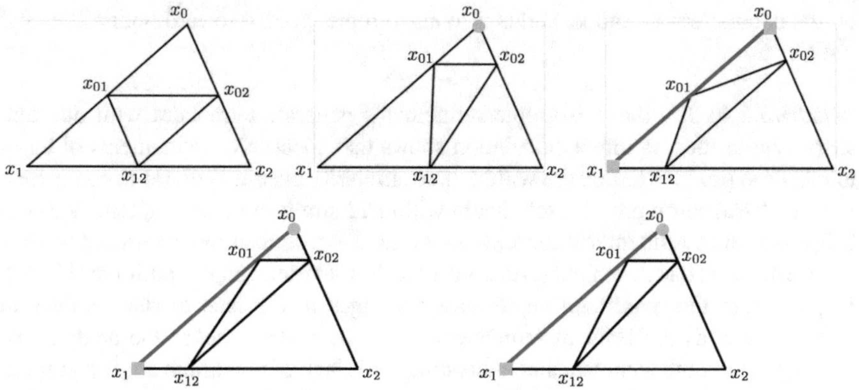

Fig. 3.5 Refinement of a triangle (green dot = v-singular vertex, green box = e-singular vertex, red line segment = singular edge). Top row (left–right): o-element, v-element, e-element. Bottom row (ev-elements, assuming v-singular vertex $x_0 = c \in \mathcal{C}$, singular edge $x_0x_1 \subset \bar{e} \in \mathcal{E}$, and $\kappa_{e_r} = 0.5$ for any $e_r \in \mathcal{E}_c \setminus \{e\}$): $\kappa_e = \kappa_c$ (left); $\kappa_e > \kappa_c$ (right).

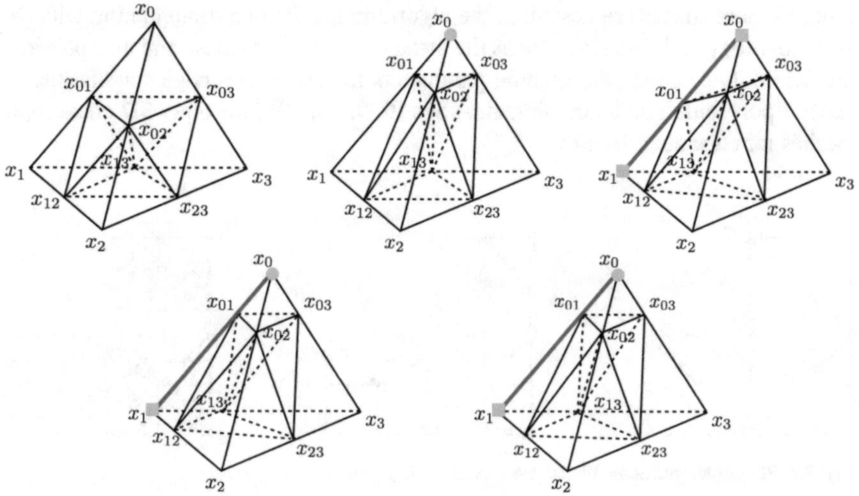

Fig. 3.6 Refinement of a tetrahedron (green dot = v-singular vertex, green box = e-singular vertex, red line segment = singular edge). Top row (left–right): o-element, v-element, e-element. Bottom row (ev-elements, assuming v-singular vertex $x_0 = c \in \mathcal{C}$, singular edge $x_0x_1 \subset \bar{e} \in \mathcal{E}$, and $\kappa_{e_r} = 0.5$ for any $e_r \in \mathcal{E}_c \setminus \{e\}$): $\kappa_e = \kappa_c$ (left); $\kappa_e > \kappa_c$ (right).

while the value $\kappa_p = 0.5$ corresponds to a quasi-uniform refinement. It is apparent that this refinement method results in very different mesh geometries. In a region away from the sets $\mathcal{C}$ and $\mathcal{E}$, the mesh is isotropic and quasi-uniform. The local refinement for a v- or v_e-element in fact follows the same rule: the mesh is isotropic and graded toward the singular vertex. In a region close to an edge (or side) $e \in \mathcal{E}$ with $\kappa_e < 0.5$ that is away from the vertices, the resulting mesh in general is anisotropic and graded toward the edge (or side). The mesh refinement in an ev-element depends on the parameters κ_c and κ_e and is also anisotropic, graded toward both $e \in \mathcal{E}$ and $c \in C$.

Remark 3.8 In 3D, the anisotropic refinements generate tetrahedra with different shape regularities. A direct calculation shows that successive refinements of an o-tetrahedron produce tetrahedra within three similarity classes [29]; refinements for a v- or v_e-tetrahedron produce tetrahedra within 22 similarity classes (Remark 3.4 in [67]). However, refinements for an e- or ev-tetrahedron lead to anisotropic meshes toward the edge that in general do not preserve the maximum angle condition [17, 74]. In particular, the maximum angle between edges in the face of the tetrahedron approaches π as the level of refinement n increases. This makes the analysis for this algorithm both technical and interesting. This lack of maximum angle condition can also occur in 2D when singular edges are defined in the initial mesh. See Figs. 3.8, 3.10, and 3.11.

Remark 3.9 Algorithm 3.9 has a few notable properties: 1. it is simple, explicit, and defined recursively; 2. the meshes $\mathcal{T}_j$, $j \leq n$, are conforming and the associated finite element spaces are nested; 3. the algorithm results in a triangulation with the same topology and data structure as the usual quasi-uniform mesh and also provides the flexibility to adjust the grading parameters for vertex and edge singularities in general polygonal/polyhedral domains. See [6, 9, 26, 106] for other 3D anisotropic meshes for edge singularities.

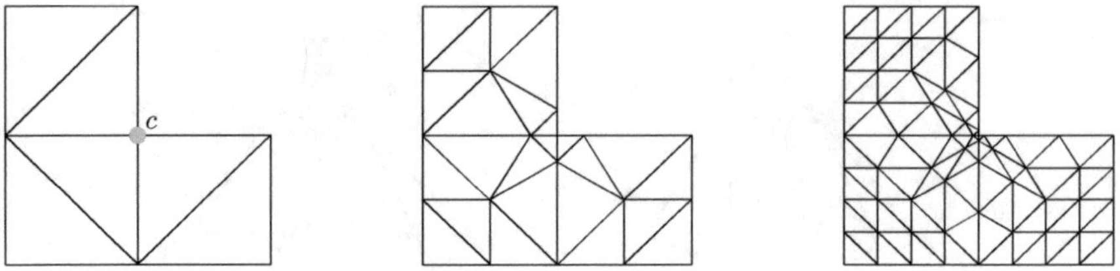

Fig. 3.7 2D graded refinements toward a vertex c ($\kappa_c = 0.2$).

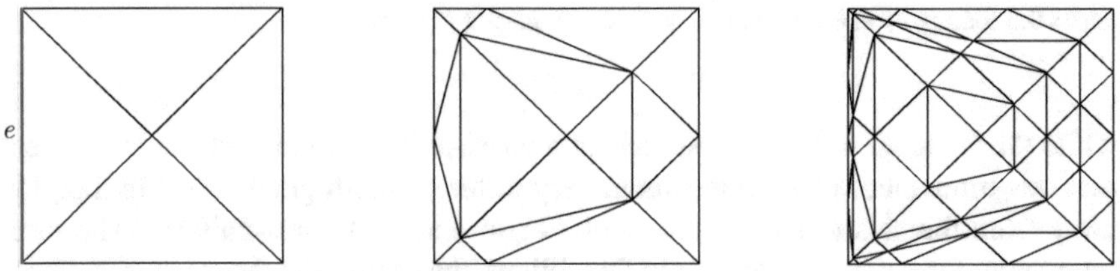

Fig. 3.8 2D graded refinements toward an edge e ($\kappa_e = 0.2$).

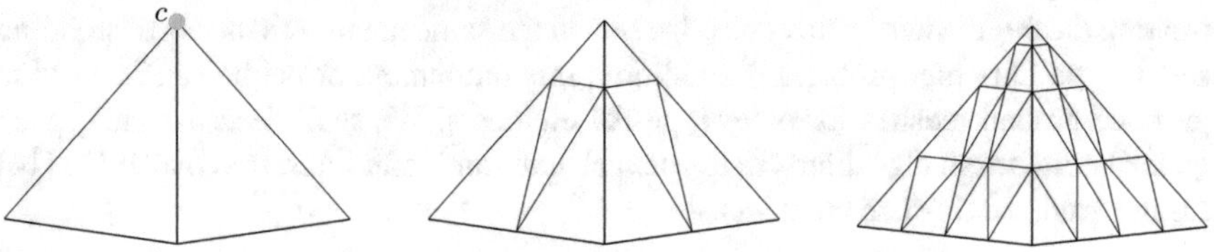

Fig. 3.9 3D graded refinements toward a vertex c ($\kappa_c = 0.3$).

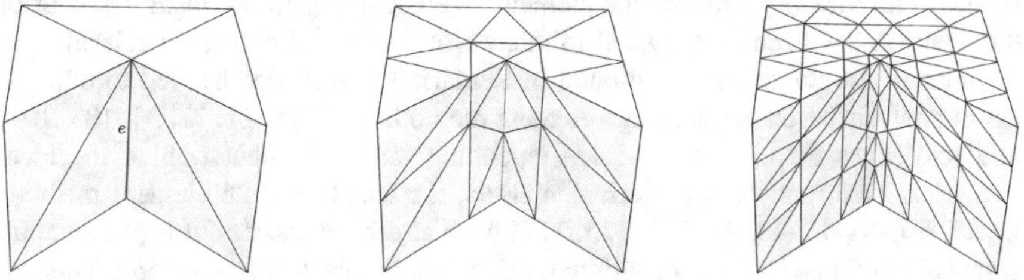

Fig. 3.10 3D graded refinements toward an edge e ($\kappa_e = 0.3$).

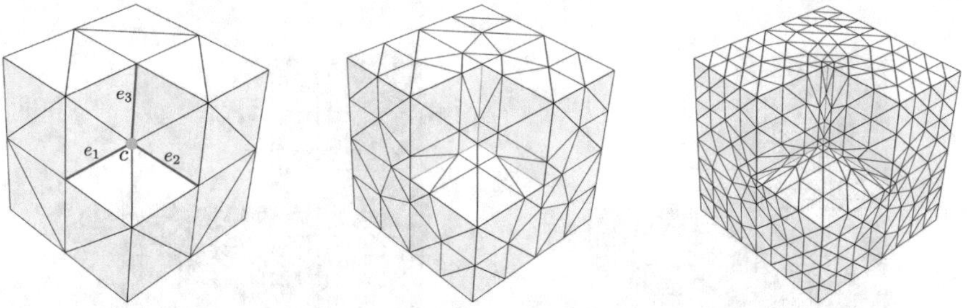

Fig. 3.11 3D graded refinements toward a vertex c and its adjacent edges ($\kappa_c = \kappa_{e_i} = 0.3$, $1 \leq i \leq 3$).

Remark 3.10 Another difference between the v-singular vertex and the e-singular vertex is that as the refinement progresses, the number of v-singular vertices does not change, while the number of e-singular vertices grows. The latter case occurs when there is a singular edge in the initial mesh. For example, in Figs. 3.7 and 3.9, there is always one v-singular vertex regardless of the level of refinements. In Figs. 3.8, 3.10, and 3.11, the new nodes on the singular edge generated in each refinement are e-singular vertices. These e-singular vertices lead to anisotropic decompositions toward the singular edge. Nevertheless, all the singular points in the triangulation are on the boundary of the domain. The singular vertices and singular edges in Definition 3.1 are special points and line segments on the boundary that are defined for the Dirichlet problem (3.32). Recall that in (3.3) where the more general mixed boundary condition is imposed, the set $\mathcal{E}$ (resp. C) also includes extra "edges" that are line segments on the boundary when $d = 3$ (resp. extra "vertices" that are isolated boundary points when $d = 2$ or intersection points of edges in $\mathcal{E}$ when $d = 3$) where the boundary condition changes. In this case, Algorithm 3.9 can also be applied by treating the augmented vertex and edge sets in the same manner as the corresponding sets for the Dirichlet problem. In addition, this refinement principle can be used to produce graded meshes for other types of elements. We shall describe the graded quadrilateral mesh algorithms in the next chapter and refer interested readers to [14] for the graded 3D prism elements.

Although often technically involved, the development of effective finite element methods for singular problems has been a central focus in computational mathematics. The idea of changing the element size according to the local behavior of the solution has been investigated in many works. For 2D corner singularities, in addition to the references mentioned in Remark 3.6, this idea has led to other a-priori analysis based graded finite element methods. See [18, 21, 22, 34, 103, 104] and references therein. Meanwhile, depending on the computation of the local numerical error and on a-posteriori analysis, the adaptive finite element methods [3, 15, 30, 40, 41, 48, 96, 114, 115, 117] have also been successful approximating singular solutions. Despite the different principles (*a-priori* or *a-posteriori*) applied, these 2D locally refined elements are *isotropic* and shape regular (see Fig. 3.12), which is aligned with the behavior of corner singularities near the vertices.

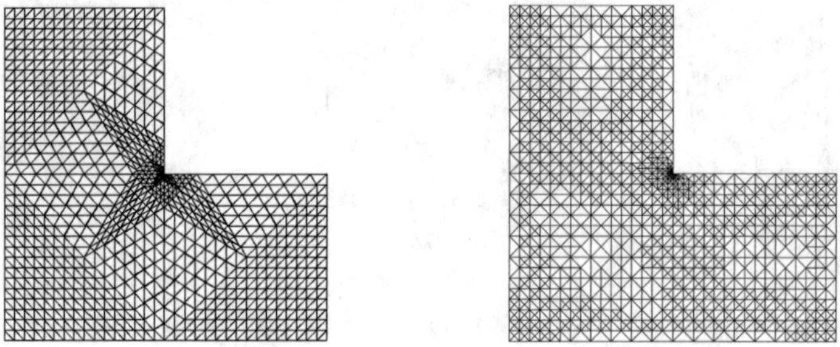

Fig. 3.12 Locally refined meshes for 2D singularities around the reentrant corner: a graded mesh derived from a-priori analysis (left); an adaptive mesh based on a-posteriori estimates (right).

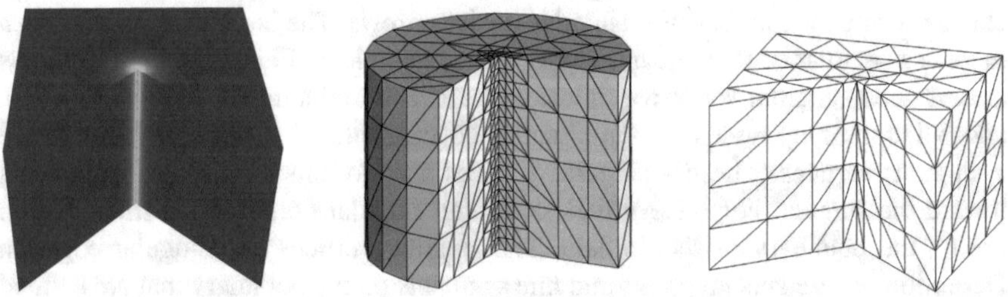

Fig. 3.13 A 3D anisotropic edge singularity (left). Other graded meshes for singularities near an edge: an isotropic mesh [6] (center); an anisotropic mesh (right).

The development of effective 3D finite element methods for singular solutions is much more challenging, especially when the solution possesses anisotropic edge singularities. See the first picture in Fig. 3.13 for an illustration of such functions. According to the aspect ratio of the element, the 3D graded mesh algorithms can be divided into two categories: isotropic and anisotropic. See the last two pictures in Fig. 3.13 for example. The *isotropic* meshes include most of the adaptive meshes from a-posteriori estimates [51, 53, 117] and the dyadic-partitioning meshes [8, 57, 92] based on a-priori analysis. These meshes are shape regular but the associated finite element methods lose the optimal convergence when the edge singularity is strong or when high-order finite elements are used. After all, using isotropic meshes to approximate anisotropic solutions can be a mismatch. For *anisotropic* meshes, tetrahedral elements are usually used to take into account the rather general geometry of the domain. Besides the graded mesh algorithms presented in this book, some other anisotropic mesh algorithms in the literature include the one in [6, 9, 10, 13] that is based on the coordinate transformation from a quasi-uniform mesh, and the one in [23, 26] that involves extra steps for prism refinements to maintain the angle condition of the simplex. For both meshes, the finite element methods obtain the optimal convergence. However, these algorithms are complicated to implement in general polyhedral domains and may not result in well-structured (nested) finite element spaces. This is because these algorithms have different geometric restrictions on simplexes in order to keep the maximum angle condition of the mesh. Recall that according to the works of Babuška and Aziz [17] and of Křížek [74], the maximum angle condition is often a rule of thumb to start with in developing numerical schemes. For prism domains, there are also tensor-product anisotropic meshes based on 2D graded algorithms [14, 106] available. Hexahedral meshes can be constructed in this case by allowing hanging nodes.

Chapter 4
Error Estimates in Polygonal Domains

Consider the second-order elliptic equations with mixed boundary conditions in a polygonal domain. This chapter is concerned with the regularity analysis in a class of weighted Sobolev spaces and the effective graded finite element approximations for possible singular solutions due to the nonsmoothness of the domain or to the change of boundary conditions. In particular, we show that the Laplace operator with the associated boundary condition is a Fredholm operator between these weighted spaces. The regularity estimates for singular solutions in weighted spaces resemble the *full regularity* estimates for smooth domains in the usual Sobolev space. Based on these regularity results and the graded mesh algorithm in Chap. 3, 2D graded mesh algorithms are formulated that can recover the optimal convergence rate in the finite element approximation, even when the solution is singular. Rigorous error analysis and various numerical examples are presented. This chapter is suitable to readers who are interested in 2D graded meshes and interested in regularity and error analysis in weighted Sobolev spaces for solutions with singularities.

Let $\Omega \subset \mathbb{R}^2$ be a bounded polygonal domain with boundary $\partial\Omega$. Given $f \in L^2(\Omega)$, consider the elliptic boundary value problem

$$-\Delta u = f \quad \text{in} \quad \Omega, \qquad u = 0 \quad \text{on} \quad \Gamma_D \quad \text{and} \quad \partial_n u = 0 \quad \text{on} \quad \Gamma_N, \tag{4.1}$$

where Γ_D and Γ_N consist of open subsets of the boundary such that $\overline{\Gamma_D} \cup \overline{\Gamma_N} = \partial\Omega$ and $\Gamma_D \neq \emptyset$. See (3.3) for detailed description of Γ_D, Γ_N, and the vertex set C. Then for (4.1), the vertex set $C = \{c_i\}_{i=1}^I$ consists of all the nonsmooth boundary points and the points where the boundary condition changes. According to the regularity estimates in Sect. 2.3 and in Sect. 3.2, the solution may possess singularities near a vertex that can slow down the convergence of the numerical solution.

H. Li, *Graded Finite Element Methods for Elliptic Problems in Nonsmooth Domains*, Surveys and Tutorials in the Applied Mathematical Sciences 10,
https://doi.org/10.1007/978-3-031-05821-9_4

4.1 Regularity Analysis in Weighted Sobolev Spaces

From now on, to simplify the presentation, we assume no two adjacent sides of Ω in (4.1) are both assigned the Neumann boundary condition. The *full regularity* estimate

$$\|u\|_{H^{m+2}(\Omega)} \leq C\|f\|_{H^m(\Omega)}, \quad m \geq 0$$

in general does not hold for (4.1) in the presence of nonsmooth boundary points. See Sect. 3.2 for discussions on the corner singularity that does not belong to H^2. For the Dirichlet problem, these singularities are associated with reentrant corners. Meanwhile, for $m > 0$, corners with smaller angles can also result in singular components of the solution that do not belong to H^{m+2}. A useful tool for the study of singular solutions is the analysis in weighted Sobolev spaces. In this section, we review regularity results for (4.1) in weighted spaces. Many of the results can be found in Kondratiev's seminal paper [71] along with the works [4, 25, 49, 61, 66, 72, 73] and references therein.

Definition 4.1 (Weighted Sobolev Space) Recall $C = \{c_i\}_{i=1}^I$. Let $r_i(x)$ be the distance from $x \in \Omega$ to $c_i \in C$. Recall the multi-index $\alpha = (\alpha_1, \alpha_2) \in \mathbb{Z}^2_{\geq 0}$ with nonnegative integer components from Definition 2.6, such that $\partial^\alpha = \partial_x^{\alpha_1}\partial_y^{\alpha_2}$ and $|\alpha| = \alpha_1 + \alpha_2$. Let $\boldsymbol{a} = (a_1, \ldots, a_I) \in \mathbb{R}^I$ be a vector. For $b \in \mathbb{R}$, set

$$\vartheta := \prod_i r_i \quad \text{and} \quad \vartheta^{b+\boldsymbol{a}} := \prod_i r_i^{b+a_i} = \vartheta^b \vartheta^{\boldsymbol{a}}. \tag{4.2}$$

Then for an integer $m \geq 0$, define the weighted Sobolev space

$$\mathcal{K}^m_{\boldsymbol{a}}(\Omega) := \{v \in H^m_{loc}(\Omega) : \vartheta^{|\alpha|-\boldsymbol{a}}\partial^\alpha v \in L^2(\Omega), \ \forall\, |\alpha| \leq m\}. \tag{4.3}$$

For any open set $G \subseteq \Omega$ and any function v, we define the norm

$$\|v\|_{\mathcal{K}^m_{\boldsymbol{a}}(G)} := \left(\sum_{|\alpha|\leq m} \|\vartheta^{|\alpha|-\boldsymbol{a}}\partial^\alpha v\|^2_{L^2(G)}\right)^{1/2},$$

and the seminorm

$$|v|_{\mathcal{K}^m_{\boldsymbol{a}}(G)} := \left(\sum_{|\alpha|= m} \|\vartheta^{|\alpha|-\boldsymbol{a}}\partial^\alpha v\|^2_{L^2(G)}\right)^{1/2}.$$

The use of the extra weight function ϑ shall allow in $\mathcal{K}^m_{\boldsymbol{a}}(\Omega)$ singular functions that do not belong to the usual Sobolev space $H^m(\Omega)$. It turns out that with a proper selection of the index $\boldsymbol{a}$, the singular solution in (4.1) can be well characterized by the weighted space $\mathcal{K}^m_{\boldsymbol{a}}(\Omega)$. Let ω_i be the interior angle at the vertex $c_i \in C$. We use the local polar coordinates (r_i, θ_i) such that r_i is the distance to c_i, and $\theta_i = 0$ and $\theta_i = \omega_i$ correspond to the two adjacent sides of the boundary at c_i. See also Fig. 2.1. We denote the neighborhood of c_i by

$$N_{\delta,i} := \{(r_i, \theta_i) : 0 < r_i < \delta, 0 < \theta_i < \omega_i\} \subset \Omega, \tag{4.4}$$

where $\delta > 0$ is sufficiently small such that $N_{\delta,i} \cap N_{\delta,j} = \emptyset$ if $i \neq j$. Note that it is the same definition as in (3.7). Based on (4.2), there exist constants $C_2 > C_1 > 0$ independent of i, such that $C_1 r_i \leq \vartheta \leq C_2 r_i$ in $N_{\delta,i}$. The vector exponent notation (4.2) ensures that in $N_{\delta,i}$, the weight $\vartheta^{|\alpha|-\boldsymbol{a}}$ in the space $\mathcal{K}^m_{\boldsymbol{a}}$ (4.3) is equivalent to $r_i^{|\alpha|-a_i}$, where a_i is the ith component of $\boldsymbol{a}$. To simplify the exposition in the text below, we will omit the index i in r_i, θ_i, $N_{\delta,i}$, and a_i when there is no ambiguity on the underlying vertex.

Then we have the following properties for functions in $\mathcal{K}^m_{\boldsymbol{a}}(\Omega)$.

Lemma 4.1 *Define m and $\boldsymbol{a}$ as in Definition 4.1. Then the weighted Sobolev space $\mathcal{K}^m_{\boldsymbol{a}}(G)$ and the Sobolev space $H^m(G)$ are equivalent, where $G \subset \Omega$ is a subset whose closure is away from the vertices. In addition, in the neighborhood N_δ of a vertex, we have*

$$\mathcal{K}^m_{\boldsymbol{a}}(N_\delta) = \{v : r^{-a}(r\partial_r)^i \partial^j_\theta v \in L^2(N_\delta), \ i + j \leq m\},$$

where a is the component of $\boldsymbol{a}$ corresponding to the vertex.

Proof According to Definition 4.1, the function ϑ is smooth and bounded from above and away from 0 in G. Therefore, the H^m-norm and the $\mathcal{K}^m_{\boldsymbol{a}}$-norm are equivalent in this region.

In the neighborhood N_δ, the differential operators ∂_x and ∂_y can be written in terms of the polar coordinates (r, θ) centered at the vertex:

$$\partial_x = (\cos\theta)\,\partial_r - \frac{\sin\theta}{r}\partial_\theta, \qquad \partial_y = (\sin\theta)\,\partial_r + \frac{\cos\theta}{r}\partial_\theta.$$

Since ϑ is equivalent to r in N_δ, we have

$$\begin{aligned}
&\sum_{j+k\leq m} \|\vartheta^{i+j-a}\partial^i_x \partial^j_y u\|_{L^2(N_\delta)} \\
&\quad \leq C \sum_{i+j\leq m} \left\| r^{i+j-a}\left(\cos\theta\,\partial_r - \frac{\sin\theta}{r}\partial_\theta\right)^i \left(\sin\theta\,\partial_r + \frac{\cos\theta}{r}\partial_\theta\right)^j u\right\|_{L^2(N_\delta)} \\
&\qquad \leq C \sum_{i+j\leq m} \left\| r^{-a}(r\partial_r)^i \partial^j_\theta u\right\|_{L^2(N_\delta)}.
\end{aligned}$$

There is a similar formula expressing $r\partial_r$ and ∂_θ in terms of $r\partial_x$ and $r\partial_y$, which provides the opposite inequality and completes the proof. □

According to Lemma 4.1, the spaces $\mathcal{K}^m_{\boldsymbol{a}}$ and H^m are equivalent in regions away from the vertices. In the neighborhood of a vertex $c_i \in C$, we have the observation below.

Lemma 4.2 *Define m and $\boldsymbol{a}$ as in Definition 4.1. In the neighborhood N_δ of a vertex, let $a \in \boldsymbol{a}$ be the parameter associated with the vertex. Then for $a \geq m$, we have*

$$\mathcal{K}^m_{\boldsymbol{a}}(N_\delta) \subset H^m(N_\delta).$$

Proof Let r be the distance to the vertex. Then for any function $v \in \mathcal{K}^m_{\boldsymbol{a}}(N_\delta)$, by (4.2) and (4.3), we have

$$r^{|\alpha|-a}\partial^\alpha v \in L^2(N_\delta), \quad \forall\, |\alpha| \le m.$$

Note that for a multi-index α with $|\alpha| \le m$, $|\alpha| - a \le 0$. Therefore, we obtain

$$\int_{N_\delta} |\partial^\alpha v|^2\, dx \le C \int_{N_\delta} |r^{|\alpha|-a}\partial^\alpha v|^2\, dx < \infty,$$

which implies $v \in H^m(N_\delta)$ and hence completes the proof. □

Meanwhile, we summarize several other important properties of the weighted space.

Lemma 4.3 *The function $\vartheta^{i+j-a}\partial_x^i\partial_y^j\vartheta^a$ is bounded in Ω.*

Proof For any interior subset $G \subset \Omega$ whose closure is away from the vertices, the function $\vartheta^{j+k-a}\partial_x^j\partial_y^k\vartheta^a$ is bounded in G because ϑ is smooth and bounded from 0.

It remains to verify that this function is bounded in the neighborhood N_δ of a vertex as well. By changing x, y into r, θ in the polar coordinates, we have

$$\begin{aligned}
|\vartheta^{i+j-a}\partial_x^i\partial_y^j\vartheta^a| &\le C\left| r^{i+j-a}\left(\cos\theta\,\partial_r - \frac{\sin\theta}{r}\partial_\theta\right)^i\left(\sin\theta\,\partial_r + \frac{\cos\theta}{r}\partial_\theta\right)^j r^a\right| \\
&\le C\left| r^{-a}\sum_{s+t\le i+j}(r\partial_r)^s\partial_\theta^t r^a\right| \\
&\le C\left| r^{-a}\sum_{s\le i+j}(r\partial_r)^s r^a\right| \le C\left|\sum_{s\le i+j} a^s\right|.
\end{aligned}$$

Therefore, $\vartheta^{i+j-a}\partial_x^i\partial_y^j\vartheta^a$ is bounded in Ω, as stated. □

Lemma 4.4 *Let $\boldsymbol{b} = (b_1, \ldots, b_I) \in \mathbb{R}^I$ be a vector. Then the multiplication by $\vartheta^{\boldsymbol{b}}$ defines an isomorphism $\mathcal{K}^m_{\boldsymbol{a}}(\Omega) \to \mathcal{K}^m_{\boldsymbol{a}+\boldsymbol{b}}(\Omega)$. Namely, $\vartheta^{\boldsymbol{b}}\mathcal{K}^m_{\boldsymbol{a}}(\Omega) = \mathcal{K}^m_{\boldsymbol{a}+\boldsymbol{b}}(\Omega)$, where $\vartheta^{\boldsymbol{b}}\mathcal{K}^m_{\boldsymbol{a}}(\Omega) = \{\vartheta^{\boldsymbol{b}}v :\ v \in \mathcal{K}^m_{\boldsymbol{a}}(\Omega)\}$.*

Proof Let $v \in \mathcal{K}^m_{\boldsymbol{a}}(\Omega)$ and $w = \vartheta^{\boldsymbol{b}}v$. Then $|\vartheta^{i+j-\boldsymbol{a}}\partial_x^i\partial_y^j v| \in L^2(\Omega)$, for $i + j \le m$. Thus, we verify $w \in \mathcal{K}^m_{\boldsymbol{a}+\boldsymbol{b}}(\Omega)$ by checking the inequalities below,

$$\begin{aligned}
|\vartheta^{i+j-\boldsymbol{a}-\boldsymbol{b}}\partial_x^i\partial_y^j w| &= \left|\vartheta^{i+j-\boldsymbol{a}-\boldsymbol{b}}\sum_{s\le i,\ t\le j}\binom{i}{s}\binom{j}{t}\partial_x^s\partial_y^t\vartheta^{\boldsymbol{b}}\partial_x^{i-s}\partial_y^{j-t}v\right| \\
&\le C\sum_{s\le i,\ t\le j}|\vartheta^{(i+j-s-t)-\boldsymbol{a}}\partial_x^{i-s}\partial_y^{j-t}v| \in L^2(\Omega),
\end{aligned}$$

with the last inequality followed from Lemma 4.3. Therefore, $\vartheta^{\boldsymbol{b}}\mathcal{K}^m_{\boldsymbol{a}}(\Omega)$ is continuously embedded in $\mathcal{K}^m_{\boldsymbol{a}+\boldsymbol{b}}(\Omega)$. In other words, the map $\vartheta^{\boldsymbol{b}} : \mathcal{K}^m_{\boldsymbol{a}}(\Omega) \to \mathcal{K}^m_{\boldsymbol{a}+\boldsymbol{b}}(\Omega)$ is continuous. On the other hand, because this embedding holds for any vector $\boldsymbol{b}$, we have the opposite inclusion:

$$\mathcal{K}^m_{\boldsymbol{a}+\boldsymbol{b}}(\Omega) = \vartheta^{\boldsymbol{b}}\vartheta^{-\boldsymbol{b}}\mathcal{K}^m_{\boldsymbol{a}+\boldsymbol{b}}(\Omega) \subset \vartheta^{\boldsymbol{b}}\mathcal{K}^m_{\boldsymbol{a}}(\Omega).$$

To complete the proof, we also notice that the inverse of multiplication by $\vartheta^{\boldsymbol{b}}$ is multiplication by $\vartheta^{-\boldsymbol{b}}$, which is also continuous. □

Let $\boldsymbol{1} \in \mathbb{R}^I$ be the constant I-dimensional vector with every entry equal to 1. In addition, $\boldsymbol{0}, \boldsymbol{1/2}, \boldsymbol{3/2}$ are the constant vectors defined in the same way.

Lemma 4.5 *The map* $-\Delta : \mathcal{K}^{m+1}_{\boldsymbol{a+1}}(\Omega) \to \mathcal{K}^{m-1}_{\boldsymbol{a-1}}(\Omega)$ *is continuous for* $m \geq 1$.

Proof It is sufficient to show that there is $C > 0$ such that $\|\Delta v\|_{\mathcal{K}^{m-1}_{\boldsymbol{a-1}}(\Omega)} \leq C\|v\|_{\mathcal{K}^{m+1}_{\boldsymbol{a+1}}(\Omega)}$ for all $v \in \mathcal{K}^{m+1}_{\boldsymbol{a+1}}(\Omega)$. Let $G \subset \Omega$ be a subset whose closure is away from the vertices. In G, Δ is a second-order differential operator with bounded coefficients. Therefore, it defines a bounded operator $H^{m+1}(G) \to H^{m-1}(G)$ [56]. Lemma 4.1 then gives $\|\Delta v\|_{\mathcal{K}^{m-1}_{\boldsymbol{a-1}}(G)} \leq C\|v\|_{\mathcal{K}^{m+1}_{\boldsymbol{a+1}}(G)}$.

In the neighborhood N_δ of a vertex, write $\Delta = r^{-2}\big((r\partial_r)^2 + \partial_\theta^2\big)$. Then Lemma 4.1 and equation $r^{-\lambda}(r\partial_r)^n r^\lambda v = (r\partial_r + \lambda)^n v$ give $\|\Delta v\|_{\mathcal{K}^{m-1}_{\boldsymbol{a-1}}(N_\delta)} \leq C\|v\|_{\mathcal{K}^{m+1}_{\boldsymbol{a+1}}(N_\delta)}$. Adding all the similar inequalities completes the proof. □

Remark 4.1 We can also define weighted Sobolev spaces on the boundary:

$$\mathcal{K}^m_{\boldsymbol{a}}(S) = \oplus_{F\subset S}\mathcal{K}^m_{\boldsymbol{a}}(F),$$

where $S \subseteq \partial\Omega$ is the union of some closed sides F on the boundary $\partial\Omega$. Suppose that $c_i, c_j \in C$ are the two endpoints of F and identify F with the interval $I = [-1, 1]$ such that c_i and c_j correspond to -1 and 1, respectively. Then to describe $\mathcal{K}^m_{\boldsymbol{a}}(F)$, it is sufficient to define

$$\mathcal{K}^m_{\boldsymbol{a}}(I) = \big\{v(t) : (1+t)^{l-a_i}(1-t)^{l-a_j}v^{(l)} \in L^2(I),\ l \leq m\big\},$$

where $v^{(l)} = \partial_t^l v$. In addition, for $s \in [0, \infty)$, we define $\mathcal{K}^s_{\boldsymbol{a}}(I)$ by interpolation, and for $s < 0$, we extend this definition by duality: $\mathcal{K}^s_{\boldsymbol{a}}(I) = \big(\mathcal{K}^{-s}_{-\boldsymbol{a}}(I)\big)^*$. See [4, 5, 25, 75, 94, 113]. In these works, it was also shown that

$$\mathcal{K}^m_{\boldsymbol{a}}(\Omega) \ni v \to v|_S \in \mathcal{K}^{m-1/2}_{\boldsymbol{a-1/2}}(S), \quad m \geq 1, \tag{4.5}$$

is a continuous surjective map, which generalizes the usual property of the trace map for smooth, bounded domains. A similar result holds for the normal derivative, yielding again a continuous and surjective map

$$\mathcal{K}^m_{\boldsymbol{a}}(\Omega) \ni v \to \partial_n v|_S \in \mathcal{K}^{m-3/2}_{\boldsymbol{a-3/2}}(S), \quad m \geq 2.$$

Note that the case $S = \partial\Omega$ is included in these results. It is interesting to compare these trace results with the trace estimates in Sobolev spaces in Sect. 2.2.3.

Using the same notation as in (4.1), consider the general boundary value problem

$$-\Delta u = f \quad \text{in} \quad \Omega, \qquad u = g_D \quad \text{on} \quad \Gamma_D \quad \text{and} \quad \partial_n u = g_N \quad \text{on} \quad \Gamma_N. \tag{4.6}$$

Then the regularity result for (4.6) in weighted spaces is as follows.

Theorem 4.1 *For* $m \geq 1$*, let* $u \in \mathcal{K}^1_{\boldsymbol{a}+1}(\Omega)$ *be a solution of* (4.6) *with* $f \in \mathcal{K}^{m-1}_{\boldsymbol{a}-1}(\Omega)$*,* $g_D \in \mathcal{K}^{m+1/2}_{\boldsymbol{a}+1/2}(\Gamma_D)$*, and* $g_N \in \mathcal{K}^{m-1/2}_{\boldsymbol{a}-1/2}(\Gamma_N)$*. Then we have*

$$\|u\|_{\mathcal{K}^{m+1}_{\boldsymbol{a}+1}(\Omega)} \leq C\Big(\|f\|_{\mathcal{K}^{m-1}_{\boldsymbol{a}-1}(\Omega)} + \|g_D\|_{\mathcal{K}^{m+1/2}_{\boldsymbol{a}+1/2}(\Gamma_D)} + \|g_N\|_{\mathcal{K}^{m-1/2}_{\boldsymbol{a}-1/2}(\Gamma_N)} + \|u\|_{\mathcal{K}^{0}_{\boldsymbol{a}+1}(\Omega)}\Big),$$

for a constant $C = C(m, \boldsymbol{a}) > 0$ *independent of* f*,* g_D*,* g_N*, and* u*.*

Proof This result is standard (see [13, 25, 45, 71, 73, 94, 98]), and thus we include only a sketch. It is sufficient to study close to a vertex of the domain Ω. Assume the vertex is at the origin. A simple proof is obtained then by using a radial partition of unity of the form $\chi_n(x) := \chi_0(2^n x)$ and then applying to the functions $\chi_n u$ the usual regularity results for smooth domains. □

Theorem 4.1 extends the full regularity estimates for smooth domains (Theorem 2.16) to polygonal domains. As a consequence, it gives by induction the following regularity result for the eigenvalue problem associated with the Laplace operator.

Corollary 4.1 *Assume that* $u \in \mathcal{K}^2_{\boldsymbol{a}}(\Omega)$ *satisfies the eigenvalue problem*

$$-\Delta u = \lambda u \quad \text{in} \quad \Omega, \qquad u = 0 \quad \text{on} \quad \Gamma_D \quad \text{and} \quad \partial_n u = 0 \quad \text{on} \quad \Gamma_N.$$

Then $u \in \mathcal{K}^m_{\boldsymbol{a}}(\Omega)$ *for any* $m \in \mathbb{N}$*.*

The solution of (4.1) is well defined in $H^1_D(\Omega)$ when $\Gamma_D \neq \emptyset$ because the associated bilinear form is both continuous and coercive on $H^1_D(\Omega)$. See Example 2.2. However, the well-posedness of the solution is not obvious in weighted spaces with arbitrary sub-indices. In what follows, we present the well-posedness result in weighted spaces. Recall that two sides of the boundary $\partial\Omega$ are adjacent if they are connected by a vertex in C.

Theorem 4.2 *Use the same notation as in Theorem 4.1. Assume that in (4.6), no two adjacent sides are both assigned the Neumann boundary condition. Then there exists* $\boldsymbol{\eta} = (\eta_1, \ldots, \eta_I)$ *with* $\eta_i > 0$ *for all* $1 \leq i \leq I$*, such that for any* $f \in \mathcal{K}^{m-1}_{\boldsymbol{a}-1}(\Omega)$*,* $g_D \in \mathcal{K}^{m+1/2}_{\boldsymbol{a}+1/2}(\Gamma_D)$*, and* $g_N \in \mathcal{K}^{m-1/2}_{\boldsymbol{a}-1/2}(\Gamma_N)$*,* $m \geq 0$*, and* $|a_i| < \eta_i$ $(1 \leq i \leq I)$*, (4.6) has a unique solution* $u \in \mathcal{K}^{m+1}_{\boldsymbol{a}+1}(\Omega)$*. In addition, it holds that*

$$\|u\|_{\mathcal{K}^{m+1}_{\boldsymbol{a}+1}(\Omega)} \leq C\Big(\|f\|_{\mathcal{K}^{m-1}_{\boldsymbol{a}-1}(\Omega)} + \|g_D\|_{\mathcal{K}^{m+1/2}_{\boldsymbol{a}+1/2}(\Gamma_D)} + \|g_N\|_{\mathcal{K}^{m-1/2}_{\boldsymbol{a}-1/2}(\Gamma_N)}\Big),$$

for a constant $C > 0$ *independent of* f*,* g_D*, and* g_N*.*

Proof Assume $g_D = 0$. If $g_D \neq 0$, one can use the trace estimate (4.5) to convert (4.6) into an elliptic equation with zero Dirichlet boundary data. Then the variational formulation for (4.6) is

$$a(u, v) := (\nabla u, \nabla v) = (f, v) + \langle g_N, v \rangle_{\Gamma_N},$$

where $\langle g_N, v\rangle_{\Gamma_N}$ is the value of the distribution g_N on the function v on Γ_N. We look for a unique solution $u \in \mathcal{K}_D := \mathcal{K}_1^1(\Omega) \cap \{v|_{\Gamma_D} = 0\}$ that satisfies this formulation for any $v \in \mathcal{K}_D$.

We shall prove the theorem first for the case $m = 0$. This case will follow from the Lax–Milgram Theorem and the strict coercivity of the bilinear form $a(\cdot, \cdot)$ on the space $\mathcal{K}_D$. Note that by the Cauchy–Schwarz inequality and the definition of the space $\mathcal{K}_1^1(\Omega)$, the bilinear form $a(\cdot, \cdot)$ is continuous, namely, $a(u, v) \le C\|u\|_{\mathcal{K}_1^1(\Omega)}\|v\|_{\mathcal{K}_1^1(\Omega)}$.

For each vertex $c_i \in C$, recall its neighborhood $N_{\delta,i}$ (4.4). Then we show the coercivity of $a(\cdot, \cdot)$ in each $N_{\delta,i}$ and in $G := \Omega \setminus (\cup_i N_{\delta,i})$. In the proof below, we shall use the notation (r, θ) instead of (r_i, θ_i) for the polar coordinates in $N_{\delta,i}$. By the assumption that no two adjacent sides are both assigned the Neumann boundary condition, Γ_D is not empty. Then in G, our desired inequality is just the usual Poincaré inequality. In $N_{\delta,i}$, because ϑ is equivalent to r, it is sufficient to verify the following inequality in every $N_{\delta,i}$:

$$\int_{N_{\delta,i}} \left((\partial_x u)^2 + (\partial_y u)^2\right) dxdy \ge C \int_{N_{\delta,i}} \frac{u^2}{r^2}\, dxdy.$$

Recall that at least one side of c_i has the Dirichlet boundary condition and that ω_i is the interior angle at the vertex c_i. Assume that the positive x-semiaxis corresponds to the side with the Dirichlet boundary condition. The one-dimensional Poincaré inequality for θ in $N_{\delta,i}$ then gives, for each fixed r,

$$\int_0^{\omega_i} u^2\, d\theta \le C \int_0^{\omega_i} (\partial_\theta u)^2\, d\theta.$$

By integrating in polar coordinates, we have

$$\int_{N_{\delta,i}} \frac{u^2}{r^2}\, dxdy = \int_{N_{\delta,i}} \frac{u^2}{r}\, drd\theta \le C \int_{N_{\delta,i}} \frac{(\partial_\theta u)^2}{r}\, drd\theta.$$

Since $\int_{N_{\delta,i}} \left((\partial_x u)^2 + (\partial_y u)^2\right) dxdy = \int_{N_{\delta,i}} \left(r(\partial_r u)^2 + (\partial_\theta u)^2/r\right) drd\theta$, we get

$$\int_{N_{\delta,i}} \frac{u^2}{\vartheta^2}\, dxdy \le C \int_{N_{\delta,i}} \frac{u^2}{r^2}\, dxdy \le C \int_{N_{\delta,i}} \left((\partial_x u)^2 + (\partial_y u)^2\right) dxdy.$$

Then the strict coercivity of the bilinear form $a(\cdot, \cdot)$ on $\mathcal{K}_D$ follows by adding all these inequalities in $N_{\delta,i}$ and in G. Consequently, the Lax–Milgram Theorem proves that $-\Delta : \mathcal{K}_D \to (\mathcal{K}_D)^*$ is an isomorphism, which is our result for $m = 0$ and $\boldsymbol{a} = \boldsymbol{0}$.

We next use the continuity of the family $\vartheta^{-\boldsymbol{a}}(-\Delta)\vartheta^{\boldsymbol{a}} : \mathcal{K}_D \to (\mathcal{K}_D)^*$ (see for example the proof of Lemma 4.5) and Lemma 4.4 to prove the result for $m = 0$ and $|a_i| < \eta_i$ for some $\boldsymbol{\eta}$ that depends only on the domain and the operator $-\Delta$. Then

Theorem 4.1 shows that if the result is true for $m = 0$ and $\boldsymbol{a}$, it also holds for $m \geq 1$ and $\boldsymbol{a}$. This completes the proof. □

We now estimate the parameter $\boldsymbol{\eta}$. It turns out that in 2D, it is possible to explicitly determine the optimal value of η_i in Theorem 4.2. For a vertex $c_i \in C$, recall the small neighborhood $N_{\delta,i}$ in Theorem 4.2. By freezing the coefficients of $-\Delta$ to c_i, the behavior of the solution u in $N_{\delta,i}$ is controlled by the eigenvalue and eigenfunction of a 1D eigenvalue problem (Theorem 3.4), or more precisely, by its operator pencil's spectrum Σ_i [71, 73], which we proceed to describe.

Let $\mathrm{i} := \sqrt{-1}$. The *operator pencil* $P_i(\tau)$ associated to $-\Delta$ at c_i is defined by

$$P_i(\tau)\phi(\theta) = r^{2-\mathrm{i}\tau-a_i}(-\Delta)\big(r^{\mathrm{i}\tau+a_i}\phi(\theta)\big),$$

where $\phi(\theta)$ is an arbitrary smooth function. Using the formula $\Delta = r^{-2}\big((r\partial_r)^2 + \partial_\theta^2\big)$ and $(r\partial_r)^2 r^{\mathrm{i}\tau+a_i} = r^{\mathrm{i}\tau+a_i}(\mathrm{i}\tau + a_i)^2 = -r^{\mathrm{i}\tau+a_i}(\tau - \mathrm{i}a_i)^2$, we obtain

$$P_i(\tau) = (\tau - \mathrm{i}a_i)^2 - \partial_\theta^2. \tag{4.7}$$

For $k \geq 1$, define $\Sigma_i = \{\pm\lambda_k : \lambda_k \geq 0\}$, where λ_k^2 is an eigenvalue of $-\partial_\theta^2$ in $(0, \omega_i)$ with associated boundary conditions (Dirichlet–Dirichlet (D–D) or Dirichlet–Neumann (D–N)) inherited from (4.1). Based on (3.11), this gives

$$\Sigma_i = \Big\{\pm\frac{k\pi}{\omega_i}\ (\mathrm{D-D})\ \text{or}\ \pm\frac{(k-1/2)\pi}{\omega_i}\ (\mathrm{D-N})\Big\}. \tag{4.8}$$

Therefore, $P(\tau)$ is invertible for all $\tau \in \mathbb{R}$, as long as $a_i \notin \Sigma_i$. For $a_i \notin \Sigma_i$, using Kondratiev's method [71], one can further obtain for $m \geq 0$,

$$\Delta_{\boldsymbol{a}} = \vartheta^{-\boldsymbol{a}}(-\Delta)\vartheta^{\boldsymbol{a}} : \mathcal{K}_{\mathbf{1}}^{m+1}(\Omega) \cap \{v|_{\Gamma_D} = 0,\ \partial_n v|_{\Gamma_N} = 0\} \to \mathcal{K}_{\mathbf{-1}}^{m-1}(\Omega) \tag{4.9}$$

is a Fredholm operator.

Recall that a *continuous* operator $A : X \to Y$ between Banach spaces is Fredholm if the kernel of A (namely, the space $\ker(A) := \{x \in X : Ax = 0\}$) and Y/AX are finite-dimensional spaces. If $A : X \to Y$ is a continuous Fredholm operator, we define its index by the formula $\operatorname{ind}(A) = \dim\big(\ker(A)\big) - \dim(Y/AX)$. Thus, we have the following result.

Theorem 4.3 *Assume the same conditions as in Theorem 4.2. Then the boundary value problem* (4.6) *defines a Fredholm operator*

$$\tilde{\Delta}_{\boldsymbol{a}} := (-\Delta, \gamma, \partial_n) : \mathcal{K}_{\boldsymbol{a}+\mathbf{1}}^{m+1}(\Omega) \to \mathcal{K}_{\boldsymbol{a}-\mathbf{1}}^{m-1}(\Omega) \oplus \mathcal{K}_{\boldsymbol{a}+\mathbf{1/2}}^{m+1/2}(\Gamma_D) \oplus \mathcal{K}_{\boldsymbol{a}-\mathbf{1/2}}^{m-1/2}(\Gamma_N) \tag{4.10}$$

for all $\boldsymbol{a}$ *such that* $a_i \notin \Sigma_i$, *where* γ *is the restriction operator to* Γ_D.

We are now ready to determine the vector $\boldsymbol{\eta}$ in Theorem 4.2. Let η_i be the minimum value of $|\lambda_k|$ for $\lambda_k \in \Sigma_i$. Then $\eta_i > 0$ if no two adjacent sides are both assigned the Neumann boundary condition.

Proposition 4.1 *Assume the same conditions as in Theorem 4.2, and let $\boldsymbol{\eta}$ be defined as above. Then $\boldsymbol{\eta}$ consists of the largest values η_i for which Theorem 4.2 is valid.*

Proof As in the proof of Theorem 4.2, we assume $m = 0$. Then the operator $\Delta_{\boldsymbol{a}}$ in (4.9) is invertible iff $\tilde{\Delta}_{\boldsymbol{a}}$ in (4.10) is invertible because the trace map is continuous and surjective. Then it is sufficient to show that $\Delta_{\boldsymbol{a}}$ is invertible if $|a_i| < \eta_i$ since $\Delta_{\boldsymbol{a}}$ is not Fredholm for any $a_i = \eta_i$.

If $|a_i| < \eta_i$ for $1 \leq i \leq I$, by Theorem 4.3, $\Delta_{\boldsymbol{a}}$ is Fredholm. Theorem 4.2 states that $\tilde{\Delta}_{\boldsymbol{a}}$ is invertible for $\boldsymbol{a} = \boldsymbol{0}$. By the homotopy invariance of the index, $\tilde{\Delta}_{\boldsymbol{a}}$ is Fredholm of index zero when $|a_i| < \eta_i$. Note that the kernel of $\Delta_{\boldsymbol{a}}$ decreases as a_i increases. We therefore conclude that $\Delta_{\boldsymbol{a}}$ is injective for $0 \leq a_i < \eta_i$ with $1 \leq i \leq I$. Since the index of $\Delta_{\boldsymbol{a}}$ is zero, we have that $\Delta_{\boldsymbol{a}}$ is in fact an isomorphism. In addition, note that $(\Delta_{\boldsymbol{a}})^* = \Delta_{-\boldsymbol{a}}$. Hence, we obtain the invertibility of $\Delta_{\boldsymbol{a}}$ for $-\eta_i < a_i \leq 0$ as well. This completes the proof. □

Remark 4.2 The analysis of the singular solution in the weighted Sobolev space shares some common ground with the procedure in Sect. 3.2 to derive the singular expansion of the solution. Both of them are closely related to the 1D eigenvalue problem (3.10) that can be derived as follows. Using a smooth cut-off function, one extends the original elliptic problem near the vertex to an elliptic problem in an infinite sector. By changing the variables $(x, y) \to (r, \theta) \to (t, \theta)$, where (r, θ) are the polar coordinates centered at the vertex and $t = \ln r$, the Laplace operator $\partial_x^2 + \partial_y^2$ in the sector becomes $\partial_t^2 + \partial_\theta^2$ in an infinite strip $\mathbb{R} \times (0, \omega)$, where ω is the interior angle at the vertex. Then applying the partial Fourier transform with respect to t, one obtains a 1D elliptic operator with the parameter: $\partial_\theta^2 - \tau^2$. It turns out that the associated eigenvalue problem is critical to determine the local behavior of the solution in the neighborhood of the vertex and the Fredholm property of the elliptic operator in the weighted space. With the vector exponent notation (4.2), the solution behavior at each vertex is characterized by the corresponding component in the index vector $\boldsymbol{a}$.

Example 4.1 In the following example, we illustrate the connection between the weighted regularity estimates and the singular expansion of the solution (Sect. 3.2). Consider (4.1) with the Dirichlet boundary condition ($\Gamma_D = \partial\Omega$). Then at the vertex c_i, $\eta_i = \pi/\omega_i$. This implies that the larger the interior angle, the smaller value $|a_i| < \eta_i$ one can choose in the regularity estimate. In the case where the interior angle $\omega_i > \pi$, according to Theorem 3.4, the singular (non-H^2) component of the solution near c_i is $r^{\eta_i} \sin(\eta_i\theta)$. By a direct calculation, one can verify that with $m \geq 0$, $r^{\eta_i} \sin(\eta_i\theta) \in \mathcal{K}_{\boldsymbol{a}+\boldsymbol{1}}^{m+1}(N_{\delta,i})$ for any $|a_i| < \eta_i$ but not for $|a_i| \geq \eta_i$.

4.2 2D Graded Meshes and Mesh Layers

In this section, we discuss the finite element approximation for (4.1) on 2D graded meshes described in Chap. 3. We first revisit the rather general graded mesh algorithm (Algorithm 3.9) and extract a more explicit 2D version for solving (4.1).

Algorithm 4.4 (2D Graded Mesh) Let Ω be a polygonal domain. Recall the vertex set $C = \{c_i\}_{i=1}^{I}$ for (4.1) includes all the vertices of the domain and points where the boundary condition changes. Let $\mathcal{T}$ be a triangulation of Ω with triangles, such that each vertex in C is also a vertex of a triangle in $\mathcal{T}$, and no triangle in $\mathcal{T}$ contains more than one singular vertex in C. Let $\kappa_i \in (0, 0.5]$ be the grading parameter for c_i. Define the vector $\boldsymbol{\kappa} = (\kappa_1, \ldots, \kappa_I)$. A $\boldsymbol{\kappa}$-refinement of $\mathcal{T}$, denoted by $\boldsymbol{\kappa}(\mathcal{T})$, is obtained by dividing each edge pq of $\mathcal{T}$ into two parts as follows:

- If neither p nor q is a singular vertex, then divide pq into two equal parts.
- Otherwise, if $p = c_i \in C$ is a singular vertex, we divide pq into pr and rq such that $|pr| = \kappa_i |pq|$.

Connecting the new nodes will divide each triangle into four triangles (Fig. 4.1). Given an initial triangulation $\mathcal{T}_0$, the associated family of graded meshes $\{\mathcal{T}_n : n \geq 0\}$ is defined recursively, $\mathcal{T}_n = \boldsymbol{\kappa}(\mathcal{T}_{n-1})$. In the text below, we also use κ_p to denote the grading parameter for a point $p \in C$. Namely, $\kappa_p = \kappa_i$ when $p = c_i$. See Fig. 3.7 for the case where Ω is an L-shaped domain: the parameter corresponding to the reentrant corner (singular vertex) is $\kappa_c = 0.2$, and the parameters for other vertices are 0.5. □

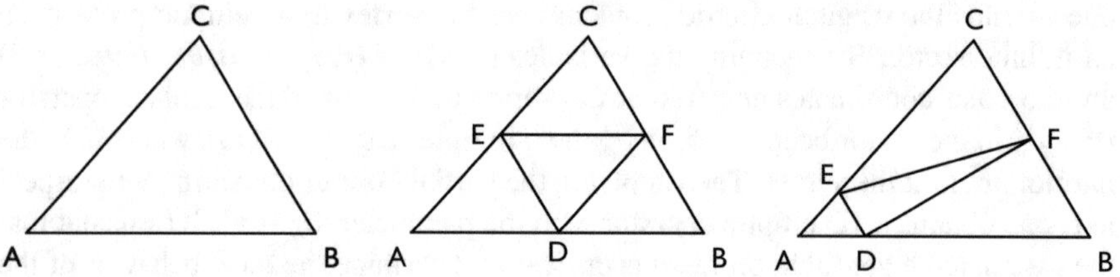

Fig. 4.1 The initial triangle $\triangle^3 ABC$ (left); the uniform refinement (center); the κ-refinement with $\kappa_A < 0.5$ for the singular vertex A (right), $\kappa_A = \frac{|AD|}{|AB|} = \frac{|AE|}{|AC|} = \frac{|DE|}{|BC|}$.

In the graded mesh $\mathcal{T}_n$, the mesh size varies according to the distance to the singular vertices. The recursive mesh construction in Algorithm 4.4 allows us to classify the triangles with comparable sizes in $\mathcal{T}_n$ via mesh layers.

Definition 4.2 (2D Mesh Layer) Let $T_{(0)} \in \mathcal{T}_0$ be an initial triangle consisting of a singular vertex $c_i \in C$. Let $T_{(j)} \subseteq T_{(0)}$ be the triangle in $\mathcal{T}_j$, $0 \leq j \leq n$, which is also attached to the vertex c_i. For $0 \leq j < n$, we define the jth mesh layer of $\mathcal{T}_n$ on $T_{(0)}$ to be the region $L_{i,j} := T_{(j)} \setminus T_{(j+1)}$; and for $j = n$, the nth layer is $L_{i,n} := T_{(n)}$. See Fig. 4.2.

Remark 4.3 The mesh layer is a group of triangles in $\mathcal{T}_n$ that are defined on initial triangles attached to the singular vertices of the domain. For example, suppose $T_{(0)} = \triangle^3 ABC$ and $A = c_i$ is the singular vertex (see Fig. 4.1). Then $L_{i,0}$ is the trapezoid $BCED$. When the initial triangle includes a singular vertex c_i ($\kappa_i < 0.5$), the resulting

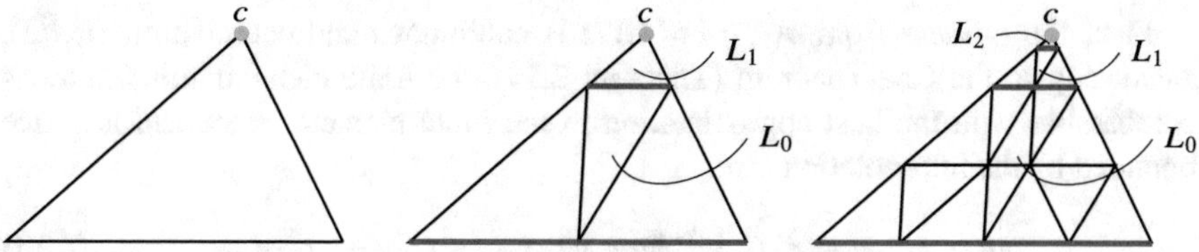

Fig. 4.2 Mesh layers (left–right): the initial triangle $T_{(0)}$ with a singular vertex $c \in C$; two layers (L_0 and L_1) after one refinement; three layers (L_0, L_1, and L_2) after two refinements. Layer boundaries are marked red, and the index i is omitted in the notation $L_{i,j}$.

triangles are smaller near the singular vertex. When the initial triangle is away from the singular vertex, Algorithm 4.4 is identical to the midpoint decomposition, and therefore, the final mesh is quasi-uniform. Recall the function ϑ in the weighted space $\mathcal{K}_a^m$ (Definition 4.1). A straightforward calculation shows that the triangles if $\mathcal{T}_n$ in the layer $L_{i,j}$ are isotropic and shape regular with mesh size $O(2^{j-n}\kappa_i^j)$. In addition, we have

$$\vartheta|_{L_{i,j}} \sim r_i|_{L_{i,j}} \sim \kappa_i^j, \text{ for } 0 \leq j < n; \quad \text{and} \quad \vartheta|_{L_{i,n}} \sim r_i|_{L_{i,n}} \leq C\kappa_i^n. \tag{4.11}$$

In comparison with Algorithm 3.9, Algorithm 4.4 involves singular vertices but no singular edges. This is due to the fact that the solution of (4.1) may possess corner singularities but rather smooth along the side of the boundary. For problems with the solution singularity in the normal direction to the boundary side, another version of Algorithm 3.9 can be developed that takes singular edges into account [79] (see also Fig. 3.8).

Recall the space $H_D^1(\Omega)$ in (2.26) with the zero trace on Γ_D. Let S_n be the Lagrange finite element space of degree $m \geq 1$ associated with the graded mesh $\mathcal{T}_n$ as defined in (2.28). Then $S_n \subset H_D^1(\Omega)$, and the finite element solution $u_n \in S_n$ of (4.1) satisfies

$$a(u_n, v) = \int_\Omega \nabla u_n \cdot \nabla v \, dx = \int_\Omega f v \, dx = (f, v), \qquad \forall\, v \in S_n. \tag{4.12}$$

4.3 Interpolation Error Estimates

We proceed with the interpolation error analysis for functions in the finite element space S_n on the graded mesh $\mathcal{T}_n$. Recall that when the solution lacks the desired regularity, the quasi-uniform mesh leads to numerical solutions with the reduced convergence rate (2.35). We shall show that with a proper selection of the grading parameter, the graded mesh gives rise to finite element approximations that converge to the solution of (4.1) in the optimal rate, even when the solution is singular.

First, the bilinear from $a(\cdot,\cdot)$ in (4.12) is continuous and coercive on $H^1_D(\Omega)$. According to the Céa Theorem (Theorem 2.19), the finite element solution u_n is comparable with the best approximation in the finite element space and is hence bounded by the interpolation error

$$\|u - u_n\|_{H^1(\Omega)} \leq C \inf_{v \in S_n} \|u - v\|_{H^1(\Omega)} \leq C\|u - u_I\|_{H^1(\Omega)}, \tag{4.13}$$

where $u_I := \mathcal{I}_{\mathcal{T}_n} u \in S_n$ is the finite element interpolation introduced in Definition 2.13. Recall the regularity estimates in Theorem 4.2. Suppose in (4.1) $f \in \mathcal{K}^{m-1}_{\boldsymbol{a-1}}(\Omega)$ for $\boldsymbol{0} < \boldsymbol{a} < \boldsymbol{\eta}$, where $\boldsymbol{\eta}$ is given in Proposition 4.1. For $\boldsymbol{a}, \boldsymbol{b} \in \mathbb{R}^I$, by $\boldsymbol{a} < \boldsymbol{b}$, we mean $a_i < b_i$ for all $1 \leq i \leq I$. Thus $u \in \mathcal{K}^{m+1}_{\boldsymbol{a+1}}(\Omega)$. The assumption $f \in \mathcal{K}^{m-1}_{\boldsymbol{a-1}}(\Omega)$ covers a large class of functions in practical models. For example, when $\boldsymbol{a} \leq \boldsymbol{1}$ and $m = 1$, we see that $L^2(\Omega) \subset \mathcal{K}^0_{\boldsymbol{a-1}}(\Omega)$. By (4.13), to obtain the finite element error estimate on $\mathcal{T}_n$, it is sufficient to derive the interpolation error on the graded mesh.

The mesh sizes are different in different regions of the domain, but they are comparable in the same mesh layer. The interpolation error estimate is based on the error analysis in each mesh layer. First, we obtain the interpolation error estimate in regions away from the singular vertices.

Lemma 4.6 *Let $\mathcal{T}_n$ be the graded mesh defined in Algorithm 4.4. Define $h := 2^{-n}$ and recall $\boldsymbol{\eta}$ in Proposition 4.1. Let $T_{(0)} \in \mathcal{T}_0$ be an initial triangle that does not consist of any singular vertex. Then there are two possible scenarios for $T_{(0)}$:*
(i) If $T_{(0)}$ is away from the vertex set C, we have

$$\|u - u_I\|_{H^1(T_{(0)})} \leq Ch^m \|u\|_{\mathcal{K}^{m+1}_{\boldsymbol{a+1}}(T_{(0)})}, \qquad \boldsymbol{0} < \boldsymbol{a} < \boldsymbol{\eta}. \tag{4.14}$$

(ii) If $T_{(0)}$ contains a regular vertex $c_i \in C$, for which the grading parameter $\kappa_i = 0.5$, we have

$$\|u - u_I\|_{H^1(T_{(0)})} \leq Ch^m \|u\|_{\mathcal{K}^{m+1}_{\boldsymbol{a+1}}(T_{(0)})}, \qquad \boldsymbol{0} < \boldsymbol{a} < \boldsymbol{\eta},\ a_i \geq m. \tag{4.15}$$

In these estimates, $C > 0$ is independent of n and u, and $a_i \in \boldsymbol{a}$ is the component corresponding to c_i.

Proof We first show (4.14). Since $T_{(0)}$ is away from the vertex set C, by Lemma 4.1, $\mathcal{K}^{m+1}_{\boldsymbol{a+1}}$ and H^{m+1} are equivalent on $T_{(0)}$. Meanwhile, according to Algorithm 4.4, the final mesh on $T_{(0)}$ is quasi-uniform with mesh size $O(2^{-n})$. Then the desired estimate follows from the interpolation error estimates in Lemma 2.5.

When $T_{(0)}$ consists of a regular vertex c_i, the final mesh on $T_{(0)}$ is also quasi-uniform with mesh size $O(2^{-n})$. Then for $a_i \geq m$, by Lemma 2.5 and Lemma 4.2, we obtain

$$\|u - u_I\|_{H^1(T_{(0)})} \leq Ch^m \|u\|_{H^{m+1}(T_{(0)})} \leq Ch^m \|u\|_{\mathcal{K}^{m+1}_{\boldsymbol{a+1}}(T_{(0)})}.$$

This completes the proof. □

Suppose that $T_{(0)} \in \mathcal{T}_0$ is an initial triangle that includes the *singular* vertex $c_i \in C$. To carry out the error analysis in the layer $L_{i,j} \subset T_{(0)}$, we need the following result regarding the dilation property in the weighted space. Recall the grading parameter $0 < \kappa_i < 0.5$ on $T_{(0)}$. Consider a new coordinate system that is a simple translation of the old xy-coordinate system with the vertex c_i now at the origin of the new coordinate system. For $0 \leq j \leq n$, we define the reference region by dilation:

$$\hat{L}_i := \kappa_i^{-j} L_{i,j}. \tag{4.16}$$

One can see that when $0 \leq j < n$, the region $\hat{L}_i$ coincides with the layer $L_{i,0}$ and is independent of j, while when $j = n$, $\hat{L}_i$ coincides with the initial triangle $T_{(0)}$. For $(x, y) \in L_{i,j}$, let $(\hat{x}, \hat{y}) := (\kappa_i^{-j} x, \kappa_i^{-j} y)$ be the image of the transformation (4.16) in $\hat{L}_i$. Meanwhile, define the dilation of a function v in $L_{i,j}$:

$$\hat{v}(\hat{x}, \hat{y}) := v(x, y), \qquad \forall\, (\hat{x}, \hat{y}) \in \hat{L}_i. \tag{4.17}$$

These definitions make sense since c_i is the origin in the new coordinate system. Meanwhile, the finite element space associated with $\mathcal{T}_n$ in $L_{i,j}$ can be transformed through (4.17) into a Lagrange finite element space on a quasi-uniform mesh in $\hat{L}_i$, which we denote by $\hat{S}_{i,j}$. Notice that $\hat{S}_{i,j}$ depends on both parameters i and j. Let $\hat{I}_j : C^0(\hat{L}_i) \to \hat{S}_{i,j}$ be the nodal interpolation operator associated with the reference region $\hat{L}_i$. Then one obtains the following scaling property in the weighted space.

Lemma 4.7 *Recall the distance function r_i to the singular vertex c_i. Then for $v \in \mathcal{K}_{\boldsymbol{a}}^m(L_{i,j})$, $m \geq 0$, and $0 \leq j \leq n$, one has*

$$\sum_{|\alpha| \leq m} \|r_i^{|\alpha|-a_i} \partial^\alpha v\|^2_{L^2(L_{i,j})} = \kappa_i^{2j(1-a_i)} \sum_{|\alpha| \leq m} \|r_i^{|\alpha|-a_i} \partial^\alpha \hat{v}\|^2_{L^2(\hat{L}_i)}. \tag{4.18}$$

Proof Since $L_{i,j}$ is in the neighborhood of the vertex c_i, according to Definition 4.1, $r_i^{|\alpha|-a_i} \sim \vartheta^{|\alpha|-a_i}$ and $r_i(x, y) = \kappa_i^j r_i(\hat{x}, \hat{y})$. Thus, for $v \in \mathcal{K}_{\boldsymbol{a}}^m(L_{i,j})$, the left hand side term of (4.18) is valid (finite). Then (4.18) follows directly from the scaling argument. □

Consequently, for $0 \leq j < n$, the error estimate in the mesh layer $L_{i,j}$ can be formulated as follows.

Lemma 4.8 *Let $\mathcal{T}_n$ be the graded mesh defined in Algorithm 4.4. Then for $0 \leq j < n$, there exists $C > 0$ independent of j and u, such that*

$$\|u - u_I\|_{H^1(L_{i,j})} \leq C\kappa_i^{ja_i} 2^{m(j-n)} \|u\|_{\mathcal{K}^{m+1}_{\boldsymbol{a}+1}(L_{i,j})}, \qquad \boldsymbol{0 < a < \eta}.$$

Proof Recall the reference region $\hat{L}_i$ from (4.16). Note that the $\mathcal{K}_{\boldsymbol{a}}^m$-norm and the H^m-norm are equivalent in $\hat{L}_i$. Therefore, by (4.18), we have

$$\|u - u_I\|^2_{H^1(L_{i,j})} \leq C \sum_{|\alpha|\leq 1} \|r_i^{|\alpha|-1} \partial^\alpha (u - u_I)\|^2_{L^2(L_{i,j})}$$
$$= C \sum_{|\alpha|\leq 1} \|r_i^{|\alpha|-1} \partial^\alpha (\hat{u} - \hat{u}_{\hat{I}})\|^2_{L^2(\hat{L}_i)} \leq C\|\hat{u} - \hat{u}_{\hat{I}}\|^2_{H^1(\hat{L}_i)},$$

where $\hat{u}_{\hat{I}} := \hat{I}_j \hat{u}$ and we used the fact $\widehat{u_I} = \hat{u}_{\hat{I}}$. Recall from Remark 4.3 that in $L_{i,j}$, the mesh size $h_{i,j} = O(\kappa_i^j 2^{j-n})$. Then by Lemma 2.5, (4.18), and (4.11), one obtains

$$\|u - u_I\|^2_{H^1(L_{i,j})} \leq C\|\hat{u} - \hat{u}_{\hat{I}}\|^2_{H^1(\hat{L}_i)} \leq C(h_{i,j}\kappa_i^{-j})^{2m} \|\hat{u}\|^2_{H^{m+1}(\hat{L}_i)}$$
$$\leq C(h_{i,j}\kappa_i^{-j})^{2m} \|\hat{u}\|^2_{\mathcal{K}^{m+1}_1(\hat{L}_i)} \leq C(h_{i,j}\kappa_i^{-j})^{2m} \sum_{|\alpha|\leq m+1} \|r_i^{|\alpha|-1} \partial^\alpha u\|^2_{L^2(L_{i,j})}$$
$$\leq C\kappa_i^{2ja_i} (h_{i,j}\kappa_i^{-j})^{2m} \sum_{|\alpha|\leq m+1} \|r_i^{|\alpha|-1-a_i} \partial^\alpha u\|^2_{L^2(L_{i,j})}$$
$$\leq C\kappa_i^{2ja_i} (h_{i,j}\kappa_i^{-j})^{2m} \|u\|^2_{\mathcal{K}^{m+1}_{a+1}(L_{i,j})} \leq C\kappa_i^{2ja_i} 2^{2m(j-n)} \|u\|^2_{\mathcal{K}^{m+1}_{a+1}(L_{i,j})}.$$

This completes the proof. □

The proof of Lemma 4.8 utilizes the fact that $\hat{L}_i = L_{i,0}$ is a region away from the singular vertex c_i. Therefore, $\hat{u} \in H^{m+1}(\hat{L}_i)$, and the usual interpolation error estimate applies in $\hat{L}_i$. However, for the last layer $L_{i,n}$, the reference region $\hat{L}_i = T_{(0)}$ includes the vertex c_i. Thus, the solution can be singular even in the reference region, and the proof has to be modified for the error analysis.

Lemma 4.9 *Let $\mathcal{T}_n$ be the graded mesh defined in Algorithm 4.4. There exists $C > 0$ independent of n and u, such that*

$$\|u - u_I\|_{H^1(L_{i,n})} \leq C\kappa_i^{na_i} \|u\|_{\mathcal{K}^{m+1}_{a+1}(L_{i,n})}, \qquad \mathbf{0} < \boldsymbol{a} < \boldsymbol{\eta}.$$

Proof For any $(x, y) \in L_{i,n}$, let $(\hat{x}, \hat{y}) = (\kappa_i^{-n}x, \kappa_i^{-n}y) \in \hat{L}_i$ be the image under the dilation (4.17) and define $\hat{u}(\hat{x}, \hat{y}) := u(x, y)$. Then according to Lemma 4.7, $\hat{u} \in \mathcal{K}^{m+1}_{a+1}(\hat{L}_i)$. Recall that the finite element basis functions in $\hat{L}_i = T_{(0)}$ are associated with a finite set of nodal points. In addition, the diameter of $\hat{L}_i$ satisfies $\text{diam}(\hat{L}_i) = O(1)$. Let $\chi : \hat{L}_i \to [0, 1]$ be a smooth function that is equal to 0 in a neighborhood of c_i but is equal to 1 at all the other nodal points in $\hat{L}_i$. Introduce the auxiliary function $v = \chi\hat{u}$ in $\hat{L}_i$. Consequently, for $m \geq 0$

$$\|v\|^2_{\mathcal{K}^m_1(\hat{L}_i)} = \|\chi\hat{u}\|^2_{\mathcal{K}^m_1(\hat{L}_i)} \leq C\|\hat{u}\|^2_{\mathcal{K}^m_1(\hat{L}_i)}, \tag{4.19}$$

where C depends on m and the smooth function χ. Moreover, since $u(c_i) = 0$, by the definition of v, one has

$$v_{\hat{I}} = \hat{u}_{\hat{I}} = \widehat{u_I} \quad \text{in } \hat{L}_i.$$

Because $v = 0$ in the neighborhood of the vertex c_i, the $\mathcal{K}^m_1$-norm and the H^m-norm are equivalent for v in $\hat{L}_i$. Then by Lemma 4.7, (4.19), and (4.11), we have

$$\begin{aligned}
\|u-u_I\|^2_{H^1(L_{i,n})} &\le C\|u-u_I\|^2_{\mathcal{K}^1_{\boldsymbol{1}}(L_{i,n})} \le C\sum_{|\alpha|\le 1}\|r_i^{|\alpha|-1}\partial^\alpha(u-u_I)\|^2_{L^2(L_{i,n})}\\
&= C\sum_{|\alpha|\le 1}\|r_i^{|\alpha|-1}\partial^\alpha(\hat u-\hat u_{\hat I})\|^2_{L^2(\hat L_i)} \le C\|\hat u - v + v - \hat u_{\hat I}\|^2_{\mathcal{K}^1_{\boldsymbol{1}}(\hat L_i)}\\
&\le C\Big(\|\hat u - v\|^2_{\mathcal{K}^1_{\boldsymbol{1}}(\hat L_i)} + \|v-\hat u_{\hat I}\|^2_{\mathcal{K}^1_{\boldsymbol{1}}(\hat L_i)}\Big)\\
&= C\Big(\|\hat u - v\|^2_{\mathcal{K}^1_{\boldsymbol{1}}(\hat L_i)} + \|v-v_{\hat I}\|^2_{\mathcal{K}^1_{\boldsymbol{1}}(\hat L_i)}\Big)\\
&\le C\Big(\|\hat u\|^2_{\mathcal{K}^1_{\boldsymbol{1}}(\hat L_i)} + \|v\|^2_{\mathcal{K}^{m+1}_{\boldsymbol{1}}(\hat L_i)}\Big) \le C\Big(\|\hat u\|^2_{\mathcal{K}^1_{\boldsymbol{1}}(\hat L_i)} + \|\hat u\|^2_{\mathcal{K}^{m+1}_{\boldsymbol{1}}(\hat L_i)}\Big)\\
&\le C\Big(\|u\|^2_{\mathcal{K}^1_{\boldsymbol{1}}(L_{i,n})} + \|u\|^2_{\mathcal{K}^{m+1}_{\boldsymbol{1}}(L_{i,n})}\Big) \le C\kappa_i^{2na_i}\|u\|^2_{\mathcal{K}^{m+1}_{\boldsymbol{a+1}}(L_{i,n})}.
\end{aligned}$$

This completes the proof. □

Consequently, Lemmas 4.6, 4.8, and 4.9 lead to the following range of the grading parameter $\boldsymbol{\kappa}$, for which one obtains the optimal convergence rate for the finite element interpolation u_I approximating $u \in \mathcal{K}^{m+1}_{\boldsymbol{a+1}}(\Omega)$.

Theorem 4.5 *Recall the parameter* $\boldsymbol{\eta}$ *in Proposition 4.1, and suppose the solution of (4.1) satisfies* $u \in \mathcal{K}^{m+1}_{\boldsymbol{a+1}}(\Omega)$ *for* $\boldsymbol{0} < \boldsymbol{a} < \boldsymbol{\eta}$. *In Algorithm 4.4, choose the grading parameter* κ_i *such that* $0 < \kappa_i \le \min\{2^{-m/a_i}, 1/2\}$. *Define* $h := 2^{-n}$. *Then*

$$\|u-u_I\|_{H^1(\Omega)} \le Ch^m\|u\|_{\mathcal{K}^{m+1}_{\boldsymbol{a+1}}(\Omega)}.$$

Proof Summing up the estimates in Lemmas 4.8, 4.9, and 4.6 for different mesh layers and regions of the domain, and using the fact $\kappa_i \le 2^{-m/a_i}$, one obtains

$$\|u-u_I\|_{H^1(\Omega)} \le C2^{-nm}\|u\|_{\mathcal{K}^{m+1}_{\boldsymbol{a+1}}(\Omega)} \le Ch^m\|u\|_{\mathcal{K}^{m+1}_{\boldsymbol{a+1}}(\Omega)},$$

which completes the proof. □

Let $\dim(S_n)$ be the dimension of the finite element space S_n. Based on the construction in Algorithm 4.4, it is clear that $\dim(S_n) = O(4^n)$. Consequently, Theorems 4.5 and (4.13) give rise to the following finite element error analysis on the graded mesh.

Corollary 4.2 *Given the same conditions as in Theorem 4.5, one has*

$$\|u-u_n\|_{H^1(\Omega)} \le C\dim(S_n)^{-m/2}\|u\|_{\mathcal{K}^{m+1}_{\boldsymbol{a+1}}(\Omega)} \le C\dim(S_n)^{-m/2}\|f\|_{\mathcal{K}^{m-1}_{\boldsymbol{a-1}}(\Omega)}.$$

Remark 4.4 Theorem 4.5 provides the range of the grading parameter κ_i such that the optimal convergence rate can be achieved for the finite element method associated with the graded mesh. The optimal grading parameter is determined by the regularity index $\boldsymbol{a}$ and the polynomial degree m in the finite element space. The extension of the graded mesh to other elliptic problems with different boundary conditions is also possible. In the next section, we will briefly discuss the numerical treatment for (4.1) when Neumann boundary conditions are allowed on adjacent sides of the domain. We conclude this section by illustrating the effectiveness of the graded finite element method.

Example 4.2 We use the linear ($m = 1$) finite element method to solve (4.1) with $f = 1$ in two cases: the L-shaped domain with the Dirichlet boundary condition and the square domain with the mixed boundary condition. See Fig. 4.3 for a more detailed description of the domains. According to Theorem 4.2 and Proposition 4.1, in both cases, the solution u belongs to $\mathcal{K}^2_{a+1}(\Omega)$ with the regularity index $a_i > 1$ for all the vertices that are associated with the convex corners where the Dirichlet boundary condition is on both adjacent sides. Consequently, the solution is in H^2 in the neighborhood of these vertices. Moreover, based on the singular solution expansion (Theorem 3.4), the solution is also in H^2 near the vertex at the bottom right corner of the square domain where the mixed boundary condition is imposed on its adjacent sides (see Fig. 4.3). Therefore, in both cases, we choose the grading parameter for each convex corner to be 0.5 (i.e., local midpoint decomposition). According to Theorem 4.5, this choice is sufficient to obtain the optimal interpolation error near these vertices. The solution is however singular ($\notin H^2$) near the reentrant corner of the L-shaped domain and near the point in the interior of the side of the square domain where the boundary condition changes.

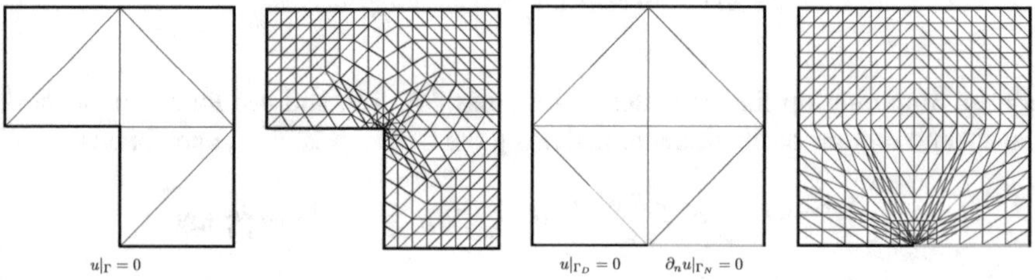

Fig. 4.3 Initial meshes vs. meshes after three graded refinements for (4.1) in two domains: Dirichlet boundary–Γ_D (black); Neumann boundary–Γ_N (green); the grading parameter to the singular vertex $\kappa = 0.2$.

For the L-shaped domain, the interior angle corresponding to the reentrant corner is $\omega = \frac{3\pi}{2}$. Therefore, the associated regularity index satisfies $a < \eta = \pi/\omega = \frac{2}{3}$. The numerical effect of the singular solution due to the reentrant corner is shown in Case 2 of Table 3.1. Based on the interpolation error analysis (Theorem 4.5), the graded mesh toward this vertex with a grading parameter $\kappa < 2^{-\frac{3}{2}} \approx 0.354$ shall recover the optimal convergence rate of the finite element method. Using the numerical convergence rate $\mathcal{R}$ in (3.2), the optimal convergence shall be achieved when $\mathcal{R} = 1$. The numerical convergence rates on meshes with different values of the grading parameter κ are displayed in Table 4.1, where j represents the number of consecutive refinements starting with the initial mesh. It is clear from Table 4.1 that the optimal rates are achieved for $\kappa < 0.354$ ($\kappa = 0.1, 0.2, 0.3$), while the convergence is not optimal for $\kappa > 0.354$ ($\kappa = 0.4, 0.5$), which is aligned with the

$j\backslash\mathcal{R}$	$\kappa = 0.1$	$\kappa = 0.2$	$\kappa = 0.3$	$\kappa = 0.4$	$\kappa = 0.5$
5	0.99	0.99	0.99	0.95	0.84
6	1.00	1.00	0.99	0.95	0.80
7	1.00	1.00	0.99	0.94	0.77
8	1.00	1.00	0.99	0.94	0.74
9	1.00	1.00	1.00	0.93	0.72
10	1.00	1.00	1.00	0.93	0.70

Table 4.1 Convergence history of the finite element method in the L-shaped domain.

theoretical prediction. Note that $\kappa = 0.4$ also corresponds to a moderately graded mesh but does not lead to the optimal convergence since it falls outside the range of κ given above.

For the case of mixed boundary conditions (Fig. 4.3), the regularity index associated with the point in the interior of the side on the boundary where the boundary condition changes is $a < \eta = \frac{\pi}{2\pi} = \frac{1}{2}$. Thus, the optimal range of the grading parameter for this point is $\kappa < 2^{-2} = 0.25$. This optimal range is verified by the numerical convergence rates $\mathcal{R}$ (3.2) in Table 4.2, which shows the optimal convergence for $\kappa = 0.1, 0.2$ and sub-optimal convergence for $\kappa = 0.3, 0.4, 0.5$.

$j\backslash\mathcal{R}$	$\kappa = 0.1$	$\kappa = 0.2$	$\kappa = 0.3$	$\kappa = 0.4$	$\kappa = 0.5$
5	0.97	0.97	0.93	0.82	0.66
6	0.99	0.98	0.92	0.78	0.60
7	0.99	0.98	0.92	0.75	0.56
8	1.00	0.99	0.91	0.72	0.53
9	1.00	0.99	0.90	0.70	0.52
10	1.00	0.99	0.90	0.95	0.51

Table 4.2 Convergence history of the finite element method for the mixed boundary condition.

In the next example, we further illustrate the use of the graded mesh in quadratic finite element methods.

Example 4.3 The quadratic ($m = 2$) finite element method is used to solve (4.1) with the Dirichlet boundary condition and $f = 1$ in a quadrilateral domain as shown in Fig. 4.4. According to the regularity theory in Chap. 3, the solution $u \in H^3$ in part of the domain that is away from the vertex A. Near A, $u \in H^{1+s}$ for $s < 1.2$, and if one uses the weighted space, $u \in \mathcal{K}^3_{a+1}(\Omega)$, where the index corresponding to A satisfies $a < \eta = 1.2$. This implies that on quasi-uniform meshes, the quadratic finite element solution will have a reduced convergence rate $\mathcal{R} \approx 1.2$, and on graded meshes toward the singular vertex A, the numerical solution will achieve the optimal convergence ($\mathcal{R} = 2$) when the grading parameter $\kappa_A < 2^{-\frac{2}{1.2}} \approx 0.315$. As in the linear finite element cases in Example 4.2, these convergence rates are confirmed by the numerical test results in Table 4.3.

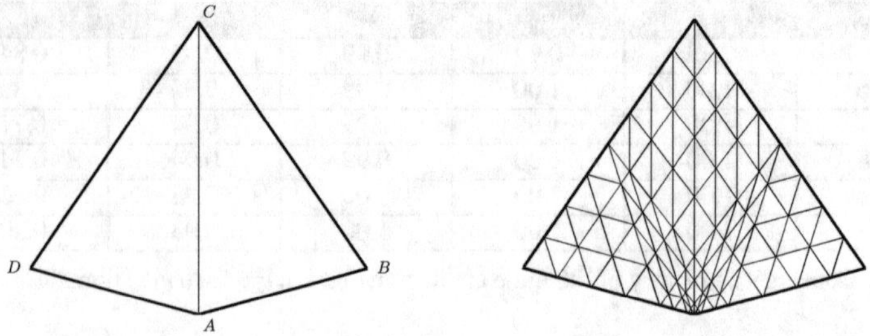

Fig. 4.4 Initial mesh vs. mesh after three graded refinements for the quadrilateral domain $ABCD$: $\angle A = 150°$, $\angle B = \angle C = \angle D = 70°$, $\kappa_A = 0.2$.

$j\backslash\mathcal{R}$	$\kappa_A = 0.1$	$\kappa_A = 0.2$	$\kappa_A = 0.3$	$\kappa_A = 0.4$	$\kappa_A = 0.5$
7	1.98	1.99	1.98	1.82	1.30
8	1.99	2.00	1.99	1.76	1.24
9	2.00	2.00	2.00	1.71	1.21
10	2.00	2.00	2.00	1.67	1.20

Table 4.3 Convergence history of the quadratic finite element method on graded meshes.

4.4 Analysis for Neumann Boundary Conditions

Consider the elliptic problem (4.1). In the previous sections, it is assumed that in any pair of two adjacent sides of the domain Ω, there is at least one side assigned the Dirichlet boundary condition. In the case where the Neumann boundary condition is imposed on both adjacent sides, the first eigenvalue associate with (3.10) is zero. This leads to a nonzero constant term in the singular expansion of the solution in the neighborhood of the common vertex of these two sides (see (3.14) and (3.15) for the Dirichlet case). Consequently, the solution may not belong to the weighted space $\mathcal{K}^m_{\boldsymbol{a+1}}$ for $\boldsymbol{a} \geq \boldsymbol{0}$. Therefore, it is necessary to revisit the regularity and interpolation error analysis for this case.

We call a vertex of the domain a Neumann–Neumann (N–N) vertex if the Neumann boundary condition $\partial_n u = 0$ is imposed on its two adjacent sides. In this section, we allow the N–N vertices in (4.1) and assume the Dirichlet boundary condition $u = 0$ on at least one side of the boundary.

The discussion on the operator pencil in Sect. 4.1 also applies to the N–N vertices. Suppose $c_i \in \mathcal{C}$ is an N–N vertex, and let ω_i be the interior angle at c_i. Then the operator $P_i(\tau) = (\tau - \mathrm{i}a_i)^2 - \partial_\theta^2$ (see (4.7)) associated with c_i is invertible for all $\tau \in \mathbb{R}$ as long as $a_i \neq \frac{k\pi}{\omega_i}$ for any $k \in \mathbb{Z}$, where $\mathbb{Z}$ is the set of all integers. This is different from a D–D vertex, for which the case $k = 0$ is not included.

For the well-posedness of (4.1), when there are N–N vertices, we consider the operator $\tilde{\Delta}_{\boldsymbol{a}}$ defined in (4.10)

$$\tilde{\Delta}_{\boldsymbol{a}} := (-\Delta, \partial_n) : \mathcal{K}^{m+1}_{\boldsymbol{a+1}}(\Omega) \cap \{v|_{\Gamma_D} = 0\} \to \mathcal{K}^{m-1}_{\boldsymbol{a-1}}(\Omega) \oplus \mathcal{K}^{m-1/2}_{\boldsymbol{a-1/2}}(\Gamma_N), \qquad (4.20)$$

which is well defined for $m \geq 1$. Define $\mathcal{H}_{\boldsymbol{a}} := \{v \in \mathcal{K}^1_{\boldsymbol{a+1}}(\Omega) : \ v = 0 \text{ on } \Gamma_D\}$. Then extend $\tilde{\Delta}_{\boldsymbol{a}}$ to the case $m = 0$ as

$$\tilde{\Delta}_a : \mathcal{H}_a \to (\mathcal{H}_{-a})^*, \quad (\Delta v, w) := -(\nabla v, \nabla w), \tag{4.21}$$

where $v \in \mathcal{H}_a$ and $w \in \mathcal{H}_{-a}$. Then we extend the set Σ_i in (4.8) to N–N vertices. Namely, for $c_i \in C$, define

$$\tilde{\Sigma}_i = \left\{ \frac{k\pi}{\omega_i} \text{ for } k \in \mathbb{Z}\,(\mathrm{N}-\mathrm{N}) \right\} \text{ or } \tilde{\Sigma}_i = \Sigma_i.$$

It can be further shown that $\tilde{\Delta}_a$ is a Fredholm operator [44, 71, 76, 84, 107].

Theorem 4.6 *Let $\tilde{\Delta}_a$ be the operator defined in* (4.20) *and* (4.21) *and $m \geq 0$. Then $\tilde{\Delta}_a$ is Fredholm if and only if $a_i \notin \tilde{\Sigma}_i$. Moreover, its index is independent of m.*

It is also possible to determine the index of the operator $\tilde{\Delta}_a$ by the following calculation. Let $\boldsymbol{a}$ and $\boldsymbol{b}$ be two vectorial weights that correspond to Fredholm operators in Theorem 4.6. Assume that there exists a vertex c_i such that $a_i < b_i$ but $a_j = b_j$ for $j \neq i$. We count the number of values in the set $(a_i, b_i) \cap \tilde{\Sigma}_i$, with the values corresponding to $k = 0$ in the definition of $\tilde{\Sigma}_i$ counted twice (because of multiplicity, which happens only in the case of N–N boundary conditions). Let N be the total number. The following result, which can be found in [44, 71, 73, 97, 98, 101]), holds.

Theorem 4.7 *Assume the conditions in Theorem 4.6. In addition, assume that $a_i < b_i$ but $a_j = b_j$ if $j \neq i$. Let N be defined as above. Then the indices of the corresponding operators satisfy*

$$\mathrm{ind}(\tilde{\Delta}_b) - \mathrm{ind}(\tilde{\Delta}_a) = -N.$$

This theorem allows to determine the index of $\tilde{\Delta}_a$. For simplicity, we compute the index only for $a_i > 0$ and small. Let η_i be the minimum value of $\tilde{\Sigma}_i \cap (0, \infty)$. Then

$$\eta_i = \begin{cases} \frac{\pi}{\omega_i}, & c_i \text{ is a D} - \text{D vertex or an N} - \text{N vertex}, \\ \frac{\pi}{2\omega_i}, & c_i \text{ is a D} - \text{N vertex}. \end{cases} \tag{4.22}$$

Furthermore, given η_i as above, let

$$\boldsymbol{\eta} := (\eta_i), \quad \text{for } 1 \leq i \leq I. \tag{4.23}$$

Theorem 4.8 *Assume the conditions in Theorem 4.6, and let N_0 be the number of N–N vertices in (4.1). Then $\tilde{\Delta}_a$ is Fredholm for $0 < a_i < \eta_i$ for all $1 \leq i \leq I$ with index*

$$\mathrm{ind}(\tilde{\Delta}_a) = -N_0.$$

Consequently, $\tilde{\Delta}_{-a}$ has index N_0 for $0 < a_i < \eta_i$.

Proof Since the index is independent of $m \geq 0$, we can assume that $m = 0$. A repeated application of Theorem 4.7 (more precisely of its generalization for $m = 0$) for each weight a_i gives that $\mathrm{ind}(\tilde{\Delta}_a) - \mathrm{ind}(\tilde{\Delta}_{-a}) = -2N_0$ (each time when we change an index from $-a_i$ to a_i, we lose a 2 in the index when c_i is an N–N vertex because the value $k = 0$ is counted twice). Since $\tilde{\Delta}_{-a} = (\tilde{\Delta}_a)^*$, we have $\mathrm{ind}(\tilde{\Delta}_{-a}) = -\,\mathrm{ind}(\tilde{\Delta}_a)$. Hence, the result follows. □

Remark 4.5 Assume the same condition $a_i < \eta_i$ as in Theorem 4.8. When the domain does not have N–N vertices, the index of the operator $\tilde{\Delta}_a$ becomes zero; this fact was used to show the well-posedness of (4.1) in Proposition 4.1. In the presence of N–N vertices, the nonzero index implies that the operator $\tilde{\Delta}_a$ is no longer a bijection between the weighted spaces in (4.20), and in consequence, the regularity result in weighted Sobolev spaces (Theorem 4.1) does not apply to this case. Nonetheless, the solution of (4.1) is uniquely defined in $H^1_D(\Omega)$ due to the Poincaré inequality since the Dirichlet boundary condition is imposed on at least one side of Ω.

We define the following functions in order to further study the invertibility properties of $\tilde{\Delta}_a$. For each N–N vertex $c_i \in C$, choose a function $\chi_i \in C^\infty(\bar{\Omega})$ that is constant equal to 1 in a neighborhood of c_i and satisfies $\partial_n \chi_i = 0$ on the boundary. In addition, we choose these functions to have disjoint supports. Let W_s be the linear span of the functions χ_i that correspond to N–N vertices. Then we have the following version of Green's formula.

Lemma 4.10 *Assume all $a_i \geq 0$ and $v, w \in \mathcal{K}^2_{a+1}(\Omega) + W_s$. Then*

$$(\Delta v, w) + (\nabla v, \nabla w) = (\partial_n v, w)_\Gamma.$$

Proof Assume first v and w are constant close to the vertices. Then we can apply the usual Green's formula without changing the terms in the formula. In general, we notice that $C(v, w) := (\Delta v, w) + (\nabla v, \nabla w) - (\partial_n v, w)_\Gamma$ depends continuously on v and w since $a_i \geq 0$ for all vertices. Then we can use a density argument to complete the proof. □

Then the following solvability result holds.

Theorem 4.9 *Recall $\boldsymbol{\eta}$ from (4.23). For $1 \leq i \leq I$, let $\boldsymbol{a} = (a_i)$ be such that $0 < a_i < \eta_i$ and $m \geq 1$. Assume the Dirichlet boundary condition is imposed on at least one side of Ω. Then for any $f \in \mathcal{K}^{m-1}_{a-1}(\Omega)$ and any $g_N \in \mathcal{K}^{m-1/2}_{a-1/2}(\Gamma_N)$, there exists a unique $u = u_{\text{reg}} + w_s$, where $u_{\text{reg}} \in \mathcal{K}^{m+1}_{a+1}(\Omega)$ and $w_s \in W_s$, satisfying $-\Delta u = f$, $u = 0$ on Γ_D, and $\partial_n u = g_N$ on Γ_N. Moreover,*

$$\|u_{\text{reg}}\|_{\mathcal{K}^{m+1}_{a+1}(\Omega)} + \|w_s\| \leq C\Big(\|f\|_{\mathcal{K}^{m-1}_{a-1}(\Omega)} + \|g_N\|_{\mathcal{K}^{m-1/2}_{a-1/2}(\Gamma_N)}\Big),$$

where the constant $C > 0$ is independent of f and g_N, and $\|\cdot\|$ can be any norm on the finite-dimensional space W_s. When $\Gamma_D = \emptyset$, the same conclusions hold if constant functions are factored out.

Proof Using the surjectivity of the trace map, we can reduce to the case $g_N = 0$. Let $V := \{v \in \mathcal{K}^{m+1}_{a+1}(\Omega) : v|_{\Gamma_D} = 0, \partial_n v|_{\Gamma_N} = 0\} + W_s$. Since $m \geq 1$, the map

$$\Delta : V \to \mathcal{K}^{m-1}_{a-1}(\Omega) \tag{4.24}$$

is well defined and continuous. Then Theorem 4.8 implies that the map (4.24) has index zero, given that the dimension of W_s is N_0. When there is at least a side in Γ_D, this map is in fact an isomorphism. Indeed, it is enough to show it is injective. This is seen as follows. Let $u \in V$ be such that $\Delta u = 0$. By Green's formula (Lemma 4.10), we have $(\nabla u, \nabla u) = (-\Delta u, u) + (\partial_n u, u)_\Gamma = 0$. Therefore, u is a constant. If there is at least one Dirichlet side, the constant must be zero, namely, $u = 0$. In the pure Neumann case, the kernel of the map (4.24) consists of constants. Another application of Green's formula shows that $(\Delta u, 1) = 0$, which identifies the range of Δ in this case as the functions with mean zero. □

Remark 4.6 According to Theorem 4.9, the solution of (4.1) allows a decomposition $u = u_{\text{reg}} + w_s$ near an N–N vertex. The function u_{reg} belongs to the weighted Sobolev space, resembling the behavior of the corner singularity near a D–D vertex, and w_s is a smooth function. In particular, w_s = constant near the N–N vertex and therefore can be approximated well by functions in the finite element space. Recall the graded mesh in Algorithm 4.4. Following the similar lines as in Sect. 4.3, one can derive the following optimal convergence of the numerical method on graded meshes [84].

Theorem 4.10 *Recall the parameter η in (4.23). For $f \in \mathcal{K}^{m-1}_{\boldsymbol{a-1}}(\Omega)$ with $\boldsymbol{0} < \boldsymbol{a} < \boldsymbol{\eta}$, let $u = u_{\text{reg}} + w_s$ be the solution of (4.1) with possible N–N vertices, where $u_{\text{reg}} \in \mathcal{K}^{m+1}_{\boldsymbol{a+1}}(\Omega)$. In Algorithm 4.4, choose the grading parameter κ_i such that $0 < \kappa_i \leq \min(2^{-m/a_i}, 1/2)$. Define $h := 2^{-n}$. Then*

$$\|u - u_n\|_{H^1(\Omega)} \leq Ch^m \|f\|_{\mathcal{K}^{m-1}_{\boldsymbol{a-1}}(\Omega)},$$

where $u_n \in S_n$ is the finite element solution associated with the graded mesh $\mathcal{T}_n$.

Example 4.4 In this example, we consider the following problem with an N–N vertex

$$-\Delta u = 1 \quad \text{in} \quad \Omega, \qquad u = r^{\frac{2}{3}} \cos(\frac{2}{3}\theta) \quad \text{on} \quad \Gamma_D \quad \text{and} \quad \partial_n u = 0 \quad \text{on} \quad \Gamma_N.$$

Here, Ω is a polygonal domain with vertices at $(0, 0)$, $(1, 0)$, $(0, 1)$, $(-1, 0)$, and $(0, -1)$, and (r, θ) are the polar coordinates centered at the origin. See Fig. 4.5 for more illustrations of the domain, the boundary condition, and the graded mesh. We use the linear finite element method to solve this problem.

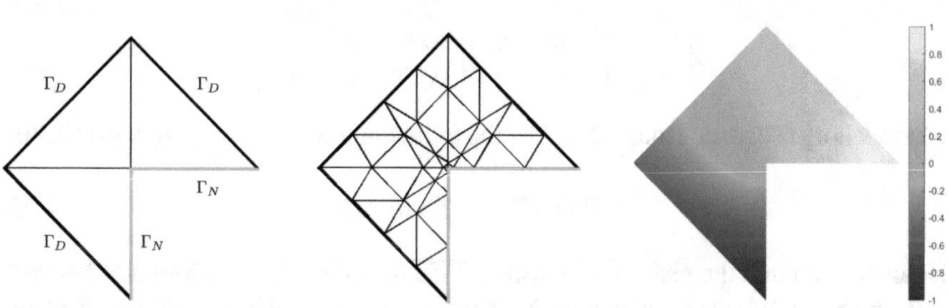

Fig. 4.5 The equation with an N–N vertex: initial mesh (left); mesh after two refinements toward the singular vertex (0, 0) with the grading parameter $\kappa = 0.2$ (center); numerical solution (right).

According to Theorem 4.10, the solution does not belong to H^2 near the N–N vertex, and therefore, the linear finite element method does not obtain the first-order convergence on quasi-uniform meshes. However, choosing a graded mesh toward the reentrant corner with the parameter $\kappa < 2^{-\frac{3}{2}} \approx 0.354$ shall give rise to the optimal convergence of the numerical method. Note that it is the same threshold value for the grading parameter as for the Dirichlet problem in the L-shaped domain (Example 4.2). This is not very surprising since based on (4.22) the upper bound η of the index a for a vertex is determined by the interior angle when the two sides of the corner have the same type of boundary condition. The numerical convergence rates $\mathcal{R}$ (3.2) corresponding to different values of the grading parameter κ toward the N–N vertex are displayed in Table 4.4, where j is the refinement level. These rates confirm the discussion above. Namely, the optimal convergence ($\mathcal{R} = 1$) is observed for $\kappa = 0.1, 0.2, 0.3$, and the convergence slows down when $\kappa = 0.4, 0.5$.

$j\backslash\mathcal{R}$	$\kappa = 0.1$	$\kappa = 0.2$	$\kappa = 0.3$	$\kappa = 0.4$	$\kappa = 0.5$
7	1.00	0.99	0.97	0.85	0.67
8	1.00	1.00	0.97	0.86	0.67
9	1.00	1.00	0.98	0.86	0.67
10	1.00	1.00	0.99	0.86	0.67
11	1.00	1.00	0.99	0.87	0.67

Table 4.4 Convergence history of the finite element method for the N–N vertex problem.

It is worth noting that the analytical tools used in this section for Neumann boundary conditions can be extended to study transmission problems that have nonsmooth interfaces. We refer interested readers to [84, 101] for more theoretical discussions. In the next example, we show how the graded mesh (Algorithm 4.4) can be extended to treat solution singularities in the interior of the domain (at the nonsmooth points of the interface) and to improve the convergence of the numerical solution for the transmission problem.

Example 4.5 Let the domain $\Omega = (-1, 1)^2$. Define

$$A = \begin{cases} R & \text{in} \quad (0,1)^2 \cup (-1,0)^2, \\ 1 & \text{in} \quad \Omega \setminus \left([0,1)^2 \cup (-1,0]^2\right), \end{cases}$$

where R is a positive constant. Consider the following Kellogg-type problem [42, 70]:

$$-\nabla \cdot (A \nabla u) = 0 \quad \text{in} \quad \Omega. \tag{4.25}$$

It can be seen that the piecewise constant diffusion coefficient A alternates between two values in subdomains surrounding the point $(0, 0)$. We are interested in the nonzero solutions of (4.25), which can be constructed as follows. Let (r, θ) be the polar coordinates centered at the origin. Then set the function

$$u(r, \theta) = r^\gamma \mu(\theta), \tag{4.26}$$

where

$$\mu(\theta) = \begin{cases} \cos\left((\pi/2-\sigma)\gamma\right)\cdot\cos\left((\theta-\pi/2+\rho)\gamma\right) & \text{if } 0 \le \theta \le \pi/2, \\ \cos(\rho\gamma)\cdot\cos\left((\theta-\pi+\sigma)\gamma\right) & \text{if } \pi/2 \le \theta \le \pi, \\ \cos(\sigma\gamma)\cdot\cos\left((\theta-\pi-\rho)\gamma\right) & \text{if } \pi \le \theta \le 3\pi/2, \\ \cos\left((\pi/2-\rho)\gamma\right)\cdot\cos\left((\theta-3\pi/2-\sigma)\gamma\right) & \text{if } 3\pi/2 \le \theta \le 2\pi. \end{cases}$$

Here, the parameters γ, ρ, σ, R satisfy

$$\begin{cases} R = -\tan\left((\pi/2-\sigma)\gamma\right)\cdot\cot(\rho\gamma), \\ 1/R = -\tan(\rho\gamma)\cdot\cot(\sigma\gamma), \\ R = -\tan(\sigma\gamma)\cdot\cot\left((\pi/2-\rho)\gamma\right), \\ 0 < \gamma < 2, \\ \max\{0, \pi\gamma-\pi\} < 2\gamma\rho < \min\{\pi\gamma, \pi\}, \\ \max\{0, \pi-\pi\gamma\} < -2\gamma\sigma < \min\{\pi, 2\pi-\pi\gamma\}. \end{cases}$$

Based on these conditions, we particularly choose the following parameters:

$$\gamma = 0.5, \ R \approx 5.828427124761, \ \rho = \pi/4, \ \sigma \approx -2.3561944901923. \quad (4.27)$$

Using the parameters in (4.27), one can verify that the function u in (4.26) is a non-trivial solution of the transmission problem (4.25). In addition, it is clear that $u \in H^{1+\delta}(\Omega)$ for $\delta < \gamma = 0.5$.

In the numerical tests, we use the linear finite element method to solve (4.25) with the Dirichlet boundary condition given by u that is defined through the parameters in (4.27). The initial mesh is designed such that the interfaces of the problem are not in the interior of any element. Unlike the other numerical examples we have discussed, the singularity of u is located at $(0, 0)$, a point not on the boundary but in the domain. Nevertheless, Algorithm 4.4 can also be used to construct graded meshes for interior singular points if the interior point is regarded as a singular vertex and is assigned a grading parameter. In this case, we let $\kappa \in (0, 0.5]$ be the grading parameter associated with $(0, 0)$ and assign 0.5 as the grading parameter for every vertex of the domain. Therefore, the origin is the only singular vertex that may be surrounded by the graded mesh. See Fig. 4.6 for more illustrations.

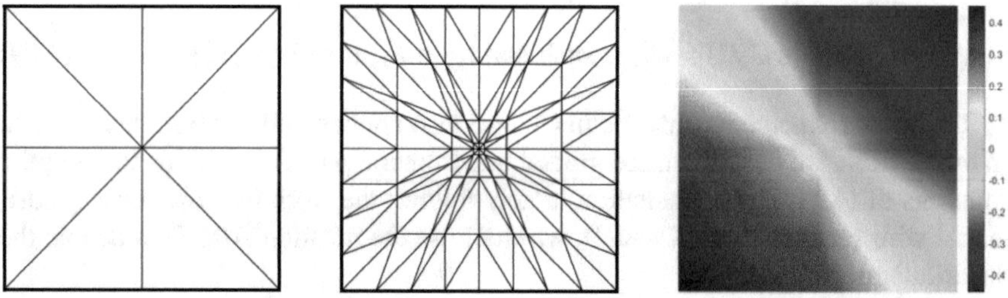

Fig. 4.6 The transmission problem: initial mesh for the domain Ω (left); mesh after two refinements toward $(0, 0)$ with the grading parameter $\kappa = 0.2$ (center); numerical solution (right).

The numerical convergence rates $\mathcal{R}$ (3.2) corresponding to different values of the grading parameter κ toward $(0, 0)$ are displayed in Table 4.5, where j is the refinement level. It can be seen that suitable graded meshes improve the convergence of the finite element solution for the Kellogg-type transmission problem.

$j\backslash\mathcal{R}$	$\kappa = 0.1$	$\kappa = 0.2$	$\kappa = 0.3$	$\kappa = 0.4$	$\kappa = 0.5$
8	0.99	0.95	0.83	0.66	0.50
9	1.00	0.97	0.84	0.66	0.50
10	1.00	0.97	0.84	0.66	0.50
11	1.00	0.98	0.85	0.66	0.50
12	1.00	0.98	0.85	0.66	0.50
13	1.00	0.99	0.85	0.66	0.50

Table 4.5 Convergence history of the finite element method for the transmission problem.

4.5 Graded Quadrilateral Meshes

As another type of grid, the quadrilateral mesh is also popular in various applications, for example, in computational fluid dynamics and in the construction of high-order finite volume methods. Let $\Omega \subset \mathbb{R}^2$ be a bounded polygonal domain and $C = \{c_i\}$, $1 \leq i \leq I$, be its vertex set. Let $\mathcal{Q}$ be a conforming mesh of Ω, consisting of convex quadrilaterals. For any quadrilateral $Q \in \mathcal{Q}$, there is a unique bilinear mapping F_Q, such that $Q = F_Q(\hat{Q})$, where $\hat{Q} = (-1, 1) \times (-1, 1)$ is the reference element. For any $v \in L^1(Q)$, we define

$$\hat{v}_Q := v \circ F_Q \in L^1(\hat{Q}). \tag{4.28}$$

Let $\hat{B}_m := \text{span}\{p_s(\hat{x})q_t(\hat{y})\}$ be the Lagrange finite element space on $\hat{Q}$, where p_s and q_t are polynomials of degree $\leq m$ and the set of nodal variables takes the function value at the nodal points. See [35, 43] for detailed construction of nodal points in the reference element. Thus, the quadrilateral finite element space of degree m associated with $\mathcal{Q}$ is

$$S_{\mathcal{Q}} = \{v \in C(\Omega) : v|_Q \text{ is such that } \hat{v}_Q \in \hat{B}_m \text{ for all } Q \in \mathcal{Q}\}. \tag{4.29}$$

The mesh refinement principle in Chap. 3 also applies to the quadrilateral mesh. In this section, we describe three graded quadrilateral mesh algorithms for singular solutions of (4.1). The presentation largely follows the work in [90]. For a quadrilateral with vertices A, B, C, and D, we shall use the notation $ABCD$ to denote the quadrilateral.

The first algorithm is based on a simple decomposition of triangles in the triangular mesh.

Algorithm 4.11 (Barycenter Refinement) Let $\mathcal{T}_j, 0 \leq j \leq n$, be the graded triangular mesh defined in Algorithm 4.4. Then we divide each triangle $T \in \mathcal{T}_j$ into three quadrilaterals using the midpoint of each edge and the barycenter of the triangle (Fig. 4.7). Thus, the quadrilateral mesh $\mathcal{Q}_j$ of Ω, $0 \leq j \leq n$, consists of all the quadrilaterals from the decomposition. See also Fig. 4.8 for example. □

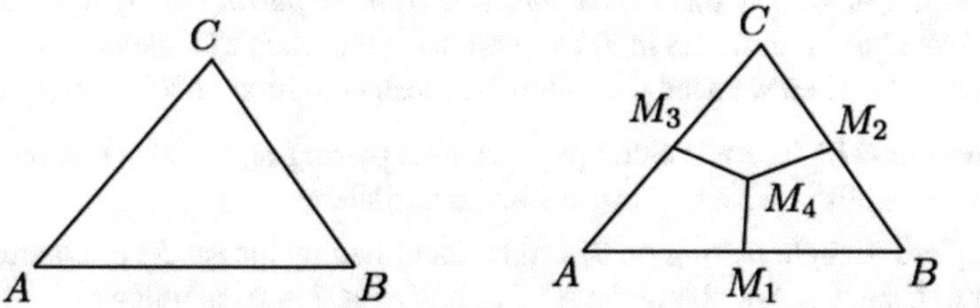

Fig. 4.7 Quadrilaterals from a triangle, left–right: a triangle in the mesh $\mathcal{T}_j$; the resulting quadrilaterals (M_4 is the barycenter).

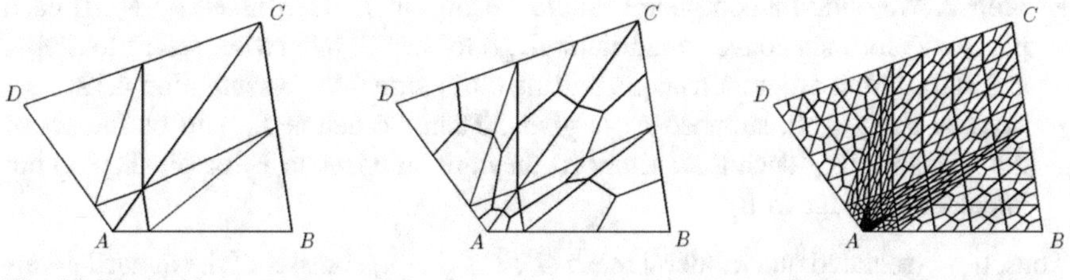

Fig. 4.8 Mesh refinements for a polygonal domain based on Algorithm 4.11 ($\kappa_A = 0.2$, $\kappa_B = \kappa_C = \kappa_D = 0.5$): the initial triangular mesh $\mathcal{T}_0$ (left), the resulting quadrilateral mesh $\mathcal{Q}_0$ (center), and the quadrilateral mesh $\mathcal{Q}_2$ (right).

Proposition 4.2 *The barycenter refinement of a graded triangulation $\mathcal{T}_j$, $0 \leq j \leq n$, directly leads to a mesh $\mathcal{Q}_j$ that consists of convex and shape regular quadrilaterals.*

Proof We first show that the barycenter refinement of a triangle results in convex quadrilaterals. Consider the triangle $\triangle^3 ABC$ and the quadrilateral $M_1BM_2M_4$ in Fig. 4.7. The barycenter M_4 is the intersection of two medians AM_2 and CM_1. Therefore, $\angle M_1M_4M_2 = \angle AM_4C$. Note $\angle AM_4C < \pi$ since $\angle AM_4C$ is an interior angle of $\triangle^3 AM_4C$. Therefore, $\angle M_1M_4M_2 < \pi$. Meanwhile, since M_4 is always an interior point of $\triangle^3 ABC$, it is straightforward to see that other interior angles of $M_1BM_2M_4$ are less than π. Hence, the quadrilateral $M_1BM_2M_4$ is convex. For the same reason, quadrilaterals $M_3M_4M_2C$ and $AM_1M_4M_3$ are both convex.

Note that successive graded refinements (Algorithm 4.4) for a triangle $T \in \mathcal{T}_0$ generate smaller triangles within at most four similarity classes. Therefore, the triangles in $\mathcal{T}_j$ are within at most $4N_0$ similarity classes, where N_0 is the number of initial triangles in $\mathcal{T}_0$. Thus, the barycentric refinement of the triangular mesh $\mathcal{T}_j$ leads to quadrilaterals within at most $12N_0$ similarity classes.

Hence, we have shown that the resulting mesh Q_j consists of convex and shape regular quadrilaterals. □

Instead of dividing a triangle, the next algorithm produces quadrilaterals by combining adjacent triangles. Let $\mathcal{T}_j$, $0 \leq j \leq n$, be the graded triangular mesh from Algorithm 4.4. We say that it is a *complete triangle pairing* on $\mathcal{T}_j$ if we can group every two adjacent triangles in $\mathcal{T}_j$ as a pair, such that each triangle is in one and only one such pair. Then we need the following mesh condition on $\mathcal{T}_1$ to proceed.

Assumption 4.12 There is a complete triangle pairing on $\mathcal{T}_1$, such that the union of the two triangles in each pair is a convex quadrilateral. □

A complete triangle pairing on $\mathcal{T}_j$ can be identified by the set E_j of common edges of paired triangles. We obtain the set E_j on $\mathcal{T}_j$, $1 \leq j \leq n$, as follows.

Algorithm 4.13 (Triangle Pairing) E_1 is given by Assumption 4.12.

- **Step 1.** We define E_2 to be the set of the edges in $\mathcal{T}_2$, such that: either (1) they are on edges in E_1 or (2) they do not intersect any edge in E_1. See Fig. 4.9 for example.
- **Step 2.** We obtain a complete triangle pairing on $\mathcal{T}_2$ determined by E_2. If each pair on $\mathcal{T}_2$ forms a convex quadrilateral, go to Step 3; otherwise, return to Step 1 and start with a different triangle pairing on $\mathcal{T}_1$ satisfying Assumption 4.12.
- **Step 3.** For $j \geq 2$, suppose E_j is given. Then we define E_{j+1} to be the set of the edges in $\mathcal{T}_{j+1}$, such that: either (i) they are on edges in E_j or (ii) they do not intersect any edge in E_j.

Thus, the associated quadrilateral mesh Q_j, $1 \leq j \leq n$, consists of the quadrilaterals determined by the triangle pairing on $\mathcal{T}_j$. □

Remark 4.7 Based on how the two triangles in a pair will be refined in the next triangular refinement (Algorithm 4.4), there are three possible types of pairings for triangles in $\mathcal{T}_j$ ($j \geq 1$): (1) both triangles will be uniformly refined (Fig. 4.9); (2) one triangle will be specially refined toward a vertex, and the other triangle will be uniformly refined (Fig. 4.10); (3) both triangles will be specially refined toward a common vertex (Fig. 4.11). See Fig. 4.12 as an example for the triangle pairing in a polygonal domain.

For Algorithm 4.13 to proceed on triangular meshes $\mathcal{T}_j$, $1 \leq j \leq n$, the initial triangulation $\mathcal{T}_0$ should be able to lead to a triangle pairing on $\mathcal{T}_1$ that satisfies Assumption 4.12 and consequently lead to a triangle pairing on $\mathcal{T}_2$ that enables Step 3 in the algorithm. This is in fact not a restrictive condition on $\mathcal{T}_0$. As indicated in the algorithm, we only need to ensure that the paired triangles in $\mathcal{T}_2$ form convex quadrilaterals. We give a concrete proof of this argument below.

Proposition 4.3 *Algorithm 4.13 gives rise to the sets E_j of edges, $1 \leq j \leq n$, provided that each pair of triangles on $\mathcal{T}_2$ forms a convex quadrilateral. In turn, each*

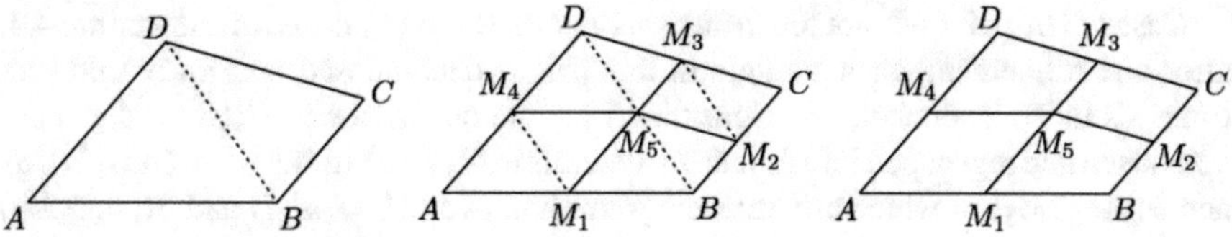

Fig. 4.9 Triangle pairing I (uniform refinement): a pair in $\mathcal{T}_j$ (left), the child pairs in $\mathcal{T}_{j+1}$ (center), and the resulting quadrilaterals in $\mathcal{Q}_{j+1}$ (right). Each pair is identified by the common edge (the dotted line).

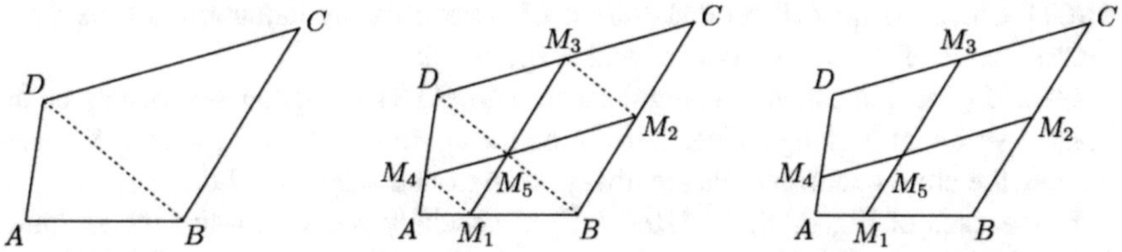

Fig. 4.10 Triangle pairing II (the graded refinement toward an isolated singular vertex): a pair in $\mathcal{T}_j$ (left), the child pairs in $\mathcal{T}_{j+1}$ (center), and the resulting quadrilaterals in $\mathcal{Q}_{j+1}$ (right). Each pair is identified by the common edge (the dotted line).

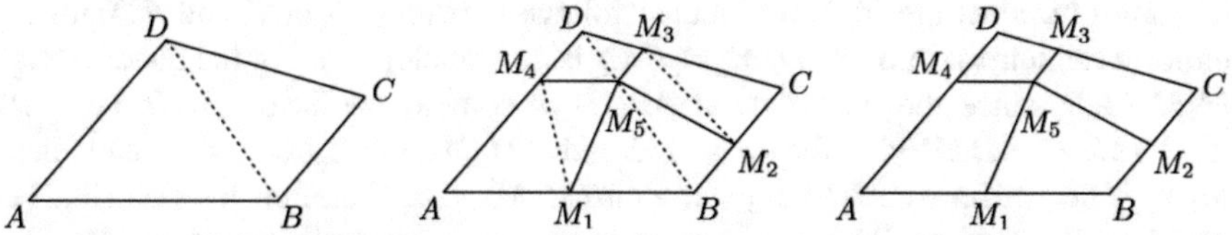

Fig. 4.11 Triangle pairing III (the graded refinement toward a common singular vertex): a pair in $\mathcal{T}_j$ (left), the child pairs in $\mathcal{T}_{j+1}$ (center), and the resulting quadrilaterals in $\mathcal{Q}_{j+1}$ (right). Each pair is identified by the common edge (the dotted line).

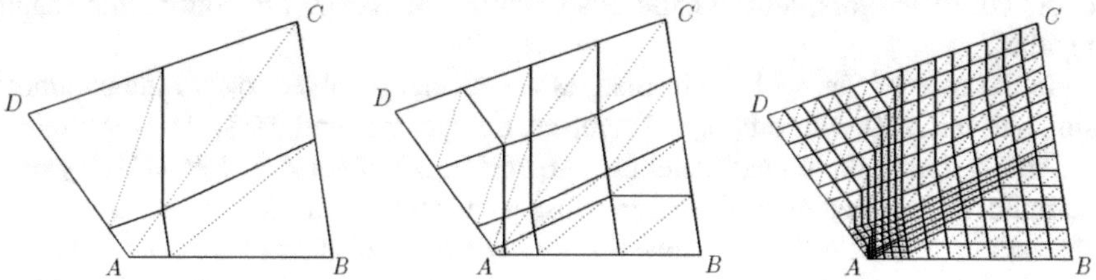

Fig. 4.12 Mesh refinements for a polygonal domain based on Algorithm 4.13 ($\kappa_A = 0.2, \kappa_B = \kappa_C = \kappa_D = 0.5$): the resulting quadrilateral mesh $\mathcal{Q}_1$ from a triangular mesh $\mathcal{T}_1$ (left), the quadrilateral mesh $\mathcal{Q}_2$ (center), and the quadrilateral mesh $\mathcal{Q}_4$ (right). The dotted lines are the common edges for the pairs.

edge set E_j determines a complete triangle pairing on $\mathcal{T}_j$ that leads to convex and shape regular quadrilaterals.

Proof It suffices to consider the following two cases for a triangle pair (a quadrilateral $ABCD$) in $\mathcal{T}_2$.

Case I (The pair does not touch any singular vertex in C.) Based on Algorithm 4.4, after one refinement, each triangle in this pair is decomposed uniformly into four triangles in $\mathcal{T}_3$. Therefore, by Algorithm 4.13, the quadrilateral $ABCD \in Q_2$ is decomposed and regrouped into four quadrilaterals in Q_3 (see Fig. 4.9), two ($AM_1M_5M_4$ and $M_5M_2CM_3$) of which are parallelograms and two ($M_4M_5M_3D$ and $M_1BM_2M_5$) of which are similar to the original quadrilateral $ABCD$. Note that by Algorithm 4.13, the subsequent uniform refinements of parallelograms lead to similar parallelograms; and for quadrilaterals $M_4M_5M_3D$ and $M_1BM_2M_5$, since they are similar to $ABCD$, the subsequent uniform refinements lead to child quadrilaterals similar to those of $ABCD$. Since the quadrilateral $ABCD \in Q_2$ is convex, by induction, all the child quadrilaterals of $ABCD$ are convex and shape regular.

Case II (The pair touches a singular vertex in C.) The singular vertex may be an isolated vertex of a triangle (Fig. 4.10) or on the common edge (Fig. 4.11). We need to show the child quadrilaterals are always convex and shape regular.

In the case of Fig. 4.10, $\triangle^3 ABD \in \mathcal{T}_2$ is specially refined toward the singular vertex A with ratio κ_A in the next step, and $\triangle^3 BCD \in \mathcal{T}_2$ is uniformly refined. Recall that $\mathcal{T}_2$ is the triangular mesh obtained from the initial triangulation after two graded refinements. Let $T \in \mathcal{T}_1$ be the parent triangle of $\triangle^3 ABD \in \mathcal{T}_2$. Thus, $\triangle^3 ABD$ is similar to T with ratio κ_A, and $AM_1M_5M_4 \subset \triangle^3 ABD$ is similar to $ABCD \subset T$ with the same ratio since the graded refinement follows the same rule on $\mathcal{T}_1$ and $\mathcal{T}_2$. With the uniform refinement in $\triangle^3 BCD$, $M_5M_2CM_3$ is a parallelogram. For the quadrilateral $M_4M_5M_3D$, since the quadrilateral $ABCD$ is convex, we have $\angle M_3DM_4 < \pi$, $\angle DM_4M_5 \leq \angle DAB < \pi$ for $\kappa_A \leq 1/2$ and $\angle M_5M_3D = \angle BCD < \pi$, and since $M_1M_5//BC$, we have $\angle M_4M_5M_3 = \pi - \angle M_1M_5M_4 < \pi$. Therefore, the quadrilateral $M_4M_5M_3D$ is convex. With a similar argument, so is the quadrilateral $M_1BM_2M_5$. Further subsequent refinements of the quadrilaterals $M_1BM_2M_5$, $M_5M_2CM_3$, and $M_4M_5M_3D$ will fall into Case I. The subsequent refinements of the pair $AM_1M_5M_4$ create child quadrilaterals similar to those of $ABCD$ since $AM_1M_5M_4$ is similar to $ABCD$. Therefore, all the child quadrilaterals of $ABCD$ are convex and shape regular.

In the case of Fig. 4.11, both triangles are specially refined toward the common singular vertex D with ratio κ_D. Therefore, the quadrilateral $M_4M_5M_3D$ is similar to $ABCD$. Other three quadrilaterals $AM_1M_5M_4$, $M_1BM_2M_5$, and $M_2CM_3M_5$ will be uniformly refined as in Case I in subsequent refinements. Therefore, as for the case in Fig. 4.10, it suffices to show that these three quadrilaterals are convex. Given that $ABCD$ is convex, following a similar argument as above, it is straightforward to see that all the interior angles in these three quadrilaterals are less than π, except for $\angle M_1M_5M_2$, for which additional calculations are needed. Recall $\kappa_D \leq 1/2$. Thus, $\angle M_1M_5B \leq \angle ADB$ and $\angle BM_5M_2 \leq \angle BDC$. Hence, $\angle M_1M_5M_2 \leq \angle ADC < \pi$. Therefore, all the child quadrilaterals of $ABCD$ are convex and shape regular, which completes the proof. □

Remark 4.8 Algorithms 4.11 and 4.13 are based on additional modifications on the existing graded triangular meshes, and therefore, both of them preserve the local mesh size in the graded triangular mesh. Meanwhile, they are also apparently different. The barycenter refinement has a simple formulation, but the resulting quadrilaterals in general do not converge to parallelograms. The triangle pairing is more complicated and requires moderate initial mesh conditions on $\mathcal{T}_0$ to proceed. Recall that the triangles in $\mathcal{T}_j$, if away from the singular vertex, are uniformly refined. Therefore, most of quadrilaterals in $\mathcal{Q}_j$ from Algorithm 4.13 are parallelograms, except for those that are either covered by more than one initial triangle in $\mathcal{T}_0$ or close to the singular vertex (see the mesh $\mathcal{Q}_4$ in Fig. 4.12). We also note that with the same number of refinements, Algorithm 4.11 leads to more quadrilaterals than Algorithm 4.13, and hence a larger finite element space.

The next algorithm directly refines a quadrilateral mesh, without requiring auxiliary triangular meshes.

Algorithm 4.14 (Graded 2-Refinement) Let $\mathcal{Q}$ be a conforming quadrilateral mesh of Ω with convex quadrilaterals. Suppose that the vertices in $\mathcal{C}$ are included in the vertices in $\mathcal{Q}$, and each quadrilateral contains at most one singular vertex of the domain. Define the vector $\boldsymbol{\kappa} = (\kappa_1, \ldots, \kappa_I)$, for $\kappa_i \in (0, 1/2]$. A $\boldsymbol{\kappa}$-refinement of $\mathcal{Q}$, denoted by $\boldsymbol{\kappa}(\mathcal{Q})$, is obtained by dividing each quadrilateral $Q \in \mathcal{Q}$ into four small quadrilaterals as follows:

- **Step 1.** (Edge Point Selection) Let AB be an edge in the quadrilateral mesh $\mathcal{Q}$.
  - If neither A nor B is a singular vertex, then we mark the midpoint M of AB.
  - Otherwise, if $A = c_i \in \mathcal{C}$ is a singular vertex, we mark the point M on AB, such that $|AM| = \kappa_i |AB|$.
- **Step 2.** (Interior Point Selection) Let $Q \in \mathcal{Q}$ be a quadrilateral.
  - If Q does not contain a singular vertex, then we use two straight lines to connect the points on the opposite edges of Q that are marked in Step 1. Then mark the intersection of these two lines.
  - If a singular vertex $c_i \in \mathcal{C}$ is a vertex of Q, let AC be the diagonal of Q, such that $A = c_i$. Then we mark the point M on AC, such that $|AM| = \kappa_i |AC|$.

Then we divide each quadrilateral Q into four quadrilaterals by connecting the marked interior point to the four marked edge points (see Fig. 4.13). Given an initial mesh $\mathcal{Q}_0$ consisting of convex quadrilaterals, the associated family of graded quadrilateral meshes $\{\mathcal{Q}_j : j \geq 0\}$ is defined recursively, $\mathcal{Q}_{j+1} = \boldsymbol{\kappa}(\mathcal{Q}_j)$. □

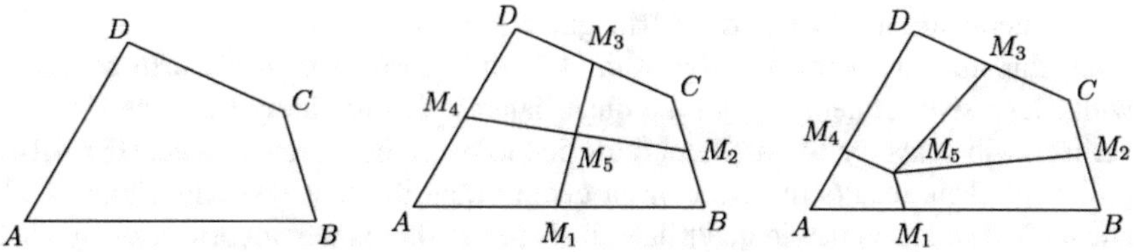

Fig. 4.13 Direct quadrilateral refinements, left–right: an initial quadrilateral; a uniform 2-refinement; a graded refinement toward A, $\kappa_A = \frac{|AM_1|}{|AB|} = \frac{|AM_4|}{|AD|} = \frac{|AM_5|}{|AC|}$.

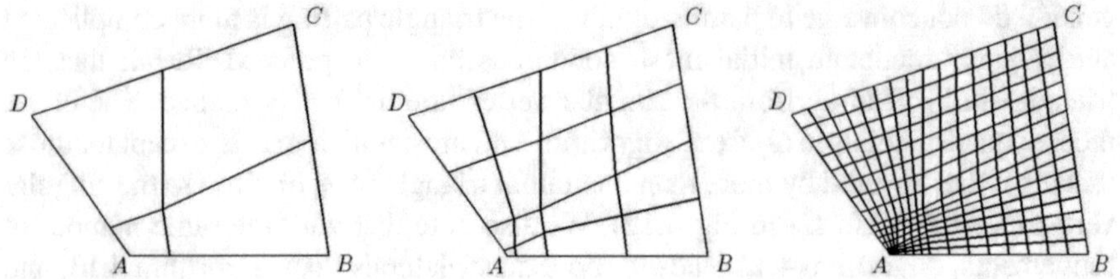

Fig. 4.14 Mesh refinements for a polygonal domain based on Algorithm 4.14 ($\kappa_A = 0.2, \kappa_B = \kappa_C = \kappa_D = 0.5$): the initial quadrilateral mesh $\mathcal{Q}_0$ (left), $\mathcal{Q}_1$ (center), and $\mathcal{Q}_3$ (right).

The graded 2-refinement also results in meshes of good quality.

Proposition 4.4 *For $0 \le j \le n$, Algorithm 4.14 leads to a sequence of meshes $\mathcal{Q}_j$ consisting of convex and shape regular quadrilaterals.*

Proof It suffices to consider the graded 2-refinements of a single quadrilateral in $\mathcal{Q}_0$. For a convex quadrilateral that is away from the singular vertex, it is known [16] that the uniform refinements (the second picture in Fig. 4.13) lead to convex and shape regular quadrilaterals. Thus, we need to show this is the case for a quadrilateral touching a singular vertex.

For a convex quadrilateral $ABCD \in \mathcal{Q}_0$ with one singular vertex $A \in \mathcal{C}$ (the third picture in Fig. 4.13), we can divide the quadrilateral into two triangles using its diagonal AC. Then for this quadrilateral, Algorithm 4.14 and a special case of Algorithm 4.13 (the triangle pairing in Fig. 4.11) produce the same child quadrilaterals after one refinement. Therefore, the four child quadrilaterals of $ABCD$ are convex and shape regular. For the three child quadrilaterals, $M_1BM_2M_5$, $M_5M_2CM_3$, and $M_5M_3DM_4$, which are away from the singular vertex, they will be uniformly refined in the subsequent refinements and hence have child quadrilaterals with desired quality (see the discussions in the first paragraph of the proof). Note that the quadrilateral $AM_1M_5M_4$ is similar to $ABCD$. Further refinements on $AM_1M_5M_4$ will repeat the pattern as for $ABCD$. Namely, $AM_1M_5M_4$ will have child quadrilaterals similar to those of $ABCD$. Thus, all the child quadrilaterals of $ABCD$ from the subsequent refinements are convex and shape regular. This completes the proof. □

Remark 4.9 We have described three graded quadrilateral mesh algorithms for the finite element approximation of (4.1). The graded 2-refinement directly refines a quadrilateral mesh and needs no triangular mesh information. It generalizes the triangular mesh refinement (Algorithm 4.4) to quadrilateral meshes. In addition, with successive refinements, all the quadrilaterals that are away from the singular vertices will converge to parallelograms due to the uniform refinements (Fig. 4.14) [16, 120]. This results in better mesh quality than the first two Algorithms 4.11 and 4.13. The difference in quadrilateral shapes in these three algorithms may lead to different approximation properties in the case where the mapped finite element space (4.28) on the reference element $\hat{Q}$ is only a proper subspace of the polynomial space $\hat{B}_m$, such as the serendipity elements [16].

Let Q_n be a quadrilateral mesh given by one of the aforementioned algorithms (Algorithms 4.11, 4.13, and 4.14). Let S_{Q_n} be the quadrilateral finite element space (4.29) associated with Q_n. Then the quadrilateral finite element solution of (4.1) is $u_n \in S_n$ such that (4.12) holds for all $v \in S_n$, where $S_n := S_{Q_n} \cap H^1_D(\Omega)$. The following error analysis can be obtained by using the similar lines as in Sect. 4.3. The readers are referred to [90] for the detailed proofs.

Theorem 4.15 *Recall the parameter* $\boldsymbol{\eta}$ *in (4.23). For* $f \in \mathcal{K}^{m-1}_{\boldsymbol{a}-\boldsymbol{1}}(\Omega)$ *with* $\boldsymbol{0} < \boldsymbol{a} < \boldsymbol{\eta}$, *let* $u = u_{\text{reg}} + w_s$ *be the solution of (4.1) with possible N–N vertices, where* $u_{\text{reg}} \in \mathcal{K}^{m+1}_{\boldsymbol{a}+\boldsymbol{1}}(\Omega)$. *In Algorithms 4.11, 4.13, and 4.14, for a vertex* $c_i \in \mathcal{C}$, *choose the grading parameter* κ_i *such that* $0 < \kappa_i \leq \min(2^{-m/a_i}, 1/2)$. *Define* $h := 2^{-n}$. *Then*

$$\|u - u_n\|_{H^1(\Omega)} \leq Ch^m \|f\|_{\mathcal{K}^{m-1}_{\boldsymbol{a}-\boldsymbol{1}}(\Omega)}.$$

Example 4.6 We illustrate the graded quadrilateral finite element method in the following example. Consider the equation

$$-\Delta u = f \quad \text{in} \quad \Omega, \qquad u = g \quad \text{on} \quad \partial\Omega,$$

where Ω is an L-shaped domain (Fig. 4.15). The functions f and g are chosen such that the solution is

$$u = r^{\frac{2}{3}} \sin\left(\frac{2}{3}\left(\theta - \frac{\pi}{2}\right)\right),$$

where (r, θ) are the polar coordinates with O as the origin. Let a_1 and η_1 be the parameters in Theorem 4.15 corresponding to the vertex O. Therefore, $u \in \mathcal{K}^2_{\boldsymbol{a}+\boldsymbol{1}}(\Omega)$, where $0 < a_i < \eta_i$, $\eta_1 = \frac{2}{3}$, and $\eta_i > 1$ for $i > 1$. We use bilinear elements ($m = 1$) in the numerical approximation of this problem. Then to achieve the optimal convergence, it is sufficient to choose the grading parameters $0 < \kappa_1 < 2^{-\eta_1} \approx 0.354$ and $\kappa_i = 0.5$ ($i > 1$) for other vertices. In Tables 4.6, 4.7, and 4.8, we list the degrees of freedom (DOFs) of the finite element space and the errors $|u - u_j|_{H^1(\Omega)}$ for each value of κ_1, where u_j is the finite element solution on the mesh after j refinements. Next to the error, displayed in the parentheses is the error reduction rate that is calculated by

$$\log_2\left(\frac{|u - u_{j-1}|_{H^1(\Omega)}}{|u - u_j|_{H^1(\Omega)}}\right).$$

Given $m = 1$, the optimal convergence corresponds to the error reduction rate 1. It is clear that the optimal convergence rate is achieved for $\kappa_1 < 0.35$ in all three meshes as predicated in Theorem 4.15. Another interesting observation is that although simpler to formulate, Algorithm 4.11 leads to a larger finite element space than the other two algorithms at the same level of refinement, while the actual errors are comparable.

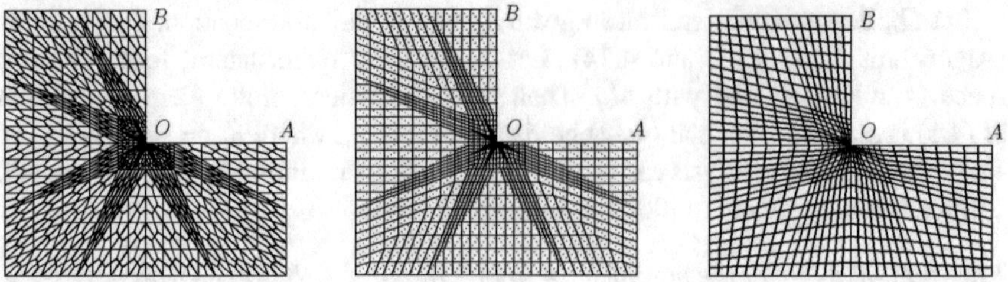

Fig. 4.15 The L-shaped domain and meshes Q_2 from the three algorithms, $\kappa_1 = 0.2$, (left–right): Algorithms 4.11, 4.13, and 4.14.

j	DOFs	$\lvert u - u_j \rvert_{H^1(\Omega)}$				
		κ_1=0.1	κ_1=0.2	κ_1=0.3	κ_1=0.4	κ_1=0.5
2	1217	4.13E-02(0.95)	2.78E-02(0.98)	2.57E-02(0.98)	3.24E-02(0.86)	4.56E-02(0.68)
3	4737	2.10E-02(0.99)	1.42E-02(0.99)	1.32E-02(0.98)	1.81E-02(0.86)	2.89E-02(0.67)
4	18689	1.05E-02(1.02)	7.12E-03(1.00)	6.74E-03(0.98)	1.01E-02(0.86)	1.83E-02(0.67)
5	74241	5.18E-03(1.02)	3.57E-03(1.00)	3.43E-03(0.98)	5.57E-03(0.86)	1.15E-02(0.67)
6	295937	2.57E-03(1.01)	1.78E-03(1.00)	1.73E-03(0.98)	3.07E-03(0.86)	7.27E-03(0.67)

Table 4.6 Convergence history for bilinear elements with meshes from Algorithm 4.11.

j	DOFs	$\lvert u - u_j \rvert_{H^1(\Omega)}$				
		κ_1=0.1	κ_1=0.2	κ_1=0.3	κ_1=0.4	κ_1=0.5
2	225	4.76E-02(0.98)	3.75E-02(1.03)	3.70E-02(1.01)	4.53E-02(0.89)	6.13E-02(0.72)
3	833	2.47E-02(1.01)	1.93E-02(1.02)	1.95E-02(0.98)	2.57E-02(0.87)	3.90E-02(0.69)
4	3201	1.25E-02(1.01)	9.76E-03(1.01)	1.009E-02(0.98)	1.44E-02(0.86)	2.47E-02(0.68)
5	12545	6.23E-03(1.01)	4.90E-03(1.01)	5.19E-03(0.98)	8.06E-03(0.85)	1.56E-02(0.67)
6	49665	3.11E-03(1.01)	2.46E-03(1.01)	2.65E-03(0.98)	4.48E-03(0.85)	9.88E-03(0.67)

Table 4.7 Convergence history for bilinear elements with meshes from Algorithm 4.13.

j	DOFs	$\lvert u - u_j \rvert_{H^1(\Omega)}$				
		κ_1=0.1	κ_1=0.2	κ_1=0.3	κ_1=0.4	κ_1=0.5
2	225	4.72E-02(0.97)	3.85E-02(0.99)	3.80E-02(0.98)	4.56E-02(0.88)	6.13E-02(0.72)
3	833	2.48E-02(0.99)	2.01E-02(1.00)	2.02E-02(0.97)	2.60E-02(0.86)	3.90E-02(0.69)
4	3201	1.27E-02(1.00)	1.03E-02(1.00)	1.06E-02(0.96)	1.47E-02(0.85)	2.47E-02(0.68)
5	12545	6.39E-03(1.00)	5.19E-03(1.00)	5.47E-03(0.97)	8.20E-03(0.85)	1.56E-02(0.67)
6	49665	3.21E-03(1.00)	2.61E-03(1.00)	2.81E-03(0.97)	4.56E-03(0.85)	9.88E-03(0.67)

Table 4.8 Convergence history for bilinear elements with meshes from Algorithm 4.14.

Chapter 5
Regularity Estimates and Graded Meshes in Polyhedral Domains

Consider elliptic equations with mixed boundary conditions in a polyhedral domain. The solution can possess vertex singularities and anisotropic edge singularities according to the nature of the nonsmooth points on the boundary. The solution may also have singularities owing to the change of boundary conditions. The aim of this chapter is twofold. First, it presents regularity estimates for these singular solutions in various anisotropic weighted Sobolev spaces. Second, it lays out important geometric properties of the 3D anisotropic mesh described in Chap. 3. This chapter is suitable for readers who are interested in regularity analysis for 3D anisotropic singular solutions and who plan to study the anisotropic graded mesh in detail.

We study the problem associated with the Laplace operator in a bounded polyhedral domain $\Omega \subset \mathbb{R}^3$ with the mixed boundary condition

$$-\Delta u = f \quad \text{in} \quad \Omega, \qquad u = 0 \quad \text{on} \quad \Gamma_D \qquad \partial_n u = 0 \quad \text{on} \quad \Gamma_N, \tag{5.1}$$

where Γ_D and Γ_N are open subsets of the boundary $\Gamma := \partial\Omega$ such that $\overline{\Gamma_D} \cup \overline{\Gamma_N} = \Gamma$. For simplicity, we suppose that each face of Γ is included either in Γ_D or in Γ_N and $\Gamma_D \neq \emptyset$. Nonetheless, the results discussed here shall apply to the more general setting of Γ_D and Γ_N as in (3.3). According to the Poincaré inequality and the Lax–Milgram Theorem, the solution of (5.1) is uniquely defined in $H^1_D(\Omega)$ (2.26) for $f \in \left(H^1_D(\Omega)\right)^*$.

It is generally a difficult task to approximate 3D problems due to the intensive computations involved. In addition, based on the discussions in Sect. 3.3, the solution can have different types of singularities: the vertex singularity and the anisotropic edge singularity. The combination of different types of singularities, together with the complexity in the 3D geometry, can significantly complicate the analysis and numerical algorithm design. In this chapter, we explore the regularity analysis and study the properties of the anisotropic mesh for (5.1).

H. Li, *Graded Finite Element Methods for Elliptic Problems in Nonsmooth Domains*,
Surveys and Tutorials in the Applied Mathematical Sciences 10,
https://doi.org/10.1007/978-3-031-05821-9_5

5.1 Regularity Analysis in Anisotropic Weighted Sobolev Spaces

For a polyhedral domain Ω, denote by $\mathcal{E}$ the finite set of open edges and by C the finite set of vertices. Let $\mathcal{S} := C \cup \mathcal{E}$. We also denote by $\mathcal{E}_c \subset \mathcal{E}$ the set of edges joining at $c \in C$ and by $C_e \subset C$ the set of endpoints of $e \in \mathcal{E}$. For any $c \in C$ (resp. $e \in \mathcal{E}$), we define $r_c(x)$ (resp. $r_e(x)$) to be the distance from $x \in \Omega$ to c (resp. to e). We further define $r_{c,e}(x) := r_e(x)/r_c(x)$ as the angular distance from x to the edge e near c. Then for any vertex $c \in C$ and edge $e \in \mathcal{E}$, we define the following subsets of Ω:

$$\begin{cases} \mathcal{V}_c = \{x \in \Omega : \ r_c(x) < \varepsilon\}, \\ \mathcal{V}_c^e = \{x \in \mathcal{V}_c : \ r_{c,e}(x) < \varepsilon\}, \\ \mathcal{V}_c^o = \{x \in \mathcal{V}_c : \ r_{c,e}(x) \geq \varepsilon, \ \forall\, e \in \mathcal{E}_c\}, \\ \mathcal{V}_e^o = \{x \in \Omega : \ r_c(x) \geq \varepsilon, \ r_{c,e}(x) < \varepsilon, \ \forall\, c \in C_e\}, \end{cases} \tag{5.2}$$

with $\varepsilon > 0$ small enough, such that all these sets are disjoint. We further define

$$\mathcal{V}^o = \Omega \setminus \left((\cup_{c \in C} \mathcal{V}_c) \cup (\cup_{e \in \mathcal{E}} \mathcal{V}_e^o) \right). \tag{5.3}$$

See Fig. 5.1 for an illustration of the domain decomposition near c and e.

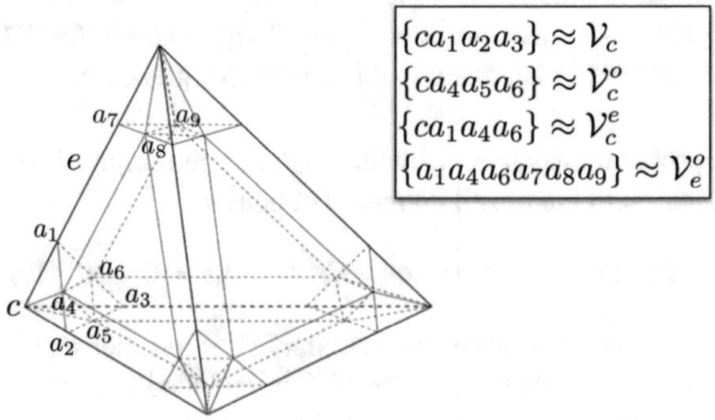

Fig. 5.1 A domain decomposition with $c \in C$ and $e \in \mathcal{E}$. Each polyhedral region is denoted by its vertices in the text box.

We shall also use the following notation in the study of (5.1). The subsets in (5.2) are neighborhoods of different nonsmooth points on the boundary. In the neighborhoods $\mathcal{V}_e^o$ and $\mathcal{V}_c^e$, we choose a local Cartesian coordinate system in which the edge $e \in \mathcal{E}$ lies on the z-axis. Let $\alpha_\perp = (\alpha_1, \alpha_2)$ consist of the first two entries of the multi-index $\alpha = (\alpha_1, \alpha_2, \alpha_3) \in \mathbb{Z}^3_{\geq 0}$. Therefore, in $\mathcal{V}_e^o$ and $\mathcal{V}_c^e$, $\partial^{\alpha_\perp} = \partial_x^{\alpha_1}\partial_y^{\alpha_2}$ is the derivative in a plane perpendicular to the edge e. In addition, both of the following terms are used to represent the same directional derivative: $\partial_1 = \partial_x$, $\partial_2 = \partial_y$, and $\partial_3 = \partial_z$. Then we shall study the solution regularity for (5.1) in two cases: (i) the Dirichlet boundary condition ($\overline{\Gamma_D} = \partial\Omega$) and (ii) the mixed boundary condition ($\Gamma_D, \Gamma_N \neq \emptyset$) in which the Neumann boundary condition is allowed on adjacent faces of the boundary.

5.1.1 Dirichlet Boundary Conditions and Weighted Spaces

Assume $\overline{\Gamma_D} = \partial\Omega$ in (5.1). We first define the following weighted Sobolev space.

Definition 5.1 (Anisotropic Weighted Space) Let $H^m_{loc}(\Omega) := \{v : v \in H^m(G)\}$, where G is any open subset with compact closure such that $\bar{G} \subset \Omega$. For $c \in C$ and $e \in \mathcal{E}$, let $\mu_c \in \mathbb{R}$ and $\mu_e \in \mathbb{R}$ be the associated parameters, respectively. Recall $\mathcal{S} = C \cup \mathcal{E}$. Define $\boldsymbol{\mu} = (\mu_s)_{s \in \mathcal{S}}$ to be the vector that consists of all these parameters. Then we define the anisotropic weighted space [38, 46]

$$\mathcal{M}^m_{\boldsymbol{\mu}}(\Omega) := \Big\{v \in H^m_{loc}(\Omega) : r_c^{|\alpha|-\mu_c}\partial^\alpha v \in L^2(\mathcal{V}^o_c),\ r_e^{|\alpha_\perp|-\mu_e}\partial^\alpha v \in L^2(\mathcal{V}^o_e), \quad (5.4)$$
$$r_c^{|\alpha|-\mu_c} r_{c,e}^{|\alpha_\perp|-\mu_e}\partial^\alpha v \in L^2(\mathcal{V}^e_c),\ \forall\, |\alpha| \le m\Big\}.$$

For any $v \in \mathcal{M}^m_{\boldsymbol{\mu}}(\Omega)$, the associated norm is

$$\|v\|^2_{\mathcal{M}^m_{\boldsymbol{\mu}}(\Omega)} := \|v\|^2_{H^m(\mathcal{V}^o)} + \sum_{|\alpha| \le m}\Bigg(\sum_{c \in C}\|r_c^{|\alpha|-\mu_c}\partial^\alpha v\|^2_{L^2(\mathcal{V}^o_c)}$$
$$+ \sum_{e \in \mathcal{E}, c \in C_e}\|r_c^{|\alpha|-\mu_c} r_{c,e}^{|\alpha_\perp|-\mu_e}\partial^\alpha v\|^2_{L^2(\mathcal{V}^e_c)} + \sum_{e \in \mathcal{E}}\|r_e^{|\alpha_\perp|-\mu_e}\partial^\alpha v\|^2_{L^2(\mathcal{V}^o_e)}\Bigg).$$

For the polyhedral domain Ω, let N_s be the number of entries in $\boldsymbol{\mu}$. For any two N_s-dimensional vectors $\boldsymbol{a}$ and $\boldsymbol{b}$, we write $\boldsymbol{a} < (\le, >, \ge)\, \boldsymbol{b}$ if each entry $a_\ell < (\le, >, \ge)\, b_\ell$. We denote by $\boldsymbol{1}$ (resp. $\boldsymbol{0}$) the constant N_s-dimensional vectors with all entries being 1 (resp. 0).

Remark 5.1 Recall the weighted space $\mathcal{K}^m_{\boldsymbol{a}}$ (Definition 4.1) for polygonal domains. Its 3D counterpart for a polyhedral domain Ω can be defined as

$$\mathcal{K}^m_{\boldsymbol{\mu}}(\Omega) := \Big\{v \in H^m_{loc}(\Omega) : r_c^{|\alpha|-\mu_c}\partial^\alpha v \in L^2(\mathcal{V}^o_c),\ r_e^{|\alpha|-\mu_e}\partial^\alpha v \in L^2(\mathcal{V}^o_e),$$
$$r_c^{|\alpha|-\mu_c} r_{c,e}^{|\alpha|-\mu_e}\partial^\alpha v \in L^2(\mathcal{V}^e_c),\ \forall\, |\alpha| \le m\Big\}.$$

The space $\mathcal{K}^m_{\boldsymbol{\mu}}$ is isotropic, in which the exponent of the weight function depends on the total order of the derivative but not on the direction of the derivative. The space $\mathcal{K}^m_{\boldsymbol{\mu}}$ can be used to formulate isotropic regularity estimates (see [38]) but hardly characterize the anisotropic edge singularity that may occur in the solution. In contrast, the space $\mathcal{M}^m_{\boldsymbol{\mu}}$ is anisotropic in the sense that the transverse derivatives $\partial^{\alpha_\perp}$ and the longitudinal derivatives along the edge play different roles in the formulation. Therefore, the space $\mathcal{M}^m_{\boldsymbol{\mu}}$ can better describe the anisotropic behavior of the singular solution, especially the additional regularity in the edge direction. For example, let $\boldsymbol{\eta} = (\eta_s)_{s \in \mathcal{S}}$ be the vector such that

$$\eta_c = \sqrt{\lambda_c + 1/4} \ \text{ for } s = c \in C \ \text{ and } \ \eta_e = \lambda_e = \pi/\omega_e \ \text{ for } s = e \in \mathcal{E}, \qquad (5.5)$$

where $\lambda_c > 0$ is the smallest positive eigenvalue of the Laplace–Beltrami operator with the zero Dirichlet boundary condition in the spherical polygon G on the unit sphere S^2 centered at c (see (3.26)), and ω_e is the interior dihedral angle of the edge e. Then for $m \geq 0$, the solution $u \in H_0^1(\Omega)$ of equation (5.1) satisfies [38, 46]

$$\|u\|_{\mathcal{M}^m_{\mu+1}(\Omega)} \leq C\|f\|_{\mathcal{M}^m_{\mu-1}(\Omega)}, \qquad 0 \leq \mu < \eta. \tag{5.6}$$

According to the discussion in Sect. 3.3, when the parameter $\eta_e < 1$ (resp. $\eta_c < 1$), the solution has a singular (non-H^2) component near the edge e (resp. near the vertex $c \in C$). Thus, for the Dirichlet problem in the next subsection (Sect. 5.1.2), we shall call an edge $e \in \mathcal{E}$ (resp. a vertex $c \in C$) a *singular edge* (resp. a *singular vertex*) of the domain if $\eta_e = \lambda_e < 1$ (resp. $\eta_c < 1$). Therefore, an edge is singular when the associated angle $\omega_e > \pi$. Note that this definition of singular edge and singular vertex is different than Definition 3.1, in which the singular edges and singular vertices were some special edges and vertices of the elements in the triangulation.

Remark 5.2 The regularity estimate (5.6) establishes the continuous dependence of the solution on the given data in weighted spaces, despite the lack of regularity in usual Sobolev spaces. Compared with the regularity results in the usual Sobolev space in Sect. 2.3, the estimate (5.6) does not give a shifting in the index m. Consequently, a smoother f (i.e., $f \in H^m_{loc}(\Omega)$) is expected in order for u to be in $\mathcal{M}^m_{\mu+1}(\Omega)$. For example, the condition $u \in \mathcal{M}^2_{\mu+1}(\Omega)$ needs $f \in H^2_{loc}(\Omega)$ (i.e., $f \in \mathcal{M}^2_{\mu-1}(\Omega)$), while when the domain is smooth, $f \in L^2(\Omega)$ will lead to $u \in H^2(\Omega)$. Meanwhile, the convergence analysis of numerical schemes usually requires the solution to be in a high-order space such as $\mathcal{M}^m_{\mu+1}(\Omega)$, $m \geq 1$. In this case, a rough given function f (e.g., $f \in L^2_{loc}(\Omega)$) will be prohibited. In what follows, we present a different regularity result in anisotropic weighted spaces that merely needs f to be in a weighted L^2 space and that gives rise to $u \in H^2_{loc}(\Omega)$. The main result is summarized in Corollary 5.2.

5.1.2 Dirichlet Boundary Conditions and Rough Given Data

Assume $\overline{\Gamma_D} = \partial\Omega$ in (5.1). Define the following weighted L^2 space.

Definition 5.2 (Weighted L^2 Space) For $\boldsymbol{\mu} = (\mu_e)_{e\in\mathcal{E}}$ with $\mu_e \in \mathbb{R}$, introduce the function

$$w_{\boldsymbol{\mu}}(x) = \begin{cases} 1 & \text{if } x \in \mathcal{V}^o \cup \left(\bigcup_{c\in C} \mathcal{V}^o_c\right), \\ r_{c,e}^{-\mu_e} & \text{if } x \in \mathcal{V}^e_c, \quad \forall\, c \in C, e \in \mathcal{E}_c, \\ r_e^{-\mu_e} & \text{if } x \in \mathcal{V}^o_e, \quad \forall\, e \in \mathcal{E}. \end{cases} \tag{5.7}$$

For any $G \subseteq \Omega$, define the space

$$L^2_{\boldsymbol{\mu}}(G) = \{v \in L^2_{loc}(\Omega) : \; w_{\boldsymbol{\mu}} v \in L^2(G)\}.$$

This space is a Hilbert space with its natural inner product

$$(v, z)_{\mu} = \int_{\Omega} w_{\mu}(x)^2 v(x) z(x)\, dx, \quad \forall\, v, z \in L^2_{\mu}(\Omega),$$

and the associated norm $\|v\|_{\mu} = (v, v)^{\frac{1}{2}}_{\mu}$, for $v \in L^2_{\mu}(\Omega)$.

From now on, we fix the parameter $\boldsymbol{\mu}^* = (\mu^*_e)_{e \in \mathcal{E}}$, such that

$$\mu^*_e \in [1/2, \lambda_e) \quad \text{if } \omega_e > \pi, \qquad \text{and} \qquad \mu^*_e = 0 \quad \text{otherwise.} \tag{5.8}$$

Based on (5.5), $\lambda_e > 1/2$. Therefore, the parameter μ^*_e is well defined. Then we study equation (5.1) with $f \in L^2_{\boldsymbol{\mu}^*}(\Omega)$.

The following analysis is based on regularity estimates in different sub-regions (see (5.2)) of the domain near the vertices and edges. To better understand the dependence of the solution on the domain and on the given data, we begin with a slightly different assumption on f: we assume $f \in L^2_{\boldsymbol{\mu}}(\Omega)$, where $\boldsymbol{\mu} = (\mu_e)_{e \in \mathcal{E}}$ satisfies

$$\mu_e \in (0, \lambda_e) \quad \text{if } \omega_e > \pi, \qquad \text{and} \qquad \mu_e = 0 \quad \text{otherwise.} \tag{5.9}$$

It is clear that $L^2_{\boldsymbol{\mu}^*}(\Omega) \subseteq L^2_{\boldsymbol{\mu}}(\Omega) \subset L^2(\Omega)$ if $\mu_e \leq \mu^*_e$ for all $e \in \mathcal{E}$. Therefore, the regularity results obtained for $f \in L^2_{\boldsymbol{\mu}}(\Omega)$ shall still hold for $f \in L^2_{\boldsymbol{\mu}^*}(\Omega)$, but it will simplify the exposition.

Regularity Estimates in $\mathcal{V}^o_e$ We start with an improved regularity result for the solution along the edges for $f \in L^2_{\boldsymbol{\mu}}(\Omega)$.

Theorem 5.1 *Let $u \in H^1_0(\Omega)$ be the solution of* (5.1) *with $f \in L^2_{\boldsymbol{\mu}}(\Omega)$ as defined in (5.9). Then for any $e \in \mathcal{E}$, we have*

$$u \in L^2_{\boldsymbol{\mu}}(\mathcal{V}^o_e), \qquad \nabla u \in [L^2_{\boldsymbol{\mu}}(\mathcal{V}^o_e)]^3, \tag{5.10}$$

$$\partial^2_3 u \in L^2_{\boldsymbol{\mu}}(\mathcal{V}^o_e), \tag{5.11}$$

where ∂_3 is the derivative in the direction of the edge e.

Proof For an edge e with $\omega_e < \pi$, the results are immediate since u belongs to $H^2(\mathcal{V}^o_e)$ (see [49] or [9, Theorem 2.4]). Hence it remains to prove the results for a singular edge e for which $\omega_e > \pi$.

Let ζ be a fixed interior point of a singular edge e and let χ be a cut-off function such that $\chi \equiv 1$ in a neighborhood of ζ and $\chi \equiv 0$ in a neighborhood of the vertices and the other edges. Assume the edge e is on the z-axis. Without loss of generality, we can assume that ζ is at the origin and we drop the index e. Denote by $D = \Sigma \times \mathbb{R}$ the dihedral truncated cone that coincides with Ω near ζ, where Σ is the truncated two-dimensional cone of opening ω, namely

$$\Sigma = \{(r\cos\theta, r\sin\theta) \in \mathbb{R}^2 \,:\, 0 < r < 1, 0 < \theta < \omega\}.$$

We assume that the support of χ is included in D.

Note that $\tilde{u} := \chi u$ is a weak solution of

$$-\Delta\tilde{u} = \tilde{f} \quad \text{in } D, \tag{5.12}$$

where $\tilde{f}$ is given by

$$\tilde{f} = \chi f - 2\nabla u \cdot \nabla\chi - u\Delta\chi \in L^2(D).$$

To have the regularity $\tilde{u} \in L^2_\mu(D) = \{v \in L^2(D) : r^{-\mu} v \in L^2(D)\}$, where r is the distance to the edge $\{0\} \times \mathbb{R}$ of D, we use Theorem 2.4 of [9] that shows

$$r^{\delta-2}\tilde{\chi}u \in L^2(D) \quad \text{and} \quad r^{\delta-1}\nabla(\tilde{\chi}u) \in [L^2(D)]^3, \tag{5.13}$$

for any $\delta > 1 - \lambda_e$, where $\tilde{\chi}$ is similar to χ, except that $\tilde{\chi} \equiv 1$ on the support of χ. This result directly leads to (5.10) and, in particular, to $\tilde{u} \in L^2_\mu(D)$.

Then we apply a Fourier transform technique (see also [63]). Namely, performing a partial Fourier transform in z, we see that $V = \mathfrak{F}_z(\tilde{u})(\zeta)$ is the solution of

$$\begin{cases} -\Delta V + \zeta^2 V = \mathfrak{F}_z(\tilde{f})(\zeta), & \text{in } \Sigma, \\ V = 0, & \text{on } \partial\Sigma. \end{cases}$$

By Parseval's identity, we have

$$\|\tilde{f}\|^2_{L^2_\mu(D)} = \int_{\mathbb{R}} \|\mathfrak{F}_z(\tilde{f})(\zeta)\|^2_{L^2_\mu(\Sigma)}\, d\zeta,$$

where $\|v\|^2_{L^2_\mu(\Sigma)} = \int_\Sigma |r(x)^{-\mu} v(x)|^2\, dx$. Then we apply Corollary 2.12 of [50] that furnishes

$$\zeta^2\|V\|_{L^2_\mu(\Sigma)} \le C\|\mathfrak{F}_z(\tilde{f})(\zeta)\|_{L^2_\mu(\Sigma)}.$$

Taking the square of this estimate, using the Fourier transform back and again Parseval's identity, we find that

$$\|\partial_3^2\tilde{u}\|^2_{L^2_\mu(D)} \le C\|\tilde{f}\|^2_{L^2_\mu(D)}. \tag{5.14}$$

This proves (5.11). □

Regularity Estimates in $\mathcal{V}_c^e$ Now we describe the extra regularity in a neighborhood of a vertex $c \in \mathcal{C}$ that is an endpoint of the singular edge e. Let Λ_c be the cone that coincides with Ω at c. Recall that r_c is the distance to c. Then for any $\beta \in \mathbb{R}$, $k \in \mathbb{Z}_{\ge 0}$, we define the space

$$V^k_\beta(\Lambda_c) = \{v \in L^2_{loc}(\Lambda_c) : r_c^{\beta+|\alpha|-k} D^\alpha v \in L^2(\Lambda_c),\ \forall\, |\alpha| \le k\}. \tag{5.15}$$

We fix a cut-off function χ such that $\chi \equiv 1$ in a neighborhood of c and $\chi \equiv 0$ in a neighborhood of the other vertices of Ω. Note that for (5.1), the exponent of the vertex singularity near c is given by $-\frac{1}{2} \pm \sqrt{\lambda_{c,k} + \frac{1}{4}}$, where $\{\lambda_{c,k}\}_{k=1}^\infty$ is the spectrum (repeated according to their multiplicity) of the positive Laplace–Beltrami operator Δ' with the Dirichlet boundary condition on the intersection G between Λ_c and the unit sphere. See [9, 61] and also Theorem 3.7. Since u is in H^1, we are only interested in exponents larger than $-\frac{1}{2}$. Therefore, let

$$\tau_{c,k} = -\frac{1}{2} + \sqrt{\lambda_{c,k} + \frac{1}{4}}. \tag{5.16}$$

The associated singular function $\sigma_{c,k}$ is given by

$$\sigma_{c,k} = r_c^{\tau_{c,k}} \phi_{c,k},$$

where $\phi_{c,k}$ is the eigenvector of Δ' associated with $\lambda_{c,k}$. Then we have the regularity estimate for the regular part of the solution near the vertex.

Lemma 5.1 *Assume $\tau_{c,k} \neq \frac{1}{2}$ for all $k \in \mathbb{N}$. Recall the cut-off function χ defined above. Let $u \in H_0^1(\Omega)$ be the solution of* (5.1) *with $f \in L_\mu^2(\Omega)$ as defined in (5.9). Then χu admits the splitting*

$$\chi u = u_0 + \sum_{0<\tau_{c,k}<\frac{1}{2}} c_k \sigma_{c,k}, \tag{5.17}$$

where $u_0 \in V_{-1}^1(\Lambda_c)$, $c_k \in \mathbb{C}$ and $\Delta u_0 \in L_\mu^2(\Lambda_c)$. Furthermore we have

$$r_c^{-2} w_\mu u_0 \in L^2(\Lambda_c) \quad \text{and} \quad r_c^{-1} w_\mu \nabla u_0 \in [L^2(\Lambda_c)]^3, \tag{5.18}$$

where w_μ is defined as in (5.7)*:*

$$w_\mu(x) = \begin{cases} r_{c,e}^{-\mu_e} & \text{if } r_{c,e}(x) < \varepsilon,\ \forall\, e \in \mathcal{E}_c, \\ 1 & \text{else.} \end{cases}$$

Proof We first apply Theorem 2.6 of [9] (see also Lemma 17.4 of [49]) that yields the decomposition (5.17) with $c_k \in \mathbb{C}$ and $u_0 \in V_{-1}^1(\Lambda_c)$ that can be split up in the form

$$u_0 = u_r + u_{\text{edge}},$$

with $u_r \in V_0^2(\Lambda_c) \cap H_0^1(\Lambda_c)$ and $u_{\text{edge}} \in V_{-1}^1(\Lambda_c)$. Note that $\sigma_{c,k}$ are harmonic functions. Thus $\Delta u_0 = \Delta(\chi u)$, and, therefore, Δu_0 belongs to $L_\mu^2(\Lambda_c)$.

Note that $u_r \in V_0^2(\Lambda_c)$ and $u_{\text{edge}} \in V_{-1}^1(\Lambda_c)$ yields (5.18) far from the edges. Hence we only need to show extra regularities along the edge. Now, we fix one edge $e \in \mathcal{E}_c$. Without loss of generality we assume that the edge e is contained in the z-axis and that c is at the origin. Fix further the spherical coordinates (ρ, θ, φ) such that $\varphi = 0$ corresponds to the z-axis (hence $\rho = r_c$ near c and $\varphi = r_{c,e}$ near e).

To prove the extra regularity of u_0 along e, we notice that Theorem 2.7 of [9] shows that

$$\rho^{-1} \varphi^{\delta_e - 1} \nabla u_{\text{edge}} \in [L^2(\Lambda_c^e)]^3,$$

for any $\delta_e \geq 0$ such that $\delta_e > 1 - \lambda_e$ if $\omega_e > \pi$, and $\delta_e = 0$ else, where

$$\Lambda_c^e = \{x \in \Lambda_c : r_{c,e} < \varepsilon\}.$$

As $\mu_e < \lambda_e$, we can always pick δ_e, such that $1 - \mu_e \geq \delta_e > 1 - \lambda_e$ and, therefore,

$$\rho^{-1}\varphi^{-\mu_e}\nabla u_{\text{edge}} \in [L^2(\Lambda_c^e)]^3. \tag{5.19}$$

For simplicity, we write μ instead of μ_e. Now we take advantage of the fact that $u_{\text{edge}}(0) = 0$ to write

$$u_{\text{edge}}(\rho, \theta, \varphi) = \int_0^\rho \frac{\partial u_{\text{edge}}}{\partial \rho}(s, \theta, \varphi)\, ds.$$

Using the Hardy operator H defined in [61, p. 28], we have

$$\rho^{-1}u_{\text{edge}}(\rho, \theta, \varphi) = H\left(\frac{\partial u_{\text{edge}}}{\partial \rho}(\cdot, \theta, \varphi)\right).$$

Since for almost all (θ, φ), we have

$$\int_0^\infty \left|\frac{\partial u_{\text{edge}}}{\partial \rho}(\cdot, \theta, \varphi)\right|^2 d\rho < \infty,$$

we can apply Hardy's inequality [61, p. 28] to get

$$\int_0^\infty |\rho^{-1}u_{\text{edge}}(\rho, \theta, \varphi)|^2\, d\rho \lesssim \int_0^\infty \left|\frac{\partial u_{\text{edge}}}{\partial \rho}(\rho, \theta, \varphi)\right|^2 d\rho.$$

Multiplying this estimate by $\varphi^{-2\mu}\sin\varphi$ and integrating in θ and φ, we find that

$$\int_{\Lambda_c} |\rho^{-2}\varphi^{-\mu}u_{\text{edge}}(x)|^2\, dx \lesssim \int_{\Lambda_c} \left|\rho^{-1}\varphi^{-\mu}\frac{\partial u_{\text{edge}}}{\partial \rho}(x)\right|^2 dx < \infty.$$

This shows that

$$\rho^{-2}\varphi^{-\mu}u_{\text{edge}} \in L^2(\Lambda_c) \tag{5.20}$$

and yields the requested regularity of u_{edge} near the edge.

For the regular part u_r, in view of its regularity $V_0^2(\Lambda_c)$, we only need to show extra regularity along the edge e. Therefore, we fix a cut-off function χ depending only on φ that is equal to 1 in a neighborhood of $\varphi = 0$ and equal to 0 outside a larger neighborhood of $\varphi = 0$. Then the regularity $u_r \in V_0^2(\Lambda_c)$ implies that

$$\rho^{-1}\partial_j(\chi u_r) \in L^2(\Lambda_c), \quad \forall\, j = 1, 2, 3.$$

Now for $j \in \{1, 2, 3\}$, we write

$$\partial_j(\chi u_r)(\rho,\theta,\varphi) = -\int_\varphi^\infty \frac{\partial}{\partial\varphi}\partial_j(\chi u_r)(\rho,\theta,\tilde{\varphi})\, d\tilde{\varphi}.$$

Hence using the Hardy operator L defined in [61, p. 28], we have

$$\varphi^{-1}\partial_j(\chi u_r)(\rho,\theta,\varphi) = -L\left(\frac{\partial}{\partial\varphi}\partial_j(\chi u_r)(\rho,\theta,\cdot)\right).$$

Since for almost all (ρ,θ),

$$\int_0^\infty \left|\frac{\partial}{\partial\varphi}\partial_j(\chi u_r)(\rho,\theta,\tilde{\varphi})\right|^2 \tilde{\varphi}^{1+2\epsilon}\, d\tilde{\varphi} < \infty,$$

for all $\epsilon > 0$, we can apply Hardy's inequality (see [61, p. 28]) to get

$$\int_0^\infty |\tilde{\varphi}^{\epsilon-1}\partial_j(\chi u_r)(\rho,\theta,\tilde{\varphi})|^2\tilde{\varphi}\, d\tilde{\varphi} \lesssim \int_0^\infty \left|\tilde{\varphi}^\epsilon \frac{\partial}{\partial\varphi}\partial_j(\chi u_r)(\rho,\theta,\tilde{\varphi})\right|^2 \tilde{\varphi}\, d\tilde{\varphi},$$

for all $\epsilon > 0$. Integrating in ρ and θ, we find that for all $\epsilon > 0$,

$$\int_{\Lambda_c} |\rho^{-1}\varphi^{\epsilon-1}\partial_j(\chi u_r)(x)|^2\, dx \lesssim \int_{\Lambda_c} \left|\rho^{-1}\frac{\partial}{\partial\varphi}\partial_j(\chi u_r)(x)\right|^2\, dx < \infty.$$

By choosing ϵ small enough, we will have $\varphi^{-\mu} \lesssim \varphi^{\epsilon-1}$, and therefore, we deduce that

$$\rho^{-1}\varphi^{-\mu}\partial_j(\chi u_r) \in L^2(\Lambda_c), \quad \forall\, j = 1,2,3. \tag{5.21}$$

Now we use the fact that $u_r(0) = 0$ to write

$$\chi u_r(\rho,\theta,\varphi) = \int_0^\rho \frac{\partial(\chi u_r)}{\partial\rho}(s,\theta,\varphi)\, ds,$$

and with the help of Hardy's inequality and (5.21), we deduce that

$$\rho^{-2}\varphi^{-\mu}\chi u_r \in L^2(\Lambda_c). \tag{5.22}$$

The conclusion (5.18) follows from (5.19), (5.20) (5.21), and (5.22). □

Besides the estimates for low-order derivatives of u_0 in Lemma 5.1, we now derive the regularity estimate for the second derivative of u_0 along the edge.

Lemma 5.2 *Under the assumption of Lemma 5.1, let $u \in H_0^1(\Omega)$ be the solution of* (5.1) *with $f \in L^2_\mu(\Omega)$ as defined in (5.9). Then u_0 from* (5.17) *satisfies*

$$r_{c,e}^{-\mu_e}\partial_3^2 u_0 \in L^2(\mathcal{V}_c^e), \qquad \forall\, c \in \mathcal{C}, e \in \mathcal{E}_c, \tag{5.23}$$

where ∂_3 is the derivative in the direction of the edge e.

Proof We assume that the edge e is contained in the z-axis and that c is at the origin. Fix the spherical coordinates (ρ, θ, φ), such that $\varphi = 0$ corresponds to the z-axis (hence $\rho = r_c$ near c and $\varphi = r_{c,e}$ near e). To simplify the notation, we shall drop the index e in μ_e. We use a dyadic covering technique. Namely, for all $j \in \mathbb{N}$, we define

$$\Sigma_{0j} := \{x \in \Lambda_c : 1 < 2^j|x| < 2\} \quad \text{and} \quad \Sigma_{1j} := \{x \in \Lambda_c : 1/2 < 2^j|x| < 4\},$$

which are, respectively, homothetic to

$$\hat{\Sigma}_0 := \{x \in \Lambda_c : 1 < |x| < 2\} \quad \text{and} \quad \hat{\Sigma}_1 := \{x \in \Lambda_c : 1/2 < |x| < 4\},$$

via the mapping

$$h_j : \Lambda_c \to \Lambda_c : x \to 2^j x.$$

For a fixed $j \in \mathbb{N}$, set $\hat{u}_0(\hat{x}) = u_0(h_j^{-1}\hat{x})$. We fix a cut-off function $\hat{\chi}$ equal to 1 on $\hat{\Sigma}_0$ and equal to 0 outside $\hat{\Sigma}_1$. Then applying the estimate (5.14) to $\hat{\chi}\hat{u}_0$, we find that

$$\|\varphi(\hat{x})^{-\mu}\partial_3^2\hat{u}_0\|^2_{\hat{\Sigma}_{0,e}} \lesssim \|\varphi(\hat{x})^{-\mu}\Delta\hat{u}_0\|^2_{\hat{\Sigma}_{1,e}} + \|\varphi(\hat{x})^{-\mu}\nabla\hat{u}_0\|^2_{\hat{\Sigma}_{1,e}} + \|\varphi(\hat{x})^{-\mu}\hat{u}_0\|^2_{\hat{\Sigma}_{1,e}},$$

where

$$\hat{\Sigma}_{0,e} = \{\hat{x} \in \hat{\Sigma}_0 : \varphi(\hat{x}) < \varepsilon\} \quad \text{and} \quad \hat{\Sigma}_{1,e} = \{\hat{x} \in \hat{\Sigma}_1 : \varphi(\hat{x}) < 2\varepsilon\}.$$

As $\hat{\rho}$ is equivalent to 1 on $\hat{\Sigma}_1$, this estimate implies that

$$\|\varphi^{-\mu}\partial_3^2\hat{u}_0\|^2_{\hat{\Sigma}_{0,e}} \lesssim \|\varphi^{-\mu}\Delta\hat{u}_0\|^2_{\hat{\Sigma}_{1,e}} + \|\hat{\rho}^{-1}\varphi^{-\mu}\nabla\hat{u}_0\|^2_{\hat{\Sigma}_{1,e}} + \|\hat{\rho}^{-2}\varphi^{-\mu}\hat{u}_0\|^2_{\hat{\Sigma}_{1,e}}.$$

Coming back to u_0 via the transformation h_j, we find

$$\|\varphi^{-\mu}\partial_3^2 u_0\|^2_{\Sigma_{0j,e}} \lesssim \|\varphi^{-\mu}\Delta u_0\|^2_{\Sigma_{1j,e}} + \|\rho^{-1}\varphi^{-\mu}\nabla u_0\|^2_{\Sigma_{1j,e}} + \|\rho^{-2}\varphi^{-\mu}u_0\|^2_{\Sigma_{1j,e}},$$

where

$$\Sigma_{0j,e} = \{x \in \Sigma_{0j} : \varphi(x) < \varepsilon\} \quad \text{and} \quad \Sigma_{1j,e} = \{x \in \Sigma_{1j} : \varphi(x) < 2\varepsilon\}.$$

Summing on $j \in \mathbb{N}$ and taking into account Lemma 5.1, we arrive at the estimate (5.23). □

Recall η_c in (5.5). Then we have the weighted regularity estimate for the second derivative of the solution along the edge direction in the neighborhood of the vertex.

Theorem 5.2 *Under the assumption of Lemma 5.1, let $u \in H_0^1(\Omega)$ be the solution of* (5.1) *with $f \in L^2_{\mu}(\Omega)$ as defined in (5.9). Then u satisfies*

$$r_c^{\beta_c} r_{c,e}^{-\mu_e}\partial_3^2 u \in L^2(\mathcal{V}_c^e), \qquad \forall\, c \in \mathcal{C}, e \in \mathcal{E}_c, \tag{5.24}$$

where ∂_3 is the derivative in the direction of the edge e; and $\beta_c > 1 - \eta_c$ if c is singular ($\eta_c < 1$), $\beta_c = 0$ otherwise.

Proof In view of the splitting (5.17), it suffices to show that each term satisfies the desired regularity estimate. Since $r_c^{\beta_c} \lesssim 1$, by Lemma 5.2, u_0 clearly satisfies

$$r_c^{\beta_c} r_{c,e}^{-\mu_e} \partial_3^2 u_0 \in L^2(\mathcal{V}_c^e).$$

The singular part (that is zero if c is not singular) satisfies it since by using spherical coordinates we see that $\partial_3^2 \sigma_{c,k}$ behaves like $r_c^{\tau_{c,k}-2} r_{c,e}^{\lambda_e}$. Hence direct calculations yield

$$r_c^{\beta_c} r_{c,e}^{-\mu_e} \partial_3^2 \sigma_{c,k} \in L^2(\mathcal{V}_c^e),$$

for any $\beta_c > 1 - \eta_c$. □

Now, we extend the analysis and derive regularity estimates for derivatives of the solution both along and perpendicular to the edge direction.

Corollary 5.1 *Recall η_e and η_c in (5.5). Under the assumption of Lemma 5.1, let $u \in H_0^1(\Omega)$ be the solution of (5.1) with $f \in L_\mu^2(\Omega)$ as defined in (5.9). For any $c \in C$ and $e \in \mathcal{E}_c$, let $\gamma_c, \gamma_e \in [0, 1]$ be such that $\gamma_c < \eta_c$ and $\gamma_e < \eta_e$. Then the following norms/seminorms of u are bounded by $\|f\|_{L_\mu^2(\Omega)}$:*

$$\|r_c^{-1-\gamma_c} r_{c,e}^{-1-\gamma_e} u\|_{L^2(\mathcal{V}_c^e)}, \tag{5.25}$$

$$\|r_c^{-\gamma_c} r_{c,e}^{-\gamma_e} \partial_\perp u\|_{L^2(\mathcal{V}_c^e)}, \tag{5.26}$$

$$\|r_c^{-\gamma_c} r_{c,e}^{-1} \partial_3 u\|_{L^2(\mathcal{V}_c^e)}, \tag{5.27}$$

$$\|r_c^{1-\gamma_c} r_{c,e}^{1-\gamma_e} \partial_\perp^2 u\|_{L^2(\mathcal{V}_c^e)}, \tag{5.28}$$

$$\|r_c^{1-\gamma_c} \partial_\perp \partial_3 u\|_{L^2(\mathcal{V}_c^e)}, \tag{5.29}$$

$$\|r_c^{1-\gamma_c} r_{c,e}^{-\mu_e} \partial_3^2 u\|_{L^2(\mathcal{V}_c^e)}, \tag{5.30}$$

where ∂_3 is the derivative in the direction of e and $\partial_\perp$ is either ∂_1 or ∂_2.

Proof Since we are interested in the regularity of u in $\mathcal{V}_c^e$, for a fixed vertex c and edge $e \in \mathcal{E}_c$, we use $\rho = r_c$ and $\varphi = r_{c,e}$.

With the notation from [38], $f \in L_\mu^2(\Omega)$ belongs to $\mathcal{M}_{1-\beta}^0(\Omega)$ with $\beta_{c'} = \gamma_{c'}$ and $\beta_{e'} = \gamma_{e'}$ for all vertices c' and edges e' (as $\gamma_{c'} \gamma_{e'} \in [0, 1]$). Therefore, by Theorem 3.3 in [38], we have

$$r_c^{-1-\gamma_c} r_{c,e}^{-1-\gamma_e} u \in L^2(\mathcal{V}_c^e).$$

Therefore, (5.25) is proved.

Note $f \in L_\mu^2(\Omega) \subset L^2(\Omega)$. According to Theorem 2.10 in [9], $u = u_r + u_s$ with $u_r \in H^2(\Omega) \cap H_0^1(\Omega)$ and u_s satisfying

$$\rho^{\beta-1}\varphi^{\delta-1}\partial_\perp u_s \in L^2(\mathcal{V}_c^e), \quad \rho^{\beta-1}\varphi^{-1}\partial_3 u_s \in L^2(\mathcal{V}_c^e),$$

where

$$\beta, \delta \geq 0, \quad \beta > 1 - \eta_c, \quad \delta > 1 - \eta_e. \tag{5.31}$$

Note $\beta - 1 \geq -1$ and $\beta - 1 > -\eta_c$; and $\delta - 1 \geq -1$ and $\delta - 1 > -\eta_e$. Then for the chosen γ_c and γ_e, we have

$$\rho^{-\gamma_c}\varphi^{-\gamma_e}\partial_\perp u_s \in L^2(\mathcal{V}_c^e), \qquad \rho^{-\gamma_c}\varphi^{-1}\partial_3 u_s \in L^2(\mathcal{V}_c^e). \tag{5.32}$$

To get (5.26) and (5.27), it then remains to prove a similar property for u_r. This is proved with the help of Hardy's inequalities (see [61, p. 28]). First by Lemma 7.1.1 of [72] (based on Hardy's inequalities), we have

$$\rho^{-1}\partial_j u_r \in L^2(\Omega), \quad \forall\, j = 1, 2, 3. \tag{5.33}$$

Fix any $j \in \{1, 2, 3\}$ and we again use the spherical coordinates (ρ, θ, φ) such that ρ is the distance to c and $\varphi = 0$ corresponds to the edge e. By fixing a cut-off function $\chi_e \in \mathcal{D}(\mathbb{R})$ such that $\chi_e = 1$ in a neighborhood of 0, we can write for almost all ρ and θ

$$\chi_e(\varphi)\partial_j u_r(\rho, \theta, \varphi) = \int_0^\varphi \frac{\partial}{\partial s}\Big(\chi_e(s)\partial_j u_r(\rho, \theta, s)\Big)\, ds,$$

since $\chi_e(\varphi = 0)\partial_j u_r(\rho, \theta, \varphi = 0) = 0$ (because $u_r = 0$ on the boundary). This identity can be written as

$$\rho^{-1}\varphi^{-1}\chi_e(\varphi)\partial_j u_r(\rho, \theta, \varphi) = H\left(\rho^{-1}\frac{\partial}{\partial s}\Big(\chi_e(s)\partial_j u_r(\rho, \theta, s)\Big)\right)(\varphi),$$

where the operator H is defined by

$$(Hv)(t) = \frac{1}{t}\int_0^t v(s)\, ds.$$

Now we show that for some $\rho_0 > 0$ and for almost all $\rho \in (0, \rho_0)$, $\theta \in (0, 2\pi)$,

$$\rho^{-1}\frac{\partial}{\partial s}\Big(\chi_e(s)\partial_j u_r(\rho, \theta, s)\Big) \in L^2(\mathbb{R}). \tag{5.34}$$

Indeed by Leibniz's rule, one has

$$\rho^{-1}\frac{\partial}{\partial s}\Big(\chi_e(s)\partial_j u_r(\rho, \theta, s)\Big) = \rho^{-1}\chi_e'(s)\partial_j u_r + \rho^{-1}\frac{\partial}{\partial s}\partial_j u_r.$$

Hence by the regularity $u_r \in H^2(\Omega)$ and (5.33), we obtain

$$\rho^{-1}\frac{\partial}{\partial s}\Big(\chi_e(s)\partial_j u_r(\rho, \theta, s)\Big) \in L^2(\Omega).$$

Since in spherical coordinates the Lebesgue measure is $\rho^2 \sin\varphi d\rho d\theta d\varphi$, we get (5.34). This regularity allows us to apply Hardy's inequality and find that for almost all $\rho \in (0, \rho_0)$ and $\theta \in (0, 2\pi)$,

$$\int_0^\infty |\rho^{-1}\varphi^{-1}\chi_e(\varphi)\partial_j u_r(\rho,\theta,\varphi)|^2\, d\varphi \leq 4\int_0^\infty \left|\rho^{-1}\frac{\partial}{\partial\varphi}\Big(\chi_e(\varphi)\partial_j u_r(\rho,\theta,\varphi)\Big)\right|^2 d\varphi.$$

Multiplying this estimate by $\rho^2 \sin\varphi$ and integrating in $\rho \in (0, \rho_0)$ and $\theta \in (0, 2\pi)$, we deduce that

$$\int_0^{\rho_0}\int_0^\infty\int_0^{2\pi} |\rho^{-1}\varphi^{-1}\chi_e(\varphi)\partial_j u_r(\rho,\theta,\varphi)|^2\, \rho^2 \sin\varphi\, d\theta d\varphi d\rho \leq 4\int_\Omega \left|\rho^{-1}\frac{\partial}{\partial\varphi}\Big(\chi_e(\varphi)\partial_j u_r(\rho,\theta,\varphi)\Big)\right|^2 dx.$$

This shows that

$$\rho^{-1}\varphi^{-1}\partial_j u_r \in L^2(\mathcal{V}_c^e),$$

and with (5.32), we deduce that (5.26) and (5.27) hold.

Meanwhile, recall that $u = u_r + u_s$. Again by Theorem 2.10 in [9] and by (5.24), we get

$$\rho^\beta\varphi^\delta\partial_\perp^2 u_s \in L^2(\mathcal{V}_c^e), \quad \rho^\beta\partial_\perp\partial_3 u_s \in L^2(\mathcal{V}_c^e) \quad \rho^\beta\varphi^{-\mu_e}\partial_3^2 u \in L^2(\mathcal{V}_c^e), \tag{5.35}$$

where β and δ are defined in (5.31). By definition, we have $1-\gamma_c \geq 0, 1-\gamma_c > 1-\eta_c$, and $1-\gamma_e \geq 0, 1-\gamma_e > 1-\eta_e$. Therefore, since $u_r \in H^2(\Omega)$, (5.35) gives rise to

$$\rho^{1-\gamma_c}\varphi^{1-\gamma_e}\partial_\perp^2 u = \rho^{1-\gamma_c}\varphi^{1-\gamma_e}\partial_\perp^2 u_r + \rho^{1-\gamma_c}\varphi^{1-\gamma_e}\partial_\perp^2 u_s \in L^2(\mathcal{V}_c^e)$$
$$\rho^{1-\gamma_c}\partial_\perp\partial_3 u = \rho^{1-\gamma_c}\partial_\perp\partial_3 u_r + \rho^{1-\gamma_c}\partial_\perp\partial_3 u_s \in L^2(\mathcal{V}_c^e)$$
$$\rho^{1-\gamma_c}\varphi^{-\mu_e}\partial_3^2 u \in L^2(\mathcal{V}_c^e).$$

Then we have proved (5.28), (5.29), and (5.30). □

Remark 5.3 From the proof of Corollary 5.1, we see that we can take $\gamma_c = 1$ if $\eta_c > 1$ and take $\gamma_e = 1$ if e is regular ($\omega_e \leq \pi$); and that the estimates (5.25)–(5.30) are also valid in $\mathcal{V}_e^o$ with r_c replaced by 1 and valid in $\mathcal{V}_c^o$ with $r_{c,e}$ replace by 1.

Consequently, we obtain the regularity estimates for (5.1) with $f \in L^2_{\mu^*}(\Omega)$.

Corollary 5.2 *Recall the interior of the domain $\mathcal{V}^o$ in (5.3). Under the assumption of Lemma 5.1, for $f \in L^2_{\mu^*}(\Omega)$ defined in (5.8), let $\gamma_e, \gamma_c \in [0, 1]$ be such that $\mu_e^* \leq \gamma_e < \eta_e = \lambda_e$, $\gamma_c < \eta_c$; and $\gamma_e = 1$ if $\lambda_e \geq 1$, $\gamma_c = 1$ if $\eta_c > 1$. Let γ be the collection of the parameters γ_c and γ_e for all the vertices $c \in C$ and edges $e \in \mathcal{E}$. Define the weighted space.*

$$\begin{aligned}\mathcal{H}^2_{\gamma}(\Omega) := \Big\{ v \in H^2_{loc}(\Omega) :\ & r_c^{-1-\gamma_c} v,\ r_c^{-\gamma_c}\partial_\perp v,\ r_c^{-\gamma_c}\partial_3 v \in L^2(\mathcal{V}_c^o),\\
& r_c^{1-\gamma_c}\partial_\perp^2 v,\ r_c^{1-\gamma_c}\partial_\perp\partial_3 v,\ r_c^{1-\gamma_c}\partial_3^2 v \in L^2(\mathcal{V}_c^o);\\
& r_c^{\gamma_e-\gamma_c} r_e^{-1-\gamma_e} v,\ r_c^{\gamma_e-\gamma_c} r_e^{-\gamma_e}\partial_\perp v,\ r_c^{1-\gamma_c} r_e^{-1}\partial_3 v \in L^2(\mathcal{V}_c^e),\\
& r_c^{\gamma_e-\gamma_c} r_e^{1-\gamma_e}\partial_\perp^2 v,\ r_c^{1-\gamma_c}\partial_\perp\partial_3 v,\ r_c^{1+\mu_e^*-\gamma_c} r_e^{-\mu_e^*}\partial_3^2 v \in L^2(\mathcal{V}_c^e);\\
& r_e^{-1-\gamma_e} v,\ r_e^{-\gamma_e}\partial_\perp v,\ r_e^{-1}\partial_3 v \in L^2(\mathcal{V}_e^o),\\
& r_e^{1-\gamma_e}\partial_\perp^2 v,\ \partial_\perp\partial_3 v,\ r_e^{-\mu_e^*}\partial_3^2 v \in L^2(\mathcal{V}_e^o)\Big\},\end{aligned}$$

with the norm

$$\begin{aligned}\|v\|^2_{\mathcal{H}^2_{\gamma}(\Omega)} :=\ & \|v\|^2_{H^2(\mathcal{V}^o)} + \sum_{c\in\mathcal{C}, e\in\mathcal{E}_c} \Big(\|r_c^{1-\gamma_c} r_{c,e}^{-\mu_e^*}\partial_3^2 v\|^2_{L^2(\mathcal{V}_c^e)}\\
& + \sum_{|\alpha_\perp|=1} \|r_c^{1-\gamma_c}\partial^{\alpha_\perp}\partial_3 v\|^2_{L^2(\mathcal{V}_c^e)} + \|r_c^{-\gamma_c} r_{c,e}^{-1}\partial_3 v\|^2_{L^2(\mathcal{V}_c^e)}\\
& + \sum_{|\alpha_\perp|\le 2} \|r_c^{|\alpha_\perp|-1-\gamma_c} r_{c,e}^{|\alpha_\perp|-1-\gamma_e}\partial^{\alpha_\perp} v\|^2_{L^2(\mathcal{V}_c^e)}\Big) + \sum_{c\in\mathcal{C}, |\alpha|\le 2} \|r_c^{|\alpha|-1-\gamma_c}\partial^{\alpha} v\|^2_{L^2(\mathcal{V}_c^o)}\\
& + \sum_{e\in\mathcal{E}} \Big(\|r_e^{-\mu_e^*}\partial_3^2 v\|^2_{L^2(\mathcal{V}_e^o)} + \sum_{|\alpha_\perp|=1} \|\partial^{\alpha_\perp}\partial_3 v\|^2_{L^2(\mathcal{V}_e^o)} + \|r_e^{-1}\partial_3 v\|^2_{L^2(\mathcal{V}_e^o)}\\
& + \sum_{|\alpha_\perp|\le 2} \|r_e^{|\alpha_\perp|-1-\gamma_e}\partial^{\alpha_\perp} v\|^2_{L^2(\mathcal{V}_e^o)}\Big),\end{aligned}$$

where ∂_3 is the derivative in the direction of e, $\partial_\perp$ is either ∂_1 or ∂_2, $\partial^{\alpha_\perp} = \partial_1^{\alpha_1}\partial_2^{\alpha_2}$ for $\alpha_\perp = (\alpha_1, \alpha_2)$, and $\alpha = (\alpha_1, \alpha_2, \alpha_3)$. Then the solution $u \in H^1_0(\Omega)$ of (5.1) satisfies

$$\|u\|_{\mathcal{H}^2_{\gamma}(\Omega)} \le C\|f\|_{L^2_{\mu^*}(\Omega)}. \tag{5.36}$$

Proof The estimate (5.36) is a direct consequence of Corollary 5.1 and Remark 5.3. Corollary 5.1 holds for μ_e in (5.9) and for all $\gamma_e \in [0, 1]$ and $\gamma_e < \lambda_e$. Thus, it still holds if we replace μ_e by μ_e^* and replace 0 by μ_e^* as the lower bound for γ_e. □

5.1.3 Mixed Boundary Conditions

In this subsection, we assume $\Gamma_N, \Gamma_D \neq \emptyset$ in (5.1) and allow the Neumann boundary condition to be imposed on two adjacent faces of the boundary $\Gamma = \partial\Omega$. Suppose that each face of Γ is included either in Γ_N or in Γ_D. Therefore, the nonsmooth boundary points (vertices and edges) are also the points where the boundary condition may change. The singular solution of (5.1) with the mixed boundary condition can be different from that in the pure Dirichlet problem. For example, near an edge or a vertex that is surrounded by Neumann faces, the solution in general does not vanish and, therefore, may not belong to the same weighted space as in the Dirichlet case.

Recall the set $\mathcal{E}$ of *open* edges and the set C of vertices of Ω. In addition, recall the set $\mathcal{E}_c \subset \mathcal{E}$ of edges joining at $c \in C$ and the set $C_e \subset C$ of endpoints of $e \in \mathcal{E}$. We say an edge $e \in \mathcal{E}$ is a Dirichlet (resp. Neumann) edge if the Dirichlet (resp. Neumann) boundary conditions are imposed on both adjacent faces of e. We say e is a DN edge if the Dirichlet condition is imposed on one adjacent face of e and the Neumann condition is on the other. Let ω_e be the opening angle between the two adjacent faces of e. For each $e \in \mathcal{E}$, define

$$\lambda_e = \begin{cases} \frac{\pi}{\omega_e}, & \text{if } e \text{ is a Dirichlet edge or a Neumann edge;} \\ \frac{\pi}{2\omega_e}, & \text{if } e \text{ is a DN edge.} \end{cases} \tag{5.37}$$

Similar to the Dirichlet problem studied in the last subsection, we define singular edges and singular vertices for (5.1) with the mixed boundary condition as follows. Let λ_e be given in (5.37). An edge $e \in \mathcal{E}$ is called *singular* if $\lambda_e < 1$; otherwise it is called *regular*. Denote by Λ_c the cone that coincides with the domain Ω at $c \in C$. Let λ_c be the first positive eigenvalue of the Laplace–Beltrami operator on the intersection of Λ_c with the unit sphere with boundary conditions inherited from (5.1). Then, if $\sqrt{\lambda_c + \frac{1}{4}} < 1$, c is called *singular*; it is *regular* otherwise. For $e \in \mathcal{E}$ and $c \in C$, we set

$$\begin{cases} \eta_e = \lambda_e \text{ if } e \text{ is singular, } \eta_e = \infty \text{ otherwise;} \\ \eta_c = \sqrt{\lambda_c + \frac{1}{4}} \text{ if } c \text{ is singular, } \eta_c = \infty \text{ otherwise.} \end{cases} \tag{5.38}$$

We shall use the same notation r_c, r_e, and $r_{c,e}$ as in the last two subsections. Let G be a subset of $\mathbb{R}^3$ such that 0 belongs to its boundary. Then we extend the space $V^k_\beta(\Lambda_c)$ in (5.15) to G as follows: for any $\beta \in \mathbb{R}$ and $k \in \mathbb{Z}_{\geq 0}$,

$$V^k_\beta(G) = \{v \in L^2_{loc}(G) : r^{\beta+|\alpha|-k} D^\alpha v \in L^2(G), \ \forall\ |\alpha| \leq k\},$$

where r is the distance to 0. Define $\lambda_{\mathcal{E}} := \max_{e \in \mathcal{E}}\{0, 1 - \lambda_e\}$. We also assume the given data in (5.1) to satisfy $f \in H^\sigma(\Omega)$ with

$$\sigma \in (\lambda_{\mathcal{E}}, 1) \text{ if } \lambda_{\mathcal{E}} > 0 \qquad \text{and} \qquad \sigma \in [0, 1) \text{ if } \lambda_{\mathcal{E}} = 0. \tag{5.39}$$

In this subsection, for a bounded domain G, the usual norm and seminorm of $H^s(G)$ $(s \geq 0)$ are denoted by $\|\cdot\|_{s,G}$ and $|\cdot|_{s,G}$, respectively. For $s = 0$, we will drop the index 0 and for $G = \Omega$, the index Ω. For two positive parameters s and t, we introduce the norm $\|\cdot\|_{s,G,t}$ on $H^s(G)$ (see [49, Definition AA.17], for instance) defined by

$$\|v\|_{s,G,t} = \left(t^{2s}\|v\|^2_G + |v|^2_{s,G}\right)^{\frac{1}{2}}, \quad \forall\ v \in H^s(G).$$

This is equivalent to the usual norm $\|\cdot\|_{s,G}$ (with constants of equivalence depending on t) if G has a Lipschitz boundary.

5.1.3.1 Regularity Estimates in a Dihedron

We proceed to develop regularity estimates for (5.1) with the mixed boundary condition. Special attention will be given to the region close to the edges where different boundary conditions are imposed.

Let $D = K \times \mathbb{R}$ be a dihedron, with K a two-dimensional cone of center 0 and opening angle ω. Let $B(0, r)$ be the ball centered at 0 with radius r. Then we consider $u \in H^1(D)$ with a support included in $(K \cap B(0, R)) \times \mathbb{R}$ for some $R > 0$ to be the solution of

$$\begin{cases} -\Delta u = f & \text{in } D, \\ u = 0 & \text{on } \Gamma_D \times \mathbb{R}, \\ \partial_n u = 0 & \text{on } \Gamma_N \times \mathbb{R}, \end{cases} \tag{5.40}$$

where $f \in H^\sigma(D)$ for some $\sigma \in [0, 1)$ and $\overline{\Gamma_D} \cup \overline{\Gamma_N} = \partial K$ such that Γ_D can be empty, a full half-plane, or $\overline{\Gamma_D} = \partial K$. In this way, we consider either the pure Dirichlet, the mixed, or the pure Neumann problem. To simplify the exposition, we use (x_1, x_2, x_3) (instead of (x, y, z)) to denote a point in D and suppose the edge of D is on the x_3-axis.

The behavior of the solution of (5.40) is well known in the case of the pure Dirichlet problem for data in L^2 (see the discussion in Sect. 3.3 and [49, 63]), but is less studied for smoother data and in the other two cases of boundary conditions. Our goal is to show that this solution can be decomposed into a regular part and a singular one with the appropriate behavior. For that purpose, we perform a partial Fourier transform in the x_3-variable that allows to reduce the study to an Helmholtz equation in K.

The Helmholtz Equation in a Cone For all $\zeta \in \mathbb{R}$, we consider the solution $v \in H^1(K)$ with a support included in $K \cap B(0, R)$ of

$$\begin{cases} -\Delta v + \zeta^2 v = g & \text{in } K, \\ v = 0 & \text{on } \Gamma_D, \\ \partial_n v = 0 & \text{on } \Gamma_N, \end{cases} \tag{5.41}$$

where $g \in H^\sigma(K)$ for some $\sigma \in [0, 1)$. We shall show that v admits a decomposition into a regular part and a singular one. Recall that the solution singularities in (5.41) are related to the singularities of the Laplace equation, namely to the singularities of (5.41) with $\zeta = 0$. Such singularities are in the form (see (3.11) and Example 3.1)

$$S_k = r^{\lambda_k} \phi_k(\theta),$$

with

$$\lambda_k = \frac{k\pi}{\omega}, \quad \forall\, k \in \mathbb{N}, \tag{5.42}$$

in the pure Dirichlet and pure Neumann cases, while

$$\lambda_k = \frac{(2k-1)\pi}{2\omega}, \quad \forall\, k \in \mathbb{N}, \tag{5.43}$$

in the mixed case. Here, r is the distance to the vertex of K and the function ϕ_k is given by (3.11). For simplicity, we denote the smallest singular exponent λ_1 by λ.

In addition, we have the following result.

Theorem 5.3 *Let $\sigma \in [0, 1)$ be such that $\sigma \neq \lambda_k - 1$ for all $k \in \mathbb{N}$. Then for all $\zeta \in \mathbb{R}$, the solution $v \in H^1(K)$ with a support included in $K \cap B(0, R)$ of (5.41) with $g \in H^\sigma(D)$ can be split up into*

$$v = v_{\text{reg}}(\zeta) + v_{\text{sing}}(\zeta), \tag{5.44}$$

with $v_{\text{reg}}(\zeta) \in H^{2+\sigma}(K)$ and $v_{\text{sing}}(\zeta) \in V^2_\delta(K) \cap H^1(K)$ for any $\delta > 1 - \lambda$ satisfying the following estimates[1]

$$\|v_{\text{reg}}(\zeta)\|_{2+\sigma,K,1+|\zeta|} \leq C\|g\|_{\sigma,K,1+|\zeta|}, \tag{5.45}$$

$$(1+\zeta^2)\|r^{-\sigma} v_{\text{reg}}(\zeta)\|_K \leq C\|g\|_{\sigma,K,1+|\zeta|}, \tag{5.46}$$

$$(1+|\zeta|^{\delta+\sigma})\|v_{\text{sing}}(\zeta)\|_{V^2_\delta(K)} \leq C\|g\|_{\sigma,K,1+|\zeta|}, \tag{5.47}$$

$$(1+|\zeta|)|v_{\text{sing}}(\zeta)|_{1,K} \leq C\|g\|_{\sigma,K,1+|\zeta|}, \tag{5.48}$$

where r is the distance to 0 (the vertex of K).

Proof We distinguish the case $|\zeta| > 1$ to the case $|\zeta| \leq 1$.
(a) For $|\zeta| > 1$, as in the proof of Theorem 16.9 in [49], we use a scaling argument. Namely, by setting $\hat{x} = |\zeta| x$ and $\hat{v}(\hat{x}) = v(x)$, we see that $\hat{v}$ is the solution of

$$\begin{cases} -\Delta\hat{v} + \hat{v} = \hat{g} & \text{in } K, \\ \hat{v} = 0 & \text{on } \Gamma_D, \\ \partial_n \hat{v} = 0 & \text{on } \Gamma_N, \end{cases} \tag{5.49}$$

where $\hat{g} \in H^\sigma(K)$ is defined by $\hat{g}(\hat{x}) = \zeta^{-2} g(x)$. The weak formulation of this problem is

$$(\hat{v}, w)_{1,K} = \int_K \hat{g} w \, dx, \quad \forall\, w \in H^1_D(K),$$

where $H^1_D(K) := \{w \in H^1(K) : w = 0 \text{ on } \Gamma_D\}$ is an Hilbert space with its natural inner product

$$(z, w)_{1,K} = \int_K (\nabla z \cdot \nabla w + zw) \, dx, \quad \forall\, z, w \in H^1_D(K).$$

Consequently, (5.49) has a unique solution $\hat{v} \in H^1_D(K)$ with the continuous dependence

$$\|\hat{v}\|_{1,K} \leq \|\hat{g}\|_{0,K}. \tag{5.50}$$

[1] Here and below the involved constants are independent of $|\zeta|$.

Now using a localization argument, by Theorem 23.7 of [49] near the origin and by the standard shift theorem far from the origin, one deduces that $\hat{v}$ admits the splitting

$$\hat{v} = \hat{v}_{\text{reg}} + \chi(\hat{r}) \sum_{k\in\mathbb{N}:\, 0<\lambda_k<1+\sigma} c_k \hat{r}^{\lambda_k} \phi_k,$$

where $\chi \in \mathcal{D}(\mathbb{R}^2)$ is a smooth cut-off function equal to 1 in a neighborhood of the origin that, without loss of generality, is assumed to have a support included in $K \cap B(0, R)$, $\hat{v}_{\text{reg}} \in H^{2+\sigma}(K)$, and $c_k \in \mathbb{R}$ with

$$\|\hat{v}_{\text{reg}}\|_{2+\sigma,K} + \sum_{k\in\mathbb{N}:\, 0<\lambda_k<1+\sigma} |c_k| \leq C\|\hat{g}\|_{\sigma,K} + \|\hat{v}\|_{1,K}.$$

Combined with (5.50), we find that

$$\|\hat{v}_{\text{reg}}\|_{2+\sigma,K} + \sum_{k\in\mathbb{N}:\, 0<\lambda_k<1+\sigma} |c_k| \leq C\|\hat{g}\|_{\sigma,K}. \tag{5.51}$$

By a transformation back, this yields (5.44) by setting $v_{\text{reg}}(x) = \hat{v}_{\text{reg}}(\hat{x})$, and[2]

$$v_{\text{sing}}(\zeta) = \chi(|\zeta| r) \sum_{k\in\mathbb{N}:\, 0<\lambda_k<1+\sigma} c_k |\zeta|^{\lambda_k} r^{\lambda_k} \phi_k.$$

Furthermore using Lemma AA.19 of [49], the estimate (5.51) is equivalent to

$$\|v_{\text{reg}}\|_{2+\sigma,K,|\zeta|} + |\zeta|^{1+\sigma} \sum_{k\in\mathbb{N}:\, 0<\lambda_k<1+\sigma} |c_k| \leq C\|g\|_{\sigma,K,|\zeta|}. \tag{5.52}$$

This estimate directly leads to the first estimate (5.45) for $|\zeta| > 1$. To prove (5.46), we first notice that the support of v_{sing} being included in $B(0, \frac{R}{|\zeta|}) \subset B(0, R)$, v_{reg} has a compact support included in $K \cap B(0, R)$. Furthermore using the estimate

$$\|v_{\text{reg}}\|_{2+\sigma,K\cap B(0,R),|\zeta|} \leq C\|g\|_{\sigma,K,|\zeta|},$$

and an interpolation inequality (see [61, Theorem 1.4.3.3]), we find that

$$\|v_{\text{reg}}\|_{\sigma,K\cap B(0,R)} \leq C|\zeta|^{-2}\|g\|_{\sigma,K,|\zeta|}.$$

Since Theorem AA.7 of [49] guarantees that $H^\sigma(K \cap B(0, R)) = V_0^\sigma(K \cap B(0, R))$, we deduce that

$$\|r^{-\sigma} v_{\text{reg}}\|_{0,K} = \|r^{-\sigma} v_{\text{reg}}\|_{0,K\cap B(0,R)} \leq C\|v_{\text{reg}}\|_{\sigma,K\cap B(0,R)} \leq C|\zeta|^{-2}\|g\|_{\sigma,K,|\zeta|},$$

which is exactly (5.46).

---

[2] Note that for $\sigma < \lambda - 1$, $v_{\text{sing}}(\zeta) = 0$.

To prove (5.47), it suffices to check that for all $k \in \mathbb{N}$ and for $0 < \lambda_k < 1+\sigma$, one has

$$|\zeta|^{\delta+\sigma+\lambda_k}|c_k|\left\|\chi(|\zeta|r)r^{\lambda_k}\phi_k\right\|_{V^2_\delta(K)} \leq C\|g\|_{\sigma,K,|\zeta|},$$

which holds, in view of (5.52), as soon as

$$|\zeta|^{\delta+\lambda_k-1}\|\chi(|\zeta|r)r^{\lambda_k}\phi_k\|_{V^2_\delta(K)} \leq C. \tag{5.53}$$

Using polar coordinates, one can show that

$$\left\|\chi(|\zeta|r)r^{\lambda_k}\phi_k\right\|^2_{V^2_\delta(K)} \leq C\int_0^\infty r^{2(\delta+\lambda_k-2)}\Big(\left|\chi(|\zeta|r)\right|^2 + \left||\zeta|r\chi'(|\zeta|r)\right|^2 + \left||\zeta|^2r^2\chi''(|\zeta|r)\right|^2\Big)r\,dr,$$

and by the change of variables $\hat{r} = |\zeta|r$, one finds

$$\|\chi(|\zeta|r)r^{\lambda_k}\phi_k\|^2_{V^2_\delta(K)} \leq C|\zeta|^{2(\delta+\lambda_k-1)}\int_0^\infty \hat{r}^{2(\delta+\lambda_k-2)}\Big(|\chi(\hat{r})|^2 + |\hat{r}\chi'(\hat{r})|^2 + |\hat{r}^2\chi''(\hat{r})|^2\Big)\hat{r}\,d\hat{r}.$$

Since the integral of this right hand side is finite as long as $\delta + \lambda_k - 1 > 0$ (which holds if $\delta + \lambda - 1 > 0$), we have found that (5.53) is valid.

The proof of (5.48) is fully similar and is left to the reader.

(b) For $|\zeta| \leq 1$, we first notice that

$$\|v\|_{1,K} \leq C|v|_{1,K}, \tag{5.54}$$

because v has a compact support included into $K \cap B(0, R)$. Since the weak formulation of problem (5.41) is

$$\int_K (\nabla v \cdot \nabla w + \zeta^2 vw)\,dx = \int_K gw\,dx, \quad \forall\, w \in H^1_D(K),$$

by taking $w = v$ in this identity and using (5.54) we find

$$\|v\|^2_{1,K} \leq C|v|^2_{1,K} \leq \int_K (|\nabla v|^2 + \zeta^2|v|^2)\,dx = \int_K gv\,dx.$$

Consequently, by the Cauchy–Schwarz inequality, we get

$$\|v\|_{1,K} \leq C\|g\|_{0,K}. \tag{5.55}$$

Now v can be seen as the solution of (compare with (5.49))

$$\begin{cases} -\Delta v + v = \tilde{g} & \text{in } K, \\ v = 0 & \text{on } \Gamma_D, \\ \partial_n v = 0 & \text{on } \Gamma_N, \end{cases}$$

where $\tilde{g} := g + v - \zeta^2 v \in H^\sigma(K)$ that, owing to (5.55), satisfies (recalling that $|\zeta| \le 1$)

$$\|\tilde{g}\|_{\sigma,K} \le C\|g\|_{\sigma,K}.$$

As in the previous discussion, we then get the decomposition

$$v = v_{\text{reg}} + \chi(r) \sum_{k\in\mathbb{N}:\, 0<\lambda_k<1+\sigma} c_k r^{\lambda_k} \phi_k,$$

where $v_{\text{reg}} \in H^{2+\sigma}(K)$ and $c_k \in \mathbb{R}$ with

$$\|v_{\text{reg}}\|_{2+\sigma,K} + \sum_{k\in\mathbb{N}:\, 0<\lambda_k<1+\sigma} |c_k| \le C\big(\|\tilde{g}\|_{\sigma,K} + \|v\|_{1,K}\big) \le C\|\tilde{g}\|_{\sigma,K}. \tag{5.56}$$

This yields (5.44) with[3]

$$v_{\text{sing}}(\zeta) = \chi(r) \sum_{k\in\mathbb{N}:\, 0<\lambda_k<1+\sigma} c_k r^{\lambda_k} \phi_k.$$

The estimate (5.45) is a direct consequence of (5.56), while the estimates (5.46)–(5.48) follow by using the previous arguments. □

Singular Decomposition in a Dihedron Recall the weighted space $\mathcal{H}^2_\gamma$ in Corollary 5.2. We define the following weighted space $\mathcal{H}^2_{\gamma,\sigma}(D)$ in the dihedron $D = K\times\mathbb{R}$ that resembles $\mathcal{H}^2_\gamma$ in the region close to the edge:

$$\mathcal{H}^2_{\gamma,\sigma}(D) := \Big\{v \in H^2_{loc}(D) : r^{-1-\gamma}v,\ r^{-\gamma}\partial_\perp v,\ r^{-1}\partial_3 v \in L^2(D), \tag{5.57}$$
$$r^{1-\gamma}\partial_\perp^2 v,\ \partial_\perp\partial_3 v,\ r^{-\sigma}\partial_3^2 v \in L^2(D)\Big\},$$

with the norm

$$\|v\|^2_{\mathcal{H}^2_{\gamma,\sigma}(D)} := \|r^{-\sigma}\partial_3^2 v\|^2_{L^2(D)} + \sum_{|\alpha_\perp|=1} \|\partial^{\alpha_\perp}\partial_3 v\|^2_{L^2(D)}$$
$$+\|r^{-1}\partial_3 v\|^2_{L^2(D)} + \sum_{|\alpha_\perp|\le 2} \|r^{|\alpha_\perp|-1-\gamma}\partial^{\alpha_\perp} v\|^2_{L^2(D)},$$

where the edge of D assumed to be on the x_3-axis, r is the distance to the edge, $\partial_\perp$ means the first-order derivative in x_1 or in x_2, $\partial_3 = \partial_{x_3}$, and $\alpha_\perp$ means that the third component of the multi-index is zero.

Then we have the following regularity estimates for (5.40).

Theorem 5.4 *Let $\sigma \in [0, 1)$ be such that for all $k \in \mathbb{N}$, $\sigma \ne \lambda_k - 1$. Recall $\lambda := \lambda_1$. Suppose $f \in H^\sigma(D)$. Then the solution $u \in H^1(D)$ of (5.40) with a support included*

[3] As before for $\sigma < \lambda - 1$, $v_{\text{sing}}(\zeta) = 0$.

in $(K \cap B(0, R)) \times \mathbb{R}$ *for some* $R > 0$ *can be split into*

$$u = u_{\text{reg}} + u_{\text{sing}}, \tag{5.58}$$

with $u_{\text{reg}} \in H^{2+\sigma}(D)$ *such that* $r^{-\sigma} \partial_3^2 u_{\text{reg}} \in L^2(D)$ *and* $u_{\text{sing}} \in \mathcal{H}^2_{\gamma,\sigma}(D)$ *for any* $\gamma < \lambda$ *satisfying the following estimates:*

$$\|u_{\text{reg}}\|_{2+\sigma,D} \leq C\ \|f\|_{\sigma,D}, \tag{5.59}$$

$$\|r^{-\sigma} \partial_3^j u_{\text{reg}}\|_{0,D} \leq C\ \|f\|_{\sigma,D}, \quad \forall\, j = 0, 1, 2, \tag{5.60}$$

$$\|u_{\text{sing}}\|_{\mathcal{H}^2_{\gamma,\sigma}(D)} \leq C\ \|f\|_{\sigma,D}. \tag{5.61}$$

Proof We perform a partial Fourier transform in x_3. Namely, let $v(\zeta) = \mathcal{F}_{x_3 \to \zeta} u$ and $g(\zeta) = \mathcal{F}_{x_3 \to \zeta} f$. Then we see that v is the solution of (5.41). Applying Theorem 5.3 to v and performing inverse Fourier transform, we find the decomposition (5.58) with

$$u_{\text{reg}} = \mathcal{F}^{-1}_{x_3 \to \zeta} v_{\text{reg}}, \quad u_{\text{sing}} = \mathcal{F}^{-1}_{x_3 \to \zeta} v_{\text{sing}}.$$

The estimate (5.59) (resp. (5.60)) follows from (5.45) (resp. (5.46)) and Proposition AA.20 from [49]. Similarly, for all $\delta > 1 - \lambda$, using the estimate (5.47) and the fact $\delta + \sigma > 0$, we get

$$\sum_{|\alpha_\perp| \leq 2} \|r^{|\alpha_\perp|-2+\delta} \partial^{\alpha_\perp} u_{\text{sing}}\|^2_{L^2(D)} \leq C\|f\|^2_{\sigma,D}.$$

Let $\gamma = 1 - \delta$. This yields

$$\sum_{|\alpha_\perp| \leq 2} \|r^{|\alpha_\perp|-1-\gamma} \partial^{\alpha_\perp} u_{\text{sing}}\|^2_{L^2(D)} \leq C\|f\|^2_{\sigma,D}. \tag{5.62}$$

Again applying (5.47) with $\delta = 2 - \sigma$ (resp. $\delta = 1 - \sigma$ if $\sigma < \lambda$ or $\delta = 1 - \lambda + \epsilon$ if $\sigma \geq \lambda$ for $\epsilon > 0$ small) that is clearly larger than $1 - \lambda$, we find

$$\|r^{-\sigma} \partial_3^2 u_{\text{sing}}\|^2_{L^2(D)} \lesssim \|f\|^2_{\sigma,D}, \tag{5.63}$$

$$\|r^{-1} \partial_3 u_{\text{sing}}\|^2_{L^2(D)} \leq C\|f\|^2_{\sigma,D}. \tag{5.64}$$

Finally applying (5.48) we obtain

$$\|\partial_3 \partial_\perp u_{\text{sing}}\|^2_{L^2(D)} \leq C\|f\|^2_{\sigma,D}. \tag{5.65}$$

The estimates (5.62)–(5.65) show that (5.61) holds. □

5.1.3.2 Regularity Results in a Dihedral Cone

We proceed to investigate the regularity of the solution of (5.1) in the region where the vertex and the edge meet. Let Λ be a cone in $\mathbb{R}^3$ of the vertex $c \in C$ (that can be identified with 0), in the sense that

$$\Lambda = \left\{x \in \mathbb{R}^3 : \frac{x}{|x|} \in G\right\},$$

where G is an open subset of the unit sphere S^2 with a piecewise smooth boundary, each smooth part being included in a great circle.

Let γ_c and γ_e be the parameters corresponding to a vertex $c \in C$ and an edge $e \in \mathcal{E}$ of Λ, respectively. Let $\boldsymbol{\gamma}$ be the collection of all the parameters γ_c and γ_e for Λ. Recall that r_c and r_e are the distances to the vertex c and to the edge e, respectively. Then given $\gamma_c, \gamma_e \geq 0$, we define the weighted space

$$\begin{aligned} \mathcal{H}^2_{\gamma,\sigma}(\Lambda) := \Big\{v \in H^2_{loc}(\Lambda) :\ & r_e^{-1-\gamma_e} v,\ r_e^{-\gamma_e}\partial_\perp v,\ r_e^{-1}\partial_3 v \in L^2(\mathcal{V}_e^o), \\ & r_e^{1-\gamma_e}\partial_\perp^2 v,\ \partial_\perp\partial_3 v,\ r_e^{-\sigma}\partial_3^2 v \in L^2(\mathcal{V}_e^o); \\ & r_c^{\gamma_e-\gamma_c} r_e^{-1-\gamma_e} v,\ r_c^{\gamma_e-\gamma_c} r_e^{-\gamma_e}\partial_\perp v,\ r_c^{1-\gamma_c} r_e^{-1}\partial_3 v \in L^2(\mathcal{V}_c^e), \\ & r_c^{\gamma_e-\gamma_c} r_e^{1-\gamma_e}\partial_\perp^2 v,\ r_c^{1-\gamma_c}\partial_\perp\partial_3 v,\ r_c^{1+\sigma-\gamma_c} r_e^{-\sigma}\partial_3^2 v \in L^2(\mathcal{V}_c^e); \\ & r_c^{-1-\gamma_c} v,\ r_c^{-\gamma_c}\partial_\perp v,\ r_c^{-\gamma_c}\partial_3 v \in L^2(\mathcal{V}_c^o), \\ & r_c^{1-\gamma_c}\partial_\perp^2 v,\ r_c^{1-\gamma_c}\partial_\perp\partial_3 v,\ r_c^{1-\gamma_c}\partial_3^2 v \in L^2(\mathcal{V}_c^o)\Big\}, \end{aligned} \tag{5.66}$$

with the norm

$$\begin{aligned} \|v\|^2_{\mathcal{H}^2_{\gamma,\sigma}(\Lambda)} := & \sum_{e\in\mathcal{E}_c}\Big(\|r_c^{1-\gamma_c} r_{c,e}^{-\sigma}\partial_3^2 v\|^2_{L^2(\mathcal{V}_c^e)} + \sum_{|\alpha_\perp|=1}\|r_c^{1-\gamma_c}\partial^{\alpha_\perp}\partial_3 v\|^2_{L^2(\mathcal{V}_c^e)} \\ & + \|r_c^{-\gamma_c} r_{c,e}^{-1}\partial_3 v\|^2_{L^2(\mathcal{V}_c^e)} + \sum_{|\alpha_\perp|\leq 2}\|r_c^{|\alpha_\perp|-1-\gamma_c} r_{c,e}^{|\alpha_\perp|-1-\gamma_e}\partial^{\alpha_\perp} v\|^2_{L^2(\mathcal{V}_c^e)}\Big) \\ & + \|v\|^2_{H^2(\mathcal{V}^o)} + \sum_{c\in C, |\alpha|\leq 2}\|r_c^{|\alpha|-1-\gamma_c}\partial^\alpha v\|^2_{L^2(\mathcal{V}_c^o)} + \sum_{e\in\mathcal{E}}\Big(\|r_e^{-\sigma}\partial_3^2 v\|^2_{L^2(\mathcal{V}_e^o)} \\ & + \sum_{|\alpha_\perp|=1}\|\partial^{\alpha_\perp}\partial_3 v\|^2_{L^2(\mathcal{V}_e^o)} + \|r_e^{-1}\partial_3 v\|^2_{L^2(\mathcal{V}_e^o)} + \sum_{|\alpha_\perp|\leq 2}\|r_e^{|\alpha_\perp|-1-\gamma_e}\partial^{\alpha_\perp} v\|^2_{L^2(\mathcal{V}_e^o)}\Big), \end{aligned}$$

where ∂_3 is the derivative in the direction of e, $\partial^{\alpha_\perp} = \partial_1^{\alpha_1}\partial_2^{\alpha_2}$ for $\alpha_\perp = (\alpha_1, \alpha_2)$, and $\alpha = (\alpha_1, \alpha_2, \alpha_3)$.

Let $B(0, R)$ be the ball in $\mathbb{R}^3$ centered at 0 with radius R. Consider $u \in H^1(\Lambda)$ with a support included in $\Lambda \cap B(0, R)$ for some $R > 0$ being the solution of

$$\begin{cases} -\Delta u = f & \text{in } \Lambda, \\ u = 0 & \text{on } \Gamma_D, \\ \partial_n u = 0 & \text{on } \Gamma_N, \end{cases} \tag{5.67}$$

where $f \in H^\sigma(\Lambda)$ for some $\sigma \in [0, 1)$ and $\overline{\Gamma_D} \cup \overline{\Gamma_N} = \partial\Lambda$ such that Γ_D (resp. Γ_N) is either empty or a finite union of plane faces. Denote by $\gamma_D = \Gamma_D \cap S^2$.

Since u is only in H^1, by a solution of (5.67) we mean that $u \in H^1_D(\Lambda) = \{v \in H^1(\Lambda) :\ v = 0 \text{ on } \Gamma_D\}$ satisfies

$$\int_\Lambda \nabla u \cdot \nabla v \, dx = \int_\Lambda f v \, dx, \quad \forall\, v \in H_D^1(\Lambda). \tag{5.68}$$

Note that the vertex singular exponent of problem (5.68) near c [9, 61] is given by $-\frac{1}{2} \pm \sqrt{\lambda_{c,k} + \frac{1}{4}}$, where $\{\lambda_{c,k}\}_{k=0}^{\infty}$ is the spectrum (enumerated in increasing order and repeated according to their multiplicity) of the non-negative Laplace–Beltrami operator Δ'_{mixed} in the intersection G between Λ and the unit sphere with the Dirichlet boundary condition on $\gamma_D = \Gamma_D \cap \partial G$ and the Neumann boundary condition on $\Gamma_N \cap \partial G$. Here we are only interested in exponents larger than $-\frac{1}{2}$. Hence for all $k \in \mathbb{N}^* := \{0, 1, 2, \ldots\}$, similar to (5.16), we set

$$\tau_{c,k} = -\frac{1}{2} + \sqrt{\lambda_{c,k} + \frac{1}{4}}.$$

The associated singular function $\sigma_{c,k}$ is given by

$$\sigma_{c,k} = r_c^{\tau_{c,k}} \phi_{c,k}, \tag{5.69}$$

where $\phi_{c,k}$ is the eigenvector of Δ'_{mixed} associated with $\lambda_{c,k}$, namely $\Delta'_{\text{mixed}}\phi_{c,k} = \lambda_{c,k}\phi_{c,k}$. Note that $\lambda_{c,0} = 0$ if and only if γ_D is empty and in that case $\sigma_{c,0} = 1$, otherwise $\lambda_{c,0} > 0$.

Note that in Λ, u may consist of singularities from the edge and singularities from the vertex. In a first step, we subtract from u its corner (vertex) singularities.

Lemma 5.3 *Let $\sigma \in [0, 1)$ be such that $\tau_{c,k} \neq \sigma + \frac{1}{2}$ for all $k \in \mathbb{N}^*$. Let $u \in H^1(\Lambda)$ with a support included in $\Lambda \cap B(0, R)$ for some $R > 0$ be the solution of* (5.68) *with $f \in H^\sigma(\Lambda)$. Then u admits the splitting*

$$u = u_0 + \sum_{-\frac{1}{2} < \tau_{c,k} < \sigma + \frac{1}{2}} c_k \sigma_{c,k}, \tag{5.70}$$

where $u_0 \in V^1_{-(1+\sigma)}(\Lambda)$, $c_k \in \mathbb{C}$ and $\Delta u_0 \in H^\sigma(\Lambda) = V_0^\sigma(\Lambda)$.

Proof By Proposition AA.27 of [49], the Mellin transform $\mathfrak{M}[u](\lambda)$ of u exists for all $\lambda \in \mathbb{C}$ with the real part $\Re\lambda = -\frac{1}{2}$ and is the variational solution of

$$(\Delta' + \lambda(\lambda + 1))\,\mathfrak{M}[u](\lambda) = \mathfrak{M}[f](\lambda - 2),$$

where Δ' is the Laplace–Beltrami operator. Since by assumption on the line $\Re\lambda = \sigma + \frac{1}{2}$, the operator $\Delta' + \lambda(\lambda + 1)$ is invertible from $H_D^1(G) = \{v \in H^1(G) : v = 0 \text{ on } \gamma_D\}$ into its dual with

$$\left\|(\Delta' + \lambda(\lambda + 1))^{-1} z\right\|_{1,G,1+|\lambda|} \lesssim \|z\|_G,$$

by the inverse Mellin transform on the line $\Re\lambda = \sigma + \frac{1}{2}$, we find the result (5.70) as in Lemma 17.4 of [49]. □

We now split up u_0 into a regular part and a singular one that contains the edge contribution.

Lemma 5.4 *Under the assumption of Lemma 5.3, u_0 admits the splitting*

$$u_0 = u_{\rm reg} + u_{\rm sing}$$

with $u_{\rm reg} \in V_0^{2+\sigma}(\Lambda)$, $r_e^{-\sigma}\partial_3^2 u_{\rm reg} \in L^2(\Lambda)$, and $u_{\rm sing} \in \mathcal{H}^2_{\gamma,\sigma}(\Lambda)$ with $\gamma_c = 1+\sigma$ and $\gamma_e < \lambda_e$, where λ_e is the smallest exponent λ_1 determined in either (5.42) *or* (5.43) *according to the associated boundary condition with ω_e instead of ω.*

Proof We start as in the proof of Proposition 17.12 of [49] by setting

$$w(t,\theta) = e^{\eta t} u_0(e^t,\theta), \quad h(t,\theta) = e^{(\eta+2)t}(\Delta u_0)(e^t,\theta)$$

with $\eta = -(\sigma+\frac{1}{2})$. These functions have the regularity $w \in H^1(\mathbb{R}\times G)$, $h \in H^\sigma(\mathbb{R}\times G)$ and w is the weak solution of (meaning that w satisfies the Dirichlet condition and Neumann condition in a weak sense)

$$\left(\Delta' + \partial_t^2 + (1-2\eta)\partial_t + \eta(\eta-1)\right) w = h.$$

Since $H^1(\mathbb{R}\times G)$ is embedded into $H^\sigma(\mathbb{R}\times G)$, this implies that

$$\left(\Delta' + \partial_t^2 + (1-2\eta)\partial_t\right) w = \tilde{h} := h - \eta(\eta-1)w \in H^\sigma(\mathbb{R}\times G).$$

Then as in Sect. 5.1.3.1, we apply a partial Fourier transform in t to find that $W = \mathcal{F}_{t\to\zeta} w$ is the weak solution of

$$\left(\Delta' - \zeta^2 + (1-2\eta)\mathrm{i}\zeta\right) W = H,$$

with $H = \mathcal{F}_{t\to\zeta}\tilde{h}$. This operator is mainly the same one as in (5.41), and therefore, we conclude that W admits a splitting similar to (5.44). Hence taking the Fourier transform back we find that (see Theorem 5.4)

$$w = w_{\rm reg} + w_{\rm sing},$$

with $w_{\rm reg} \in H^{2+\sigma}(\mathbb{R}\times G)$ such that $r_{c,e}^{-\sigma}\partial_t^j w_{\rm reg} \in L^2(\mathbb{R}\times G)$, for $j = 0, 1$, and 2 (recalling that $r_{c,e}$ is the distance to the edges in $\mathbb{R}\times G$ and hence the angular distance in Λ), and $w_{\rm sing} \in \mathcal{H}^2_{\gamma,\sigma}(\mathbb{R}\times G)$ for any $\gamma_e < \lambda_e$.

Coming back to u_0, we find the result by setting

$$u_{\rm reg}(r_c,\theta) = r_c^{-\eta} w_{\rm reg}(r_c,\theta), \quad u_{\rm sing}(r_c,\theta) = r_c^{-\eta} w_{\rm sing}(r_c,\theta).$$

Indeed, the regularity $u_{\rm reg} \in V_0^{2+\sigma}(\Lambda)$ follows from $w_{\rm reg} \in H^{2+\sigma}(\mathbb{R}\times G)$ by using Theorem AA.3 of [49], while the property $r_e^{-\sigma}\partial_3^2 u_{\rm reg} \in L^2(\Lambda)$ follows from the expression of ∂_3^2 in spherical coordinates, an Euler's change of variables, and the regularity of $w_{\rm reg}$ mentioned above (noticing that $r_e^{-\sigma} \sim r_c^{-\sigma} r_{c,e}^{-\sigma}$).

The regularity of $u_{\rm sing}$ is proved similarly. □

In summary, we have the following decomposition of the solution u of (5.68).

Corollary 5.3 *Under the assumption of Lemma 5.3, $u \in H^1(\Lambda)$ with a support included in $\Lambda \cap B(0, R)$ for some $R > 0$ being the solution of* (5.68) *with $f \in H^\sigma(\Lambda)$ for some $\sigma \in [0, 1)$ admits the splitting*

$$u = u_{\rm reg} + u_{\rm sing} + \sum_{-\frac{1}{2}<\tau_{c,k}<\sigma+\frac{1}{2}} c_k \chi \sigma_{c,k},$$

with $u_{\rm reg} \in V_0^{2+\sigma}(\Lambda)$, $r_e^{-\sigma}\partial_3^2 u_{\rm reg} \in L^2(\Lambda)$ and $u_{\rm sing} \in \mathcal{H}^2_{\gamma,\sigma}(\Lambda)$ with $\gamma_c = 1 + \sigma$ and $\gamma_e < \lambda_e$, and χ being a smooth (and radial) cut-off function with a compact support and equal to 1 on the support of u.

Proof Since $u = \chi u$, the result follows from the two previous lemmas and Leibniz's rule. □

5.1.3.3 Regularity Estimates for (5.1) with Mixed Boundary Conditions

We proceed to obtain anisotropic regularity results for (5.1) with $f \in H^\sigma(\Omega)$ for some $\sigma \in [0, 1)$. The analysis is based on regularity estimates in different subregions (5.2) of the domain near the vertices and edges and uses a localization argument.

Regularity Estimates in $\mathcal{V}_e^o$ We start with an improved regularity of the solution along the edges. For any $e \in \mathcal{E}$ and any point $\xi_e \in e$, we can fix a Cartesian system of coordinates $x_e = (x_{e,1}, x_{e,2}, x_{e,3})$ such that ξ_e corresponds to $(0, 0, 0)$. In such a situation we can fix a cut-off function χ_{ξ_e} in the form

$$\chi_{\xi_e}(x_e) = \chi_0(x_{e,1}, x_{e,2})\chi_1(x_{e,3}),$$

with χ_0, χ_1 two cut-off functions such that χ_0 (resp. χ_1) is equal to 1 near $(0, 0)$ (resp. 0) with a sufficiently small support such that the support of χ_{ξ_e} is included in $\mathcal{V}_e^o$. For all $k \in \mathbb{N}$, we further denote by $\lambda_{e,k}$ the associated singular exponents defined by (5.42) or (5.43) (according to the boundary conditions on its adjacent faces) with ω_e instead of ω.

Theorem 5.5 *Recall the space $\mathcal{H}^2_{\gamma,\sigma}$ from* (5.57) *and λ_e from Lemma 5.4. Let $u \in H_D^1(\Omega) = \{u \in H^1(\Omega) : u = 0 \text{ on } \Gamma_D\}$ be the solution of* (5.1) *with the mixed boundary condition and $f \in H^\sigma(\Omega)$ for some $\sigma \in [0, 1)$ such that $\sigma \neq \lambda_{e,k} - 1$ for all $k \in \mathbb{N}$. Then for any $e \in \mathcal{E}$ and any point $\xi_e \in e$, we have*

$$\chi_{\xi_e} u = u_{\xi_e,\rm reg} + u_{\xi_e,\rm sing}, \tag{5.71}$$

with $u_{\xi_e,\rm reg} \in H^{2+\sigma}(\mathcal{V}_e^o)$ such that $r_e^{-\sigma}\partial_3^2 u_{\xi_e,\rm reg} \in L^2(\mathcal{V}_e^o)$ and $u_{\xi_e,\rm sing} \in \mathcal{H}^2_{\gamma_e,\sigma}(\mathcal{V}_e^o)$ for any $\gamma_e \in [0, 1]$ and $\gamma_e < \lambda_e$ satisfying the estimates

$$\|u_{\xi_e,\text{reg}}\|_{2+\sigma,\mathcal{V}_e^o} \lesssim \|f\|_{\sigma,\Omega},$$
$$\|r_e^{-\sigma}\partial_3^j u_{\xi_e,\text{reg}}\|_{\mathcal{V}_e^o} \lesssim \|f\|_{\sigma,\Omega}, \quad \forall j = 0, 1, 2,$$
$$\|u_{\xi_e,\text{sing}}\|_{\mathcal{H}^2_{\gamma_e,\sigma}(\mathcal{V}_e^o)} \lesssim \|f\|_{\sigma,\Omega}. \tag{5.72}$$

Proof For simplicity we drop the indices e and ξ_e. Denote by $D = K\times\mathbb{R}$ the dihedral cone that coincides with Ω near ξ, where K is a two-dimensional cone of opening angle ω.

Now $\tilde{u} := \chi u$ is a weak solution of

$$-\Delta\tilde{u} = \tilde{f} \quad \text{in} \quad D, \tag{5.73}$$

where $\tilde{f}$ is given by

$$\tilde{f} = \chi f - 2\nabla u \cdot \nabla\chi - u\Delta\chi \in L^2(D).$$

Note that the main point is that $\tilde{f}$ actually belongs to $H^\sigma(D)$. Indeed the first term has trivially this property, and the third term is even in $H^1(D)$, hence also in $H^\sigma(D)$. For the second term, we will show that it belongs to $H^1(D)$ as well. We notice that

$$\nabla u \cdot \nabla\chi = \chi_1 \sum_{i=1,2} \partial_i\chi_0\partial_i u + \chi_0\partial_3\chi_1\partial_3 u.$$

Now the first term belongs to $H^1(D)$ because $\chi_1\partial_i\chi_0$ is zero in a neighborhood of the edge, and the H^2 regularity of u inside the domain is sufficient to get the desired regularity. On the other hand for the second term, we notice that the method of tangential differential quotients of Nirenberg (see, for instance, [61, pp. 87–90]) can be applied to ψu in the x_3 direction, with a cut-off function similar to χ such that $\psi \equiv 1$ on the support of χ, and deduce that $\partial_3(\psi u) \in H^1(\Omega)$. This leads to $\chi_0\partial_3\chi_1\partial_3 u \in H^1(D)$.

Once $\tilde{f}$ belongs to $H^\sigma(D)$, we conclude the proof by applying Theorem 5.4. □

Regularity Estimates in $\mathcal{V}_c$ Now we describe the extra regularity in a neighborhood of a vertex $c \in C$. For that purpose, we fix a cut-off function χ_c that depends only on the r_c variable and such that $\chi_c \equiv 1$ in a neighborhood of c and with a support included in $\mathcal{V}_c$ (hence $\chi_c \equiv 0$ in a neighborhood of the other vertices of Ω). We further denote by Λ_c the infinite cone that coincides with Ω near c.

Then we have the following splitting of the solution near the vertex.

Theorem 5.6 *Recall the space $\mathcal{H}^2_{\gamma,\sigma}$ from* (5.66) *and λ_e from Lemma 5.4. Under the assumption of Lemma 5.3, let $u \in H^1_D(\Omega)$ be the solution of* (5.1) *with the mixed boundary condition and $f \in H^\sigma(\Omega)$ for some $\sigma \in [0, 1)$. Then for any $c \in C$, $\chi_c u$ admits the splitting*

$$\chi_c u = u_{c,\text{reg}} + u_{c,\text{sing}} + \sum_{-\frac{1}{2}<\tau_{c,k}<\sigma+\frac{1}{2}} c_k\psi\sigma_{c,k}, \tag{5.74}$$

with $u_{c,\text{reg}} \in V_0^{2+\sigma}(\Lambda_c)$, $r_e^{-\sigma}\partial_3^2 u_{c,\text{reg}} \in L^2(\mathcal{V}_c^e)$ *for all* $e \in \mathcal{E}_c$, $u_{c,\text{sing}} \in \mathcal{H}^2_{\gamma,\sigma}(\Lambda_c)$ *with* $\gamma_c = 1 + \sigma$ *and* $\gamma_e \in [0, 1]$ *and* $\gamma_e < \lambda_e$, *and finally* ψ *being a smooth (and radial) cut-off function with a compact support and equal to 1 on the support of* χ_c.

Proof We can apply Corollary 5.3 to χu (for shortness we drop the index c) if we show that

$$\Delta(\chi u) \in H^\sigma(\Lambda).$$

As before using Leibniz's rule, $\tilde{u} := \chi u$ is a weak solution of (5.73) with

$$\tilde{f} = \chi f - 2\nabla u \cdot \nabla \chi - u\Delta \chi,$$

which actually belongs to $H^\sigma(\Lambda)$. The first and third term have trivially this property. Hence only the second term requires a careful inspection. Due to the choice of χ and using Cartesian coordinates centered at c, we have

$$\nabla u \cdot \nabla \chi = \frac{\chi'(r_c)}{r_c} \sum_{i=1,2,3} x_i \partial_i u.$$

First we may notice that χ' is zero near c. Hence the regularity of $\nabla u \cdot \nabla \chi$ is related to the regularity of u far from the corners. Recall the sub-regions

$$\mathcal{V}_c = \{x \in \Omega : r_c(x) < \varepsilon\}, \quad \mathcal{V}_c^o = \{x \in \mathcal{V}_c : r_{c,e}(x) \geq \varepsilon, \ \forall\, e \in \mathcal{E}_c\}.$$

Define a smaller set $\mathcal{V}_c' \subset \mathcal{V}_c^o$ such that

$$\mathcal{V}_c' = \{x : r_{c,e}(x) \geq \varepsilon, \ r_c(x) < \varepsilon/2, \ \forall\, e \in \mathcal{E}_c\}.$$

Let $\mathcal{V}_0 = \mathcal{V}_c^o \setminus \mathcal{V}_c'$. As u belongs to H^2 in $\mathcal{V}_0$, we get that

$$\nabla u \cdot \nabla \chi \in H^1(\mathcal{V}_0). \tag{5.75}$$

Now for a fixed edge e having c as an endpoint, we can use Cartesian coordinates such that the x_3-axis contains the edge e and can use the splitting (5.71) of Theorem 5.5. The regular part contributes to an H^1 function. Therefore, let us show that this is the same for the singular part $u_{e,\text{sing}}$. Indeed we have to show that

$$x_i \partial_i u_{e,\text{sing}} \in H^1(\mathcal{V}_e^o), \quad \forall\, i = 1, 2, 3.$$

For $i = 3$, this is a direct consequence of (5.72), while for $i = 1$ or 2, this is a consequence of (5.72) and of the bound

$$|x_i| \lesssim r_{c,e} \lesssim r_{c,e}^{1-\gamma_e}.$$

All together (5.75) is valid and the proof is complete. □

The third term of the splitting (5.74) of $\chi_c u$ is not in $\mathcal{H}^2_{\gamma,\sigma}(\Lambda \cap B(0, R))$ because (see (5.69))

$$\sigma_{c,k} = r_c^{\tau_{c,k}} \phi_{c,k}$$

and $\phi_{c,k}$ is not necessarily equal to zero at the corners of G (intersection of Λ with the unit sphere), but it can be split up into a regular part and a singular one. This allows us to show the next result.

Lemma 5.5 *Let $\tau_{c,k} > 0$ be fixed such that $\tau_{c,k} < \sigma + \frac{1}{2}$. Recall λ_e from Lemma 5.4. Then $v_{c,k} := \psi\sigma_{c,k}$ with $\sigma_{c,k}$ given by* (5.69) *can be split up into $v_{c,k} = v_1 + v_2$ such that v_1 satisfies*

$$\begin{aligned}
r_c^{\sigma_c} r_{c,e}^{-\sigma_{c,e}} \partial_3^2 v_1 &\in L^2(\mathcal{V}_c^e), \\
r_c^{\sigma_c} \partial^{\alpha_\perp} \partial_3 v_1 &\in L^2(\mathcal{V}_c^e), \quad && \forall\, |\alpha_\perp| = 1, \\
r_c^{\sigma_c} \partial^{\alpha_\perp} v_1 &\in L^2(\mathcal{V}_c^e), \quad && \forall\, |\alpha_\perp| = 2, \\
r_c^{\sigma_c - 2 + |\alpha|} \partial^{\alpha} v_1 &\in L^2(\mathcal{V}_c^e), \quad && \forall\, |\alpha| \le 1,
\end{aligned}$$

and v_2 satisfies

$$\begin{aligned}
r_c^{\sigma_c} r_{c,e}^{-\sigma_{c,e}} \partial_3^2 v_2 &\in L^2(\mathcal{V}_c^e), \\
r_c^{\sigma_c} \partial^{\alpha_\perp} \partial_3 v_2 &\in L^2(\mathcal{V}_c^e), \quad && \forall\, |\alpha_\perp| = 1, \\
r_c^{\sigma_c} r_{c,e}^{\tilde{\sigma}_{c,e}} \partial^{\alpha_\perp} v_2 &\in L^2(\mathcal{V}_c^e), \quad && \forall\, |\alpha_\perp| = 2, \\
r_c^{\sigma_c - 1} r_{c,e}^{\tilde{\sigma}_{c,e} - 1} \partial^{\alpha_\perp} v_2 &\in L^2(\mathcal{V}_c^e), \quad && \forall\, |\alpha_\perp| = 1, \\
r_c^{\sigma_c - 1} r_{c,e}^{-1} \partial_3 v_2 &\in L^2(\mathcal{V}_c^e), \\
r_c^{\sigma_c - 2} r_{c,e}^{\tilde{\sigma}_{c,e} - 2} v_2 &\in L^2(\mathcal{V}_c^e),
\end{aligned}$$

for any $\sigma_{c,e} < 1$, any $\tilde{\sigma}_{c,e} > 1 - \lambda_e$, and any $\sigma_c > \frac{1}{2} - \tau_c$, where τ_c is the smallest positive $\tau_{c,k}$ such that $-\frac{1}{2} < \tau_{c,k} < \sigma + \frac{1}{2}$, and $r_{c,e}(x) := \frac{r_e(x)}{r_c(x)}$ is the angular distance.

Proof Recall the intersection G of Λ and the unit sphere. Based on direct calculations expressing the Cartesian derivatives in spherical coordinates $(r_c, \theta_{c,e}, \varphi_{c,e})$ (where $\varphi_{c,e}$ is the angular distance to the edge e) and using the splitting

$$\phi_{c,k} = \kappa_{c,e}^{(0)} \chi_{c,e}(\varphi_{c,e}) + \kappa_{c,e} \varphi_{c,e}^{\lambda_{c,e}} \phi_{c,k,e}(\theta_{c,e}) + \phi_{c,k,R},$$

where $\kappa_{c,e}^{(0)}, \kappa_{c,e} \in \mathbb{C}$, $\chi_{c,e}$ is a cut-off function equal to 1 near $\varphi_{c,e} = 0$, $\phi_{c,k,e}$ is the eigenfunction associated with the corner singular exponent $\lambda_{c,e}$, and $\phi_{c,k,R}$ is the regular part of $\phi_{c,k}$ that either belongs to $H^2(G)$ or has to be split up into a singular part (similar to the second term of the above right hand side) and a regular part in $H^2(G)$. In this situation,

$$v_2 = \kappa_{c,e} \psi r_c^{\tau_{c,k}} \varphi_{c,e}^{\lambda_{c,e}} \phi_{c,k,e}(\theta_{c,e}),$$

while $v_1 = v_{c,k} - v_2$. □

Remark 5.4 The function $v_{c,k} = \psi\sigma_{c,k} = \psi r^{\tau_{c,k}}\phi_{c,k}$ that characterizes the vertex singularity in (5.74) satisfies the following. (i) In the case $\lambda_{c,k} = 0$, $\sigma_{c,k}$ =constant, because the eigenvector $\phi_{c,k}$ of the Laplace–Beltrami operator corresponding to the zero eigenvalue is a constant. (ii) In the case $\lambda_{c,k} > 0$, according to Lemma 5.5, the function $v_{c,k}$ admits a splitting into two functions v_1 and v_2.

Recall that $\mathcal{V}_c^o$ is part of the neighborhood of $c \in C$ that excludes the edges. Based on Theorem 5.6 and Lemma 5.5, we further obtain the regularity estimate in $\mathcal{V}_c^o$.

Corollary 5.4 *Under the assumption of Lemma 5.3, let $u \in H_D^1(\Omega)$ be the solution of* (5.1) *with the mixed boundary condition and with $f \in H^\sigma(\Omega)$ for some $\sigma \in [0, 1)$. Let $c \in C$ be a vertex and let τ_c be the smallest positive $\tau_{c,k}$. If $\tau_c < \sigma + \frac{1}{2}$, in $\mathcal{V}_c^o$, u admits the decomposition $u = u_{c,\mathrm{reg}} + u_{c,\mathrm{sing}}$ where $u_{c,\mathrm{reg}} \in H^2(\mathcal{V}_c^o)$ and $u_{c,\mathrm{sing}} \in V^2_{\sigma_c}(\mathcal{V}_c^o)$ for any $\sigma_c > \frac{1}{2} - \tau_c$. If $\tau_c \geq \sigma + \frac{1}{2}$, we have $u \in H^2(\mathcal{V}_c^o)$.*

Proof Note that the angular distance $r_{c,e}$ is bounded below from 0 in $\mathcal{V}_c^o$. Taking this into account in (5.74) and in the regularity estimates (Theorem 5.6 and Lemma 5.5), we can derive the result in this Corollary by straightforward calculations. □

5.2 3D Graded Meshes and Mesh Layers

In this section, we revisit the graded tetrahedral mesh (Algorithms 3.8 and 3.9) for 3D polyhedral domains and study its properties that are useful for the finite element error analysis. Recall the element types in Definition 3.2. Then the graded mesh for solving (5.1) can be summarized as follows.

Algorithm 5.7 (3D Graded Mesh) Let $\mathcal{T}$ be a triangulation of Ω as in Definition 3.1. To each $c \in C$ (resp. $e \in \mathcal{E}$), we associate a grading parameter κ_c (resp. κ_e) such that $\kappa_c, \kappa_e \in (0, 0.5]$. Let $T = \Delta^4 x_0x_1x_2x_3 \in \mathcal{T}$ be a tetrahedron with vertices x_0, x_1, x_2, and x_3. Assume that x_0 is the singular vertex if T is a v-, v_e-, or ev-tetrahedron, and x_0x_1 is the singular edge if T is an e- or ev-tetrahedron. Let $\boldsymbol{\kappa}$ be the collection of the parameters κ_c and κ_e for all $c \in C$ and $e \in \mathcal{E}$. Then the refinement, denoted by $\boldsymbol{\kappa}(\mathcal{T})$, proceeds as follows. We first generate new nodes x_{kl}, $0 \leq k < l \leq 3$, on each edge x_kx_l of T, based on its type.

(I) T = o-tetrahedron: $x_{kl} = (x_k + x_l)/2$.
(II) T = v-tetrahedron: Suppose $x_0 = c \in C$. Define $\kappa = \kappa_{ec} := \min_{e \in \mathcal{E}_c}\{\kappa_c, \kappa_e\}$. Then $x_{kl} = (x_k + x_l)/2$ for $1 \leq k < l \leq 3$; $x_{0l} = (1-\kappa)x_0 + \kappa x_l$ for $1 \leq l \leq 3$.
(III) T = v_e-tetrahedron: Suppose x_0 is an interior point of $e \in \mathcal{E}$. Let $\kappa = \kappa_e$. Then $x_{kl} = (x_k + x_l)/2$ for $1 \leq k < l \leq 3$; $x_{0l} = (1-\kappa)x_0 + \kappa x_l$ for $1 \leq l \leq 3$.
(IV) T = e-tetrahedron: Suppose $x_0x_1 \subset e \in \mathcal{E}$. Then $x_{kl} = (1 - \kappa_e)x_k + \kappa_e x_l$ for $0 \leq k \leq 1$ and $2 \leq l \leq 3$; $x_{01} = (x_0 + x_1)/2$, $x_{23} = (x_2 + x_3)/2$.
(V) T = ev-tetrahedron: Suppose $x_0x_1 \subset \bar{e}$ and $x_0 = c \in C$. Define $\kappa_{sc} := \min_{e_r \in \mathcal{E}_c \setminus \{e\}}\{\kappa_c, \kappa_{e_r}\}$. Then for $2 \leq l \leq 3$, $x_{0l} = (1 - \kappa_{ec})x_0 + \kappa_{ec}x_l$ and $x_{1l} = (1 - \kappa_e)x_1 + \kappa_e x_l$; $x_{01} = (1 - \kappa_{sc})x_0 + \kappa_{sc}x_1$, $x_{23} = (x_2 + x_3)/2$.

Connecting these nodes x_{kl} on all the faces of T, we obtain four sub-tetrahedra and one octahedron. The octahedron then is cut into four tetrahedra using x_{13} as the common vertex. Therefore, after one refinement, we obtain eight sub-tetrahedra for each $T \in \mathcal{T}$ denoted by their vertices:

$$\Delta^4 x_0x_{01}x_{02}x_{03},\ \Delta^4 x_{01}x_1x_{12}x_{13},\ \Delta^4 x_{02}x_{12}x_2x_{23},\ \Delta^4 x_{03}x_{13}x_{23}x_3,$$
$$\Delta^4 x_{01}x_{02}x_{03}x_{13},\ \Delta^4 x_{01}x_{02}x_{12}x_{13},\ \Delta^4 x_{02}x_{03}x_{13}x_{23},\ \Delta^4 x_{02}x_{12}x_{13}x_{23}.$$

See Fig. 5.2 for different types of decompositions. Given an initial mesh $\mathcal{T}_0$, the associated family of anisotropic meshes $\{\mathcal{T}_n : n \geq 0\}$ is defined recursively $\mathcal{T}_n = \kappa(\mathcal{T}_{n-1})$. See Fig. 5.3 for example. □

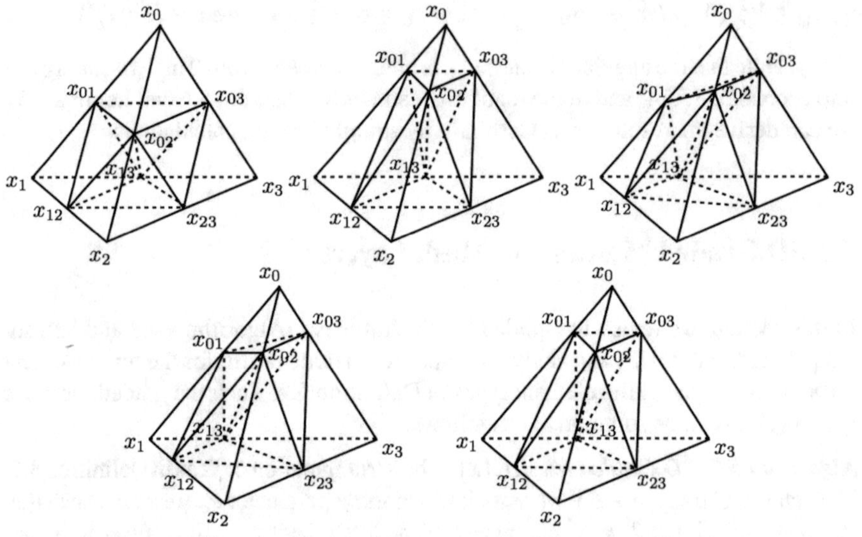

Fig. 5.2 Decompositions for a tetrahedron $\Delta^4 x_0x_1x_2x_3$, top row (left–right): o-tetrahedron, v- or v_e-tetrahedron, e-tetrahedron; bottom row (left–right): an ev-tetrahedron ($\kappa_{ec} = \kappa_e$), an ev-tetrahedron ($\kappa_{ec} < \kappa_e$).

Algorithm 5.7 is the more explicit version of Algorithm 3.9 for graded tetrahedral mesh refinements. According to Algorithm 5.7 and Algorithm 3.9, the consecutive refinements of an initial tetrahedron $T_{(0)} \in \mathcal{T}_0$ shall result in *mesh layers*. These mesh layers depend on the element type of $T_{(0)}$ and are the collections of tetrahedra in $\mathcal{T}_n$ with comparable distances to the singular points on $\partial T_{(0)}$.

Mesh Layers on Initial v- or v_e-Tetrahedra in $\mathcal{T}_0$ Let $T_{(0)} = \Delta^4 x_0x_1x_2x_3 \in \mathcal{T}_0$ be either a v- or a v_e-tetrahedron with $x_0 \in C$ or $x_0 \in e \in \mathcal{E}$. We define the mesh layer associated with $\mathcal{T}_n$ on $T_{(0)}$ as follows.

Definition 5.3 (Mesh Layer in v- and v_e-Tetrahedra) Use a local Cartesian coordinate system, such that x_0 is the origin. For $1 \leq i \leq n$, the ith refinement on $T_{(0)}$ produces a small tetrahedron with x_0 as a vertex and with one face, denoted by $P_{v,i}$, parallel to the face $\Delta^3 x_1 x_2 x_3$ of $T_{(0)}$. See Fig. 5.2 for example. Then after n refinements, we define the ith mesh layer $L_{v,i}$ of $T_{(0)}$, $1 \le i < n$, as the region in $T_{(0)}$ between $P_{v,i}$ and $P_{v,i+1}$. We denote by $L_{v,0}$ the region in $T_{(0)}$ between $\Delta^3 x_1 x_2 x_3$ and $P_{v,1}$; and let $L_{v,n}$ be the small tetrahedron with x_0 as a vertex that is bounded by $P_{v,n}$ and three faces of $T_{(0)}$. Since x_0 is the only point for the special refinement, we drop the sub-index in the grading parameter. Namely, for such $T_{(0)}$, we use κ to denote the grading parameter near x_0 ($\kappa = \kappa_{ec}$ if $x_0 = c \in C$ and $\kappa = \kappa_e$ if $x_0 \in e \in \mathcal{E}$). See the second picture in Fig. 5.3. Then by Algorithm 5.7, the dilation matrix

$$\mathbf{B}_{v,i} := \begin{pmatrix} \kappa^{-i} & 0 & 0 \\ 0 & \kappa^{-i} & 0 \\ 0 & 0 & \kappa^{-i} \end{pmatrix} \tag{5.76}$$

maps $L_{v,i}$ to $L_{v,0}$ for $0 \le i < n$ and maps $L_{v,n}$ to $T_{(0)}$ for $i = n$. We define the *initial triangulation* of $L_{v,i}$, $0 \le i < n$, to be the first decomposition of $L_{v,i}$ into tetrahedra. Thus, the initial triangulation of $L_{v,i}$ consists of those tetrahedra in $\mathcal{T}_{i+1}$ that are contained in the layer $L_{v,i}$.

Remark 5.5 Based on the refinement, on $L_{v,i}$, $0 \le i \le n$, the tetrahedra in $\mathcal{T}_n$ are isotropic with mesh size $O(\kappa^i 2^{i-n})$. In $T_{(0)}$, let r be the distance to x_0. Therefore,

$$r \sim \kappa^i \qquad \text{on } L_{v,i}, \quad 0 \le i < n. \tag{5.77}$$

Namely, if $T_{(0)}$ is a v-tetrahedron, $r \sim r_c$ for $c = x_0 \in C$; and if $T_{(0)}$ is a v_e-tetrahedron, $r \sim r_e$, where $e \in \mathcal{E}$ is the edge containing x_0.

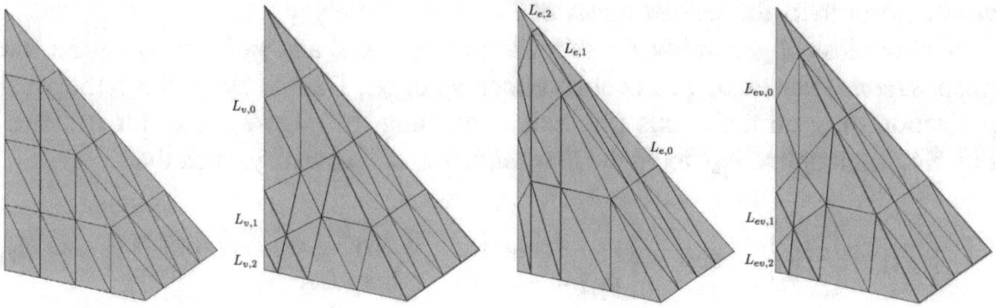

Fig. 5.3 Anisotropic triangulations after two consecutive refinements and mesh layers on an initial tetrahedron (left–right): o-tetrahedron, v- or v_e-tetrahedron ($\kappa = 0.3$), e-tetrahedron ($\kappa_e = 0.3$); ev-tetrahedron ($\kappa_{ec} = \kappa_c = 0.3$, $\kappa_e = 0.4$).

Mesh Layers on Initial e-Tetrahedra in $\mathcal{T}_0$ Now, let $T_{(0)} := \Delta^4 x_0 x_1 x_2 x_3 \in \mathcal{T}_0$ be an e-tetrahedron with $x_0 x_1$ on the edge $e \in \mathcal{E}$ and let κ_e be the associated grading parameter. Then we define the mesh layer associated with $\mathcal{T}_n$ on $T_{(0)}$ as follows.

Definition 5.4 (Mesh Layer in e-Tetrahedra) Based on Algorithm 5.7, in each refinement, an e-tetrahedron is cut by a parallelogram parallel to $x_0 x_1$. For example, in the e-tetrahedron of Fig. 5.2, the quadrilateral with vertices $x_{02}, x_{12}, x_{13}, x_{03}$ is the aforementioned parallelogram. We denote by $P_{e,i}$ the parallelogram produced in the

ith refinement, $1 \leq i \leq n$. For the mesh $\mathcal{T}_n$, let the ith layer $L_{e,i}$ on $T_{(0)}$, $0 < i < n$, be the region bounded by $P_{e,i}$, $P_{e,i+1}$, and the faces of $T_{(0)}$. Define $L_{e,0}$ to be the sub-region of $T_{(0)}$ away from e that is separated by $P_{e,1}$. Define $L_{e,n}$ to be the sub-region of $T_{(0)}$ between $P_{e,n}$ and e. See the third picture in Fig. 5.3. As in Definition 5.3, the initial triangulation of the layer $L_{e,i}$, $0 \leq i < n$, consists of the tetrahedra in $\mathcal{T}_{i+1}$ that are contained in $L_{e,i}$.

Remark 5.6 In the mesh layers, the distance r_e to the edge e satisfies

$$r_e \sim \kappa_e^i \qquad \text{on } L_{e,i}, \quad 0 \leq i < n. \tag{5.78}$$

In addition, the mesh layers of an e-tetrahedron $T_{(0)}$ also satisfy the following properties (see Fig. 5.5):

- The layer $L_{e,i}$, $2 \leq i \leq n$, is the union of two components: sub-regions from 2^{i-1} e-tetrahedra in $\mathcal{T}_{i-1}$ and sub-regions from $2^i - 2$ v_e-tetrahedra in $\mathcal{T}_{i-1}$.
- Among the aforementioned $2^i - 2$ v_e-tetrahedra in $\mathcal{T}_{i-1}$, 2^k of them are sub-regions of v_e-tetrahedra in $\mathcal{T}_k$, $1 \leq k \leq i-1$.

Now, we start to develop some estimates for the shape regularity of the mesh on $L_{e,i}$, although it is in general anisotropic and violates the maximum angle condition. These results will be useful for the interpolation error analysis.

Definition 5.5 (Relative Distance for e-Tetrahedra) Recall the initial e-tetrahedron $T_{(0)} = \triangle^4 x_0x_1x_2x_3 \in \mathcal{T}_0$. For an e-tetrahedron $T = \triangle^4 \gamma_0\gamma_1\gamma_2\gamma_3$ generated by some subsequent refinements of $T_{(0)}$ based on Algorithm 5.7, consider its two vertices on the edge x_0x_1. We call the vertex that is closer to x_0 the *first vertex* of T and call the vertex closer to x_1 the *second vertex* of T.

Without loss of generality, we suppose $\gamma_0\gamma_1 \subset e \in \mathcal{E}$ and γ_0 (resp. γ_1) is the *first* (resp. *second*) vertex of T. Let γ be either γ_2 or γ_3. Denote by γ' the orthogonal projection of γ on the z-axis (the axis containing the edge e). See, for instance, Fig. 5.4. Then define $c_{\gamma,1}$ to be the *first relative z-distance* of γ, such that

$$|c_{\gamma,1}| = \frac{|\gamma_0\gamma'|}{|\gamma_0\gamma_1|}, \quad \text{and} \quad \begin{cases} c_{\gamma,1} = |c_{\gamma,1}| & \text{if } \overrightarrow{\gamma_0\gamma'} = t(\overrightarrow{\gamma_0\gamma_1}) \text{ for } t > 0 \\ c_{\gamma,1} = -|c_{\gamma,1}| & \text{otherwise.} \end{cases} \tag{5.79}$$

Here, $\overrightarrow{\gamma_0\gamma'}$ represents the vector from γ_0 to γ'. The *second relative z-distance* of γ, denoted by $c_{\gamma,2}$, is defined by

$$|c_{\gamma,2}| = \frac{|\gamma_1\gamma'|}{|\gamma_0\gamma_1|}, \quad \text{and} \quad \begin{cases} c_{\gamma,2} = |c_{\gamma,2}| & \text{if } \overrightarrow{\gamma_1\gamma'} = t(\overrightarrow{\gamma_1\gamma_0}) \text{ for } t > 0 \\ c_{\gamma,2} = -|c_{\gamma,2}| & \text{otherwise.} \end{cases} \tag{5.80}$$

It is clear that $c_{\gamma,2} = 1 - c_{\gamma,1}$. In addition, we define the *absolute relative distance* for T, denoted by c_T, such that

$$c_T = \max\{|c_{\gamma_2,1}|, |c_{\gamma_2,2}|, |c_{\gamma_3,1}|, |c_{\gamma_3,2}|\}. \tag{5.81}$$

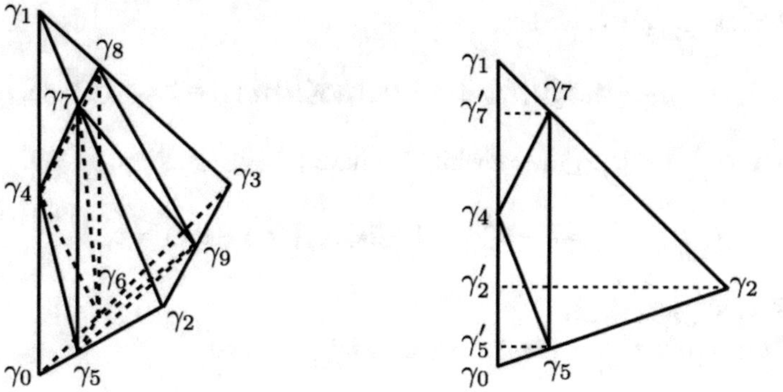

Fig. 5.4 The mesh on an e-tetrahedron $\Delta^4\gamma_0\gamma_1\gamma_2\gamma_3$ after one refinement (left); the induced triangles on an face containing the singular edge (right).

Remark 5.7 For each e-tetrahedron, there are four relative distances corresponding to the two vertices away from the z-axis. The sign of the relative distance is determined by the location of the orthogonal projection of the off-the-edge vertex. The relative distances imply, for the e-tetrahedron, how far the off-the-edge vertices shift away in the z-direction from the vertices on the z-axis.

Remark 5.8 Note that after one refinement, T is decomposed into eight sub-tetrahedra: two e-tetrahedra (denoted by T_A and T_B), two v_e-tetrahedra, and four o-tetrahedra. In this case, we call T the *parent tetrahedron* of the sub-tetrahedra and call each sub-tetrahedron the *child tetrahedron* of T. Note that Definition 5.5 is also valid for ev-tetrahedra. We shall use it later for ev-tetrahedra as well.

In what follows, we establish the connections between T and its child e-tetrahedra T_A and T_B in terms of the corresponding relative z-distances.

Lemma 5.6 *Let $T \subset T_{(0)}$ be an e-tetrahedron in $\mathcal{T}_i$, $1 \le i < n$. Let $T_A, T_B \subset T$ be the two child e-tetrahedra in $\mathcal{T}_{i+1}$. Denote by c_T, c_A, and c_B the absolute distances for T, T_A, and T_B as in (5.81). Then* $\max\{c_A, c_B\} \le \max\{c_T, 1\}$.

Proof Denote T by $T = \Delta^4\gamma_0\gamma_1\gamma_2\gamma_3$ with the *first* vertex γ_0 and the *second* vertex γ_1 on the singular edge $\gamma_0\gamma_1$. As illustrated in Fig. 5.4, we let $T_A := \Delta^4\gamma_0\gamma_4\gamma_5\gamma_6$ and $T_B := \Delta^4\gamma_1\gamma_4\gamma_7\gamma_8$. Recall the relative distance from Definition 5.5. In particular, let $c_{\gamma_2,1}, c_{\gamma_2,2}$ be the relative distances of γ_2 in T, and let $c^A_{\gamma_5,1}, c^A_{\gamma_5,2}$ (resp. $c^B_{\gamma_7,1}, c^B_{\gamma_7,2}$) be the relative distances of γ_5 (resp. γ_7) in T_A (resp. T_B). We first show $|c^A_{\gamma_5,1}|, |c^A_{\gamma_5,2}| < \max\{|c_{\gamma_2,1}|, |c_{\gamma_2,2}|, 1\}$.

Consider the triangles on the face $\Delta^3\gamma_0\gamma_1\gamma_2$ of T, induced by the sub-tetrahedra after one refinement of T (the second picture in Fig. 5.4). In addition, we have drawn three dashed line segments $\gamma_2\gamma_2'$, $\gamma_5\gamma_5'$, and $\gamma_7\gamma_7'$ that are perpendicular to $\gamma_0\gamma_1$. Then by (5.79), we have

$$|c_{\gamma_2,1}| = |\gamma_0\gamma_2'|/|\gamma_0\gamma_1| \quad \text{and} \quad |c^A_{\gamma_5,1}| = |\gamma_0\gamma_5'|/|\gamma_0\gamma_4|.$$

Note that $\Delta^3\gamma_0\gamma_5\gamma_5'$ is similar to $\Delta^3\gamma_0\gamma_2\gamma_2'$. Therefore, $|\gamma_0\gamma_5'| = \kappa_e|\gamma_0\gamma_2'|$, and $c_{\gamma_2,1}$ and $c^A_{\gamma_5,1}$ have the same sign. Recall $0 < \kappa_e < 0.5$. Then we consider all the possible cases.

In the case $c_{\gamma_2,1} < 0$, we have

$$0 > c^A_{\gamma_5,1} = -|\gamma_0\gamma_5'|/|\gamma_0\gamma_4| = -2\kappa_e|\gamma_0\gamma_2'|/|\gamma_0\gamma_1| = 2\kappa_e c_{\gamma_2,1} > c_{\gamma_2,1}.$$

Therefore, $|c^A_{\gamma_5,1}| < |c_{\gamma_2,1}|$. Meanwhile, we have

$$1 \leq c^A_{\gamma_5,2} = 1 - c^A_{\gamma_5,1} = 1 - 2\kappa_e c_{\gamma_2,1} < 1 - c_{\gamma_2,1} = c_{\gamma_2,2}.$$

Therefore, $|c^A_{\gamma_5,2}| < |c_{\gamma_2,2}|$.

In the case $0 \leq c_{\gamma_2,1} < (2\kappa_e)^{-1}$, we have $c^A_{\gamma_5,1} \geq 0$ and

$$c^A_{\gamma_5,1} = |\gamma_0\gamma_5'|/|\gamma_0\gamma_4| = 2\kappa_e|\gamma_0\gamma_2'|/|\gamma_0\gamma_1| = 2\kappa_e c_{\gamma_2,1} < 1.$$

Meanwhile, we have

$$0 \leq c^A_{\gamma_5,2} = 1 - c^A_{\gamma_5,1} = 1 - 2\kappa_e c_{\gamma_2,1} < 1.$$

In the case $c_{\gamma_2,1} \geq (2\kappa_e)^{-1}$, we have

$$1 \leq c^A_{\gamma_5,1} = |\gamma_0\gamma_5'|/|\gamma_0\gamma_4| = 2\kappa_e|\gamma_0\gamma_2'|/|\gamma_0\gamma_1| = 2\kappa_e c_{\gamma_2,1} < |c_{\gamma_2,1}|.$$

Meanwhile, we have

$$0 \geq c^A_{\gamma_5,2} = 1 - c^A_{\gamma_5,1} = 1 - 2\kappa_e c_{\gamma_2,1} > 1 - c_{\gamma_2,1} = c_{\gamma_2,2}.$$

Therefore, $|c^A_{\gamma_5,2}| < |c_{\gamma_2,2}|$. Thus, we have shown

$$|c^A_{\gamma_5,1}|, |c^A_{\gamma_5,2}| < \max\{|c_{\gamma_2,1}|, |c_{\gamma_2,2}|, 1\}.$$

With a similar calculation, we can derive the upper bounds for other relative distances in T_A and T_B, namely

$$|c^A_{\gamma_6,1}|, |c^A_{\gamma_6,2}| < \max\{|c_{\gamma_3,1}|, |c_{\gamma_3,2}|, 1\},$$
$$|c^B_{\gamma_7,1}|, |c^B_{\gamma_7,2}| < \max\{|c_{\gamma_2,1}|, |c_{\gamma_2,2}|, 1\}, \quad |c^B_{\gamma_8,1}|, |c^B_{\gamma_8,2}| < \max\{|c_{\gamma_3,1}|, |c_{\gamma_3,2}|, 1\}.$$

Hence, the proof is completed by (5.81). □

Recall that for a v- or v_e-tetrahedron in $\mathcal{T}_0$, the isotropic transformation (5.76) maps a mesh layer to a reference domain (either the tetrahedron itself or the layer $L_{v,0}$). In turn, we define the reference domain for an e-tetrahedron.

Definition 5.6 (Reference e-Tetrahedron) For the initial e-tetrahedron $\triangle^4 x_0x_1x_2x_3 = T_{(0)} \in \mathcal{T}_0$, we use a local Cartesian coordinate system, such that the z-axis contains the edge x_0x_1 with the direction of $\overrightarrow{x_0x_1}$ as the positive direction, and x_2 is in the xz-plane. We will specify the origin later. Let $l_0 := |x_0x_1|$ be the length of the singular edge. Then we define the reference tetrahedron $\hat{T} = \triangle^4\hat{x}_0\hat{x}_1\hat{x}_2\hat{x}_3$, such that

$$\hat{x}_0 = (0, 0, -l_0/2), \quad \hat{x}_1 = (0, 0, l_0/2), \quad \hat{x}_k = (\hat{\lambda}_k, \hat{\xi}_k, -l_0/2),\ k = 2, 3,$$

where $\hat{\lambda}_k, \hat{\xi}_k$ are the x- and y-components of the vertices x_2 and x_3, respectively. Therefore, $\hat{\xi}_2 = 0$ and $\hat{\lambda}_2, \hat{\lambda}_3, \hat{\xi}_3$ are constants that depend on the shape regularity of $T_{(0)}$. Thus, $\hat{T}$ is a tetrahedron with one face in the plane $z = -l_0/2$, one face in the xz-plane, such that $|\hat{x}_0\hat{x}_1| = |x_0x_1|$, $|\hat{x}_0\hat{x}_2|$ = the length of the orthogonal projection of x_0x_2 in the plane $z = -l_0/2$, and $|\hat{x}_0\hat{x}_3|$ = the length of the orthogonal projection of x_0x_3 in the plane $z = -l_0/2$. In addition, we denote by $\hat{\mathcal{T}}_1$ and $\hat{\mathcal{T}}_2$ the triangulations of $\hat{T}$ after one and two edge refinements with parameter κ_e, respectively. See Fig. 5.5 for example.

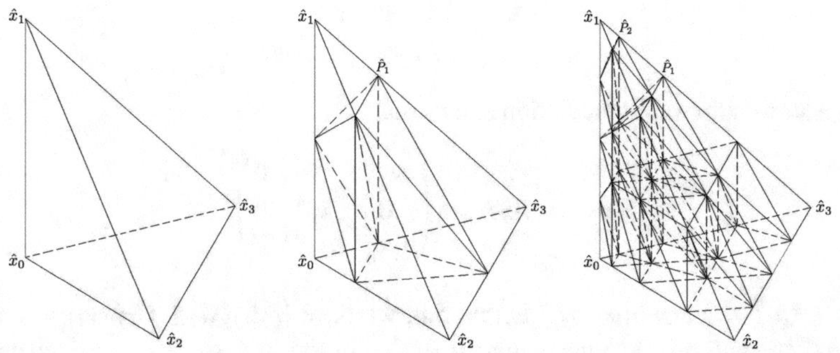

Fig. 5.5 A reference tetrahedron $\hat{T}$ (left); the triangulation $\hat{\mathcal{T}}_1$ after one edge refinement (center); the triangulation $\hat{\mathcal{T}}_2$ after two edge refinements (right).

In the following lemmas, we construct explicit linear mappings between an e-tetrahedron in the initial e-tetrahedron $T_{(0)}$ and the reference tetrahedron $\hat{T}$.

Lemma 5.7 *For an e-tetrahedron $T := \Delta^4\gamma_0\gamma_1\gamma_2\gamma_3 \subset T_{(0)}$ and $T \in \mathcal{T}_i$, $1 \le i \le n$, suppose $\gamma_0\gamma_1$ is the singular edge, and $\overrightarrow{\gamma_0\gamma_1}$ and $\overrightarrow{x_0x_1}$ share the same direction. We use the local coordinate system in Definition 5.6 and set $(\gamma_0 + \gamma_1)/2$ to be the origin. Then there exists a matrix*

$$\mathbf{B}_{e,i} = \begin{pmatrix} \kappa_e^{-i} & 0 & 0 \\ 0 & \kappa_e^{-i} & 0 \\ b_1\kappa_e^{-i} & b_2\kappa_e^{-i} & 2^i \end{pmatrix} \tag{5.82}$$

with $|b_1|, |b_2| \le C_0$, where $C_0 > 0$ depends on the initial tetrahedron $T_{(0)}$ but not on i, such that $\mathbf{B}_{e,i} : T \to \hat{T}$ is a bijection.

Proof Based on the refinement in Algorithm 5.7, and on Definition 5.6, we have $\gamma_2 = (\kappa_e^i\hat{\lambda}_2, 0, \zeta_2)$, $\gamma_3 = (\kappa_e^i\hat{\lambda}_3, \kappa_e^i\hat{\xi}_3, \zeta_3)$, and $|\gamma_0\gamma_1| = 2^{-i}l_0 = 2^{-i}|x_0x_1|$, where ζ_2 and ζ_3 are the z-coordinates of the vertices γ_2 and γ_3, respectively. Thus, the anisotropic transformation

$$\mathbf{A}_1 := \begin{pmatrix} \kappa_e^{-i} & 0 & 0 \\ 0 & \kappa_e^{-i} & 0 \\ 0 & 0 & 2^i \end{pmatrix}$$

maps T to a tetrahedron, with vertices $\mathbf{A}_1\gamma_0 = (0, 0, -l_0/2)$, $\mathbf{A}_1\gamma_1 = (0, 0, l_0/2)$, $\mathbf{A}_1\gamma_2 = (\hat{\lambda}_2, 0, 2^i\zeta_2)$, and $\mathbf{A}_1\gamma_3 = (\hat{\lambda}_3, \hat{\xi}_3, 2^i\zeta_3)$. Now, define

$$b_1 = -\frac{(2^i\zeta_2 + 2^{-1}l_0)}{\hat{\lambda}_2}, \quad b_2 = \frac{2^i(\zeta_2\hat{\lambda}_3 - \hat{\lambda}_2\zeta_3) + 2^{-1}l_0(\hat{\lambda}_3 - \hat{\lambda}_2)}{\hat{\lambda}_2\hat{\xi}_3}, \tag{5.83}$$

and let

$$\mathbf{A}_2 := \begin{pmatrix} 1 & 0 & 0 \\ 0 & 1 & 0 \\ b_1 & b_2 & 1 \end{pmatrix}.$$

Then a straightforward calculation shows that

$$\mathbf{B}_{e,i} := \mathbf{A}_2\mathbf{A}_1 = \begin{pmatrix} \kappa_e^{-i} & 0 & 0 \\ 0 & \kappa_e^{-i} & 0 \\ b_1\kappa_e^{-i} & b_2\kappa_e^{-i} & 2^i \end{pmatrix}$$

maps T to $\hat{T}$. Meanwhile, by Lemma 5.6, we have $|\zeta_2|, |\zeta_3| \le C|\gamma_0\gamma_1| = C2^{-i}l_0$, where C depends on the shape regularity of $T_{(0)}$. In addition, since $\hat{\lambda}_2, \hat{\lambda}_3, \hat{\xi}_3$ all depend on the shape regularity of $T_{(0)}$ and $\hat{\lambda}_2, \hat{\xi}_3 \neq 0$, by (5.83), we conclude $|b_1|, |b_2| \le C_0$, where $C_0 \ge 0$ depends on $T_{(0)}$ but not on i. □

Recall the parent and child tetrahedra associated with each mesh refinement in Remark 5.8. Note that for a v_e-tetrahedron $T_{(i)} \subset T_{(0)}$ in $\mathcal{T}_i$, its parent tetrahedron, which is in $\mathcal{T}_{i-1}$, can be either a v_e-tetrahedron or an e-tetrahedron. Nevertheless, there exists a v_e-tetrahedron $T_{(k)} \in \mathcal{T}_k$, $1 \le k \le i$, such that $T_{(i)} \subset T_{(k)} \subset T_{(0)}$ and $T_{(k)}$'s parent tetrahedron is an e-tetrahedron in $\mathcal{T}_{k-1}$.

Next, we construct the mapping between a v_e-tetrahedron in $\mathcal{T}_i$ and the reference domain. Recall the triangulations $\hat{\mathcal{T}}_1$ and $\hat{\mathcal{T}}_2$ of $\hat{T}$ in Definition 5.6.

Lemma 5.8 *Let $T_{(i)} \subset T_{(0)}$ be a v_e-tetrahedron in $\mathcal{T}_i$, $1 \le i \le n$. Let $T_{(k)} \in \mathcal{T}_k$, $1 \le k \le i$, be the v_e-tetrahedron, such that $T_{(i)} \subset T_{(k)}$ and $T_{(k)}$'s parent tetrahedron $T_{(k-1)} = \triangle^4\gamma_0\gamma_1\gamma_2\gamma_3 \in \mathcal{T}_{k-1}$ is an e-tetrahedron. On $T_{(k-1)}$, we use the same local coordinate system as in Lemma 5.7 with origin at $(\gamma_0 + \gamma_1)/2$. Then there is a transformation*

$$\mathbf{B}_{i,k} = \begin{pmatrix} \kappa_e^{-i+1} & 0 & 0 \\ 0 & \kappa_e^{-i+1} & 0 \\ b_1\kappa_e^{-i+1} & b_2\kappa_e^{-i+1} & 2^{k-1}\kappa_e^{k-i} \end{pmatrix} \tag{5.84}$$

that maps $T_{(i)}$ to a v_e-tetrahedron in $\hat{\mathcal{T}}_1$, where $|b_1|, |b_2| \le C_0$, for $C_0 > 0$ depending on $T_{(0)}$ but not on i or k.

Proof Based on Algorithm 5.7, the origin $(\gamma_0 + \gamma_1)/2$ is the vertex of $T_{(i)}$ on the singular edge. Then the linear mapping

$$\mathbf{A}_1 = \begin{pmatrix} \kappa_e^{k-i} & 0 & 0 \\ 0 & \kappa_e^{k-i} & 0 \\ 0 & 0 & \kappa_e^{k-i} \end{pmatrix}$$

translates $T_{(i)}$ to $T_{(k)}$. Since $T_{(k-1)}$ is an e-tetrahedron, by Lemma 5.7, the transformation

$$\mathbf{A}_2 = \begin{pmatrix} \kappa_e^{-k+1} & 0 & 0 \\ 0 & \kappa_e^{-k+1} & 0 \\ b_1\kappa_e^{-k+1} & b_2\kappa_e^{-k+1} & 2^{k-1} \end{pmatrix}$$

maps $T_{(k-1)}$ to $\hat{T}$ and also maps the restriction of $\mathcal{T}_k$ on $T_{(k-1)}$ to $\hat{\mathcal{T}}_1$, where $|b_1|, |b_2| < C_0$ for C_0 depending on $T_{(0)}$. Therefore,

$$\mathbf{B}_{i,k} := \mathbf{A}_2\mathbf{A}_1 = \begin{pmatrix} \kappa_e^{-i+1} & 0 & 0 \\ 0 & \kappa_e^{-i+1} & 0 \\ b_1\kappa_e^{-i+1} & b_2\kappa_e^{-i+1} & 2^{k-1}\kappa_e^{k-i} \end{pmatrix}$$

maps $T_{(i)}$ to one of the v_e-tetrahedra in $\hat{\mathcal{T}}_1$. This completes the proof. □

The parameters b_1 and b_2 in Lemmas 5.7 and 5.8 can be different for different tetrahedra in $\mathcal{T}_i$, but they are uniformly bounded by a constant that depends on the initial e-tetrahedron $T_{(0)}$. Now, we are ready to construct the mapping from a tetrahedron $T_{(i+1)} \in \mathcal{T}_{i+1}$ in the mesh layer $L_{e,i}$ (Definition 5.4) to the reference domain. Also recall that $T_{(i+1)}$ is a tetrahedron in the initial triangulation of $L_{e,i}$.

Lemma 5.9 *Let $T_{(i+1)} \in \mathcal{T}_{i+1}$ be a tetrahedron, such that $T_{(i+1)} \subset L_{e,i} \subset T_{(0)}$, $0 \leq i < n$.*
Case I: $T_{(i+1)}$ is a child tetrahedron of an e-tetrahedron $T_{(i)} \in \mathcal{T}_i$. Using the $T_{(i)}$-based local coordinate system as in Lemma 5.7, the transformation

$$\mathbf{B}_{e,i} = \begin{pmatrix} \kappa_e^{-i} & 0 & 0 \\ 0 & \kappa_e^{-i} & 0 \\ b_1\kappa_e^{-i} & b_2\kappa_e^{-i} & 2^{i} \end{pmatrix} \tag{5.85}$$

maps $T_{(i+1)}$ to some o-tetrahedron in $\hat{\mathcal{T}}_1$.
Case II: $T_{(i+1)}$ is a child tetrahedron of a v_e-tetrahedron $T_{(i)} \in \mathcal{T}_i$. Let $T_{(k)} \in \mathcal{T}_k$, $1 \leq k \leq i$, be the v_e-tetrahedron, such that $T_{(i)} \subset T_{(k)}$ and $T_{(k)}$'s parent tetrahedron $T_{(k-1)} \in \mathcal{T}_{k-1}$ is an e-tetrahedron. Using the $T_{(k-1)}$-based local coordinate system as in Lemma 5.8, the transformation

$$\mathbf{B}_{i,k} = \begin{pmatrix} \kappa_e^{-i+1} & 0 & 0 \\ 0 & \kappa_e^{-i+1} & 0 \\ b_1\kappa_e^{-i+1} & b_2\kappa_e^{-i+1} & 2^{k-1}\kappa_e^{k-i} \end{pmatrix} \tag{5.86}$$

maps $T_{(i+1)}$ to an o-tetrahedron in $\hat{\mathcal{T}}_2$. In both cases, $|b_1|, |b_2| \le C_0$, for $C_0 > 0$ depending on $T_{(0)}$ but not on i or k.

Proof If $T_{(i+1)}$ is a child tetrahedron of an e-tetrahedron $T_{(i)} \in \mathcal{T}_i$, the matrix in (5.82) maps $T_{(i)}$ to $\hat{T}$ and maps $P_{e,i+1} \cap T_{(i)}$ to $\hat{P}_1$, where $P_{e,i+1}$ is the parallelogram cutting $T_{(0)}$ in the $i+1$st refinement (Definition 5.4) and $\hat{P}_1$ is the parallelogram cutting $\hat{T}$ in the first edge refinement (see Fig. 5.5). Consequently, $T_{(i+1)}$ is translated to one of the four o-tetrahedra in $\hat{\mathcal{T}}_1$ by the same mapping.

For Case II, the transformation (5.84) maps $T_{(i)}$ to a v_e-tetrahedron in $\hat{\mathcal{T}}_1$. In addition, it maps $P_{e,i} \cap T_{(i)}$ to $\hat{P}_1$, and $P_{e,i+1} \cap T_{(i)}$ to $\hat{P}_2$ (see Fig. 5.5). Therefore, the same transformation maps $T_{(i+1)}$ to an o-tetrahedron in $\hat{\mathcal{T}}_2$ between $\hat{P}_1$ and $\hat{P}_2$. This completes the proof. □

Mesh Layers on Initial ev-Tetrahedra in $\mathcal{T}_0$ Denote by $T_{(0)} = \Delta^4 x_0x_1x_2x_3 \in \mathcal{T}_0$ an ev-tetrahedron, such that $x_0 = c \in \mathcal{C}$ and x_0x_1 is on the edge $e \in \mathcal{E}$. We define the mesh layer associated with $\mathcal{T}_n$ on $T_{(0)}$ as follows.

Definition 5.7 (Mesh Layer in ev-Tetrahedra) For $1 \le i \le n$, the ith refinement on $T_{(0)}$ produces a small tetrahedron with x_0 as a vertex. We denote by $P_{ev,i}$ the face of this small tetrahedron whose closure does not contain x_0 (see the last two pictures in Fig. 5.2). Then for the mesh $\mathcal{T}_n$ on $T_{(0)}$, we define the ith mesh layer $L_{ev,i}$, $1 \le i < n$, as the region in $T_{(0)}$ between $P_{ev,i}$ and $P_{ev,i+1}$. We define $L_{ev,0}$ to be the region in $T_{(0)}$ between $\Delta^3 x_1x_2x_3$ and $P_{ev,1}$ and let $L_{ev,n} \subset T_{(0)}$ be the small tetrahedron with x_0 as a vertex that is produced in the nth refinement.

For each ev-tetrahedron, one extra refinement results in one ev-tetrahedron, one e-tetrahedron, two v_e-tetrahedra, and four o-tetrahedra. Let $T = \Delta^4\gamma_0\gamma_1\gamma_2\gamma_3 \subset T_{(0)}$ be an ev-tetrahedron produced by some subsequent refinements of $T_{(0)}$, with $\gamma_0 = x_0$ and $\gamma_0\gamma_1$ on the edge $e \in \mathcal{E}$. We define the relative distances $c_{\gamma,1}$ and $c_{\gamma,2}$ for T using the same notation as in Definition 5.5 (see also Remark 5.8). In the next lemma, we show the analogue of Lemma 5.6 for ev-tetrahedra. Namely, the relative distances are bounded for ev-tetrahedra with respect to the refinement.

Lemma 5.10 *Let $T = \Delta^4\gamma_0\gamma_1\gamma_2\gamma_3 \subset T_{(0)}$ be an ev-tetrahedron in $\mathcal{T}_i$, $1 \le i < n$, with $\gamma_0 = x_0$ and $\gamma_0\gamma_1$ on the edge $e \in \mathcal{E}$. Let $T_R \subset T$ be the ev-tetrahedron in $\mathcal{T}_{i+1}$. Denote by c_T and c_R the absolute distances (5.81) for T and T_R, respectively. Then $c_R \le \max\{c_T, 1\}$.*

Proof Recall the grading parameters κ_{sc}, κ_e, and κ_{ec} for $T_{(0)}$ with $\kappa_{sc}, \kappa_e \ge \kappa_{ec}$. We use Fig. 5.4 to demonstrate the proof. Then $T_R = \Delta^4\gamma_0\gamma_4\gamma_5\gamma_6$. Consider the triangles on the face $\Delta^3\gamma_0\gamma_1\gamma_2$ of T, induced by the sub-tetrahedra after one refinement on T, where γ_5' and γ_2' are the orthogonal projections of γ_5 and γ_2 on the singular edge. However, note that instead of the mid-point of $\gamma_0\gamma_1$ for the e-tetrahedron, the location of γ_4 here is given by $|\gamma_0\gamma_4| = \kappa_{sc}|\gamma_0\gamma_1|$ for the ev-tetrahedron. Let $c_{\gamma_2,1}, c_{\gamma_2,2}$ (resp. $c^R_{\gamma_5,1}, c^R_{\gamma_5,2}$) be the relative distances of γ_2 in T (resp. γ_5 in T_R).

Based on Algorithm 5.7, $|\gamma_0\gamma_5'| = \kappa_{ec}|\gamma_0\gamma_2'|$. By (5.79), $c^R_{\gamma_5,1}$ and $c_{\gamma_2,1}$ have the same sign. We first show $|c^R_{\gamma_5,1}|, |c^R_{\gamma_5,2}| \leq \max\{|c_{\gamma_2,1}|, |c_{\gamma_2,2}|, 1\}$ by considering the following cases, in which the calculations are based on the definitions in (5.79) and (5.80).

If $c_{\gamma_2,1} < 0$, we have

$$0 > c^R_{\gamma_5,1} = -|\gamma_0\gamma_5'|/|\gamma_0\gamma_4| = -\kappa_{sc}^{-1}\kappa_{ec}|\gamma_0\gamma_2'|/|\gamma_0\gamma_1| = \kappa_{sc}^{-1}\kappa_{ec}c_{\gamma_2,1} \geq c_{\gamma_2,1}.$$

Therefore, $|c^R_{\gamma_5,1}| \leq |c_{\gamma_2,1}|$. Meanwhile, we have

$$1 \leq c^R_{\gamma_5,2} = 1 - c^R_{\gamma_5,1} = 1 - \kappa_{sc}^{-1}\kappa_{ec}c_{\gamma_2,1} < 1 - c_{\gamma_2,1} = c_{\gamma_2,2}.$$

Therefore, $|c^R_{\gamma_5,2}| \leq |c_{\gamma_2,2}|$.

If $0 \leq c_{\gamma_2,1} < \kappa_{sc}\kappa_{ec}^{-1}$, we have $c^R_{\gamma_5,1} \geq 0$ and

$$c^R_{\gamma_5,1} = |\gamma_0\gamma_5'|/|\gamma_0\gamma_4| = \kappa_{sc}^{-1}\kappa_{ec}|\gamma_0\gamma_2'|/|\gamma_0\gamma_4| = \kappa_{sc}^{-1}\kappa_{ec}c_{\gamma_2,1} < 1.$$

Meanwhile, we have

$$0 \leq c^R_{\gamma_5,2} = 1 - c^R_{\gamma_5,1} \leq 1.$$

If $c_{\gamma_2,1} \geq \kappa_{sc}\kappa_{ec}^{-1}$, we have

$$1 \leq c^R_{\gamma_5,1} = |\gamma_0\gamma_5'|/|\gamma_0\gamma_4| = \kappa_{sc}^{-1}\kappa_{ec}|\gamma_0\gamma_2'|/|\gamma_0\gamma_4| = \kappa_{sc}^{-1}\kappa_{ec}c_{\gamma_2,1} \leq |c_{\gamma_2,1}|.$$

Meanwhile, we have

$$0 \geq c^R_{\gamma_5,2} = 1 - c^R_{\gamma_5,1} = 1 - \kappa_{sc}^{-1}\kappa_{ec}c_{\gamma_2,1} \geq 1 - c_{\gamma_2,1} = c_{\gamma_2,2}.$$

Therefore, $|c^R_{\gamma_5,2}| \leq |c_{\gamma_2,2}|$.

Hence, $|c^R_{\gamma_5,1}|, |c^R_{\gamma_5,2}| \leq \max\{|c_{\gamma_2,1}|, |c_{\gamma_5,2}|, 1\}$. Using a similar calculation, we can also obtain the same estimate for relative distances of γ_3 and γ_6. Then the proof is completed by combining these estimates and by the definition of the absolute distance (5.81). □

Now, we define the reference element for the ev-tetrahedron.

Definition 5.8 (Reference ev-Tetrahedron) We use the tetrahedron $\hat{T} = \triangle^4\hat{x}_0\hat{x}_1\hat{x}_2\hat{x}_3$ in Definition 5.6 as our reference element. For $T_{(0)}$, recall the grading parameters κ_{ec}, κ_{sc}, and κ_e. For the reference ev-tetrahedron $\hat{T}$, one graded refinement using the same parameters κ_{sc}, κ_e, and κ_{ec} for $\hat{x}_0$ and $\hat{x}_0\hat{x}_1$ gives rise to a triangulation on $\hat{T}$, which we denote by $\hat{\mathcal{T}}_1$. Define the union of the seven tetrahedra in $\hat{\mathcal{T}}_1$ away from $\hat{x}_0$ to be the mesh layer $\hat{L}$ on $\hat{T}$. We denote by $\hat{\mathcal{L}}$ the initial triangulation of $\hat{L}$ that contains these seven tetrahedra.

Then we construct a mapping from an ev-tetrahedron $T \subset T_{(0)}$ in $\mathcal{T}_i$ to $\hat{T}$.

Lemma 5.11 *For an ev-tetrahedron $T := \Delta^4\gamma_0\gamma_1\gamma_2\gamma_3 \subset T_{(0)}$ in $\mathcal{T}_i$, $0 \le i \le n$, suppose $\gamma_0 = c \in C$ and $\gamma_0\gamma_1 \subset e \in \mathcal{E}$. Use a local Cartesian coordinate system, such that $(\gamma_0 + \gamma_1)/2$ is the origin, γ_1 is in the positive z-axis, and γ_2 is in the xz-plane. Then there is a mapping*

$$\mathbf{B}_{ev,i} = \begin{pmatrix} \kappa_{ec}^{-i} & 0 & 0 \\ 0 & \kappa_{ec}^{-i} & 0 \\ b_1\kappa_{ec}^{-i} & b_2\kappa_{ec}^{-i} & \kappa_{sc}^{-i} \end{pmatrix} \tag{5.87}$$

with $|b_1|, |b_2| \le C_0$, for $C_0 \ge 0$ depending on $T_{(0)}$, such that $\mathbf{B}_{ev,i} : T \to \hat{T}$ is a bijection.

Proof Recall $\hat{\lambda}_k$ and $\hat{\xi}_k$, $k = 2, 3$, in Definition 5.6. Based on Algorithm 5.7, we have $\gamma_2 = (\kappa_{ec}^i\hat{\lambda}_2, 0, \zeta_2)$, $\gamma_3 = (\kappa_{ec}^i\hat{\lambda}_3, \kappa_{ec}^i\hat{\xi}_3, \zeta_3)$, and $|\gamma_0\gamma_1| = \kappa_{sc}^i l_0 = \kappa_{sc}^i|x_0x_1|$, where ζ_2 and ζ_3 are the z-coordinates of the vertices γ_2 and γ_3, respectively. Then the transformation

$$\mathbf{A}_1 := \begin{pmatrix} \kappa_{ec}^{-i} & 0 & 0 \\ 0 & \kappa_{ec}^{-i} & 0 \\ 0 & 0 & \kappa_{sc}^{-i} \end{pmatrix}$$

maps T to a tetrahedron with vertices $\mathbf{A}_1\gamma_0 = (0, 0, -l_0/2)$, $\mathbf{A}_1\gamma_1 = (0, 0, l_0/2)$, $\mathbf{A}_1\gamma_2 = (\hat{\lambda}_2, 0, \kappa_{sc}^{-i}\zeta_2)$, and $\mathbf{A}_1\gamma_3 = (\hat{\lambda}_3, \hat{\xi}_3, \kappa_{sc}^{-i}\zeta_3)$. Let

$$b_1 = -(\kappa_{sc}^{-i}\zeta_2 + l_0/2)/\hat{\lambda}_2, \qquad b_2 = \Big(\kappa_{sc}^{-i}(\zeta_2\hat{\lambda}_3 - \hat{\lambda}_2\zeta_3) + l_0(\hat{\lambda}_3 - \hat{\lambda}_2)/2\Big)/\hat{\lambda}_2\hat{\xi}_3.$$

Define

$$\mathbf{A}_2 := \begin{pmatrix} 1 & 0 & 0 \\ 0 & 1 & 0 \\ b_1 & b_2 & 1 \end{pmatrix}.$$

By Lemma 5.10, the absolute distance for T is bounded by a constant determined by $T_{(0)}$. Therefore, we have $|\zeta_2|, |\zeta_3| \le C|\gamma_0\gamma_1| = C\kappa_{sc}^i l_0$, where C depends on the shape regularity of $T_{(0)}$. In addition, since $\hat{\lambda}_2, \hat{\lambda}_3, \hat{\xi}_3$ all depend on the shape regularity of $T^{(0)}$ and $\hat{\lambda}_2, \hat{\xi}_3 \ne 0$, the transformation

$$\mathbf{B}_{ev,i} := \mathbf{A}_2\mathbf{A}_1 = \begin{pmatrix} \kappa_{ec}^{-i} & 0 & 0 \\ 0 & \kappa_{ec}^{-i} & 0 \\ b_1\kappa_{ec}^{-i} & b_2\kappa_{ec}^{-i} & \kappa_{sc}^{-i} \end{pmatrix}$$

maps T to $\hat{T}$ with $|b_1|, |b_2| \le C_0$, where $C_0 \ge 0$ depends on $T_{(0)}$ but not on i. This completes the proof. □

Chapter 6
Anisotropic Error Estimates in Polyhedral Domains

In this chapter, we discuss the error estimate for the finite element method associated with the 3D anisotropic graded meshes in polyhedral domains. In particular, we focus on three scenarios: 1. the given data belonging to a high-order weighted space ($f \in \mathcal{M}^{m+1}_{\mu-1}(\Omega)$), 2. rough given data ($f \in L^2_{\mu^*}(\Omega)$), and 3. the mixed boundary condition with special attention to the case where the Neumann condition is imposed on adjacent faces of the boundary. Based on the error analysis, we formulate optimal graded mesh algorithms such that the finite element solution approximates the singular solution in the optimal convergence rate. Numerical examples are presented to demonstrate the effectiveness of the method.

Consider the following problem in a bounded polyhedral domain $\Omega \subset \mathbb{R}^3$ with the mixed boundary condition

$$-\Delta u = f \quad \text{in} \quad \Omega, \qquad u = 0 \quad \text{on} \quad \Gamma_D \qquad \partial_n u = 0 \quad \text{on} \quad \Gamma_N, \tag{6.1}$$

where Γ_D and Γ_N are open subsets of the boundary $\Gamma := \partial\Omega$ such that $\overline{\Gamma_D} \cup \overline{\Gamma_N} = \Gamma$. We suppose that each face of Γ is included either in Γ_D or in Γ_N and $\Gamma_D \neq \emptyset$. Note that this is the same model problem considered in Chap. 5.

The variational solution $u \in H^1_D(\Omega)$ (see (2.26)) of (6.1) satisfies

$$a(u, v) = \int_\Omega \nabla u \cdot \nabla v \, dx = \int_\Omega f v \, dx = (f, v), \quad \forall\, v \in H^1_D(\Omega).$$

Let $\mathcal{T}_n$ be a triangulation of Ω with tetrahedra. Let $S_n \subset H^1_0(\Omega)$ be the Lagrange finite element space of degree $m \geq 1$ associated with $\mathcal{T}_n$. Namely,

$$S_n = \{v \in C(\bar{\Omega}) : \ v|_{\Gamma_D} = 0, \ v|_T \in P_m(T) \text{ for any tetrahedron } T \in \mathcal{T}_n\},$$

where $P_m(T)$ is the space of polynomials of degree $\leq m$ on T. Then the finite element solution $u_n \in S_n$ for (6.1) is defined by

$$a(u_n, v_n) = (f, v_n), \quad \forall\, v_n \in S_n. \tag{6.2}$$

H. Li, *Graded Finite Element Methods for Elliptic Problems in Nonsmooth Domains*,
Surveys and Tutorials in the Applied Mathematical Sciences 10,
https://doi.org/10.1007/978-3-031-05821-9_6

Remark 6.1 By the Poincaré inequality, the bilinear form $a(\cdot, \cdot)$ is both continuous and coercive on $H^1_D(\Omega)$. The solution of (6.1) is uniquely defined in $H^1_D(\Omega)$ for $f \in (H^1_D(\Omega))^*$. Thus, the Céa Theorem (Theorem 2.19) gives rise to

$$\|u - u_n\|_{H^1(\Omega)} \leq \inf_{v_n \in S_n} \|u - v_n\|_{H^1(\Omega)}. \tag{6.3}$$

On a standard quasi-uniform triangulation $\mathcal{T}_n$, similar to the 2D scenario, the limited regularity of u in the usual Sobolev space may result in a *sub-optimal* convergence rate for the finite element approximation. Namely,

$$\|u - u_n\|_{H^1(\Omega)} \leq Ch^s \|u\|_{H^{s+1}(\Omega)}, \tag{6.4}$$

where h is the mesh size in $\mathcal{T}_n$ and $0 < s < m$ depends on the geometry of the domain. See the discussions in Chaps. 3 and 5 on the solution regularity.

We proceed to derive the finite element interpolation error estimates on the graded meshes in Algorithm 5.7. These estimates shall lead to grading parameters such that the associated finite element methods approximate the singular solution in the optimal convergence rate. To simplify the exposition, for an element in the set $\mathcal{S} = \mathcal{C} \cup \mathcal{E}$ of nonsmooth points, we define below another set of parameters corresponding to κ_c, κ_e, κ_{sc}, and κ_{ec}. When $c \in \mathcal{S}$ is a vertex, let a_c be such that

$$\kappa_c = 2^{-m/a_c}. \tag{6.5}$$

When $e \in \mathcal{S}$ is an edge, let a_e be such that

$$\kappa_e = 2^{-m/a_e}. \tag{6.6}$$

Meanwhile, we define the constant $a_{ec} = \min_{e \in \mathcal{E}_c} \{a_c, a_e\}$. Therefore,

$$\kappa_{ec} = 2^{-m/a_{ec}}. \tag{6.7}$$

In an ev-tetrahedron, suppose $pq \subset \bar{e} \in \mathcal{E}_c$ is the singular edge with $p = c \in \mathcal{C}$. For pq, define $a_{sc} = \min_{e_r \in \mathcal{E}_c \setminus \{e\}} \{a_c, a_{e_r}\}$. Thus

$$\kappa_{sc} = 2^{-m/a_{sc}}. \tag{6.8}$$

Then denote by $\boldsymbol{a}$ the collection of the parameters a_c and a_e for all $c \in \mathcal{C}$ and $e \in \mathcal{E}$.

As in Algorithm 5.7, let $\mathcal{T}_0$ be an initial triangulation of the domain Ω with tetrahedra that satisfy the condition in Definition 3.1. Recall $\mathcal{T}_n$ is the mesh obtained after n successive refinements based on the parameter $\boldsymbol{\kappa}$. Throughout this chapter, we let $h := 2^{-n}$ be the mesh parameter of $\mathcal{T}_n$. For a continuous function v, we let v_I be its Lagrange nodal interpolation associated with the underlying mesh.

Note that the tetrahedra in the initial mesh $\mathcal{T}_0 = \{T_{(0),j}\}_{j=1}^J$ are all shape regular and can be classified into five categories (Definition 3.2). Thus, with the triangulation $\mathcal{T}_n$, the interpolation error estimate on Ω is based on the interpolation error estimates on different sub-regions of Ω, each of which is represented by an initial tetrahedron $T_{(0),j} \in \mathcal{T}_0$.

6.1 Interpolation Error Estimates for $u \in \mathcal{M}^{m+1}_{\mu+1}(\Omega)$

Assume the Dirichlet boundary condition $\overline{\Gamma_D} = \partial\Omega$ in (6.1) and assume $f \in \mathcal{M}^{m+1}_{\mu-1}(\Omega)$ for $0 \leq \mu < \eta$, where η is given in (5.5). According to the regularity result (5.6), one has $u \in \mathcal{M}^{m+1}_{\mu+1}(\Omega)$.

Based on the definition, the space $\mathcal{M}^{m+1}_{\mu+1}$ $(m \geq 1)$ is equivalent to the Sobolev space H^{m+1} in any sub-region of Ω that is away from the singular set $\mathcal{S}$. Therefore, by the Sobolev embedding theorem, $u \in \mathcal{M}^{m+1}_{\mu+1}(\Omega)$ is continuous at each nodal point in the interior of the domain. On the boundary of the domain, we set $u_I = 0$ due to the boundary condition. This makes the interpolation u_I well defined. In addition, the parameters in Algorithm 5.7 are chosen as follows.

Algorithm 6.1 (Anisotropic Algorithm for $u \in \mathcal{M}^{m+1}_{\mu+1}(\Omega)$) For (6.1) with $u \in H^1_0(\Omega) \cap \mathcal{M}^{m+1}_{\mu+1}(\Omega)$, we choose the anisotropic mesh $\mathcal{T}_n$ (Algorithm 5.7) with the grading parameters (6.5)–(6.8) such that

$$0 < a_e \leq \min\{m, \mu_e\}; \tag{6.9}$$

$$0 < a_c \leq m \quad \text{and} \quad a_C := (m+1)(1 - a_c^{-1} a_{ec}) + a_{ec} \leq \mu_c, \tag{6.10}$$

where m is the degree of the piecewise polynomial space in S_n, and μ_e, μ_c are the components of μ in (5.6). Let $u_n \in S_n$ be the finite element solution (6.2). □

Recall that a vertex or an edge in the triangulation is singular when the associated grading parameter is less than 0.5, and the meshes near regular vertices and edges are quasi-uniform. In particular, we have the following observation.

Proposition 6.1 *Let $u \in \mathcal{M}^{m+1}_{\mu+1}(\Omega)$ be the solution of (6.1). If Algorithm 6.1 gives rise to the grading parameter $\kappa_c = 0.5$ for a vertex $c \in \mathcal{C}$ (resp. $\kappa_e = 0.5$ for an edge $e \in \mathcal{E}$), then we have $u \in H^{m+1}(\mathcal{V}_c^o)$ (resp. $u \in H^{m+1}(\mathcal{V}_e^o)$). Furthermore, if $c \in \bar{e}$ and $\kappa_c = \kappa_e = 0.5$ in Algorithm 6.1, then $u \in H^{m+1}(\mathcal{V}_c^e)$.*

Proof We shall show the case for an edge $e \in \mathcal{E}$. The cases for a vertex $c \in \mathcal{C}$ and for $c \in \bar{e}$ can be proved similarly. In fact, given $\kappa_e = 0.5$, by (6.6) and (6.9), we have $a_e = m$ and $\mu_e \geq m$. Therefore, by the definition of the weighted space (5.4), for $u \in \mathcal{M}^{m+1}_{\mu+1}(\mathcal{V}_e^o)$, we have

$$\begin{aligned}
\|u\|^2_{H^{m+1}(\mathcal{V}_e^o)} &= \sum_{|\alpha| \leq m+1} \|\partial^\alpha u\|^2_{L^2(\mathcal{V}_e^o)} \\
&\leq C \sum_{|\alpha| \leq m+1} \|r_e^{|\alpha_\perp| - \mu_e - 1} \partial^\alpha u\|^2_{L^2(\mathcal{V}_e^o)} \leq C\|u\|^2_{\mathcal{M}^{m+1}_{\mu+1}(\mathcal{V}_e^o)},
\end{aligned}$$

which completes the proof. □

Note that Proposition 6.1 also holds for $u \in \mathcal{M}^{m+1}_{\boldsymbol{a}+1}(\Omega)$, where $\boldsymbol{a}$ is given by (6.9)–(6.10). Now, we proceed to present interpolation error estimates for the finite element method given in Algorithm 6.1 on initial tetrahedra of different types.

6.1.1 Estimates on Initial o-, v-, and v_e-Tetrahedra in $\mathcal{T}_0$

We first have the estimate for an o-tetrahedron in the initial mesh.

Lemma 6.1 *Let $T_{(0)} \in \mathcal{T}_0$ be an o-tetrahedron. For $u \in \mathcal{M}^{m+1}_{\boldsymbol{a}+1}(\Omega)$, where $\boldsymbol{a}$ is given in (6.9) – (6.10), let u_I be its nodal interpolation on $\mathcal{T}_n$. Then we have*

$$|u - u_I|_{H^1(T_{(0)})} \leq Ch^m \|u\|_{\mathcal{M}^{m+1}_{\boldsymbol{a}+1}(T_{(0)})},$$

where $h = 2^{-n}$ and C is independent of n and u.

Proof Based on Algorithm 5.7, the restriction of $\mathcal{T}_n$ on $T_{(0)}$ is a quasi-uniform mesh with size $O(2^{-n})$. Note that $T_{(0)}$ can be either in the interior of the domain or an initial tetrahedron that includes regular vertices and edges on the boundary Γ. In either case, by the definition of the weighted space and Proposition 6.1, we have $\|u\|_{H^{m+1}(T_{(0)})} \leq C\|u\|_{\mathcal{M}^{m+1}_{\boldsymbol{a}+1}(T_{(0)})}$. Then by the standard interpolation error estimate (Lemma 2.5), we have

$$|u - u_I|_{H^1(T_{(0)})} \leq C2^{-nm}\|u\|_{H^{m+1}(T_{(0)})} \leq Ch^m\|u\|_{\mathcal{M}^{m+1}_{\boldsymbol{a}+1}(T_{(0)})}.$$

This completes the proof. □

For a v- or v_e-tetrahedron in $\mathcal{T}_0$, recall the mesh layers $L_{v,i}$ in Definition 5.3. Then we have the interpolation error estimate in $L_{v,i}$.

Lemma 6.2 *Let $T_{(0)} \in \mathcal{T}_0$ be either a v- or a v_e-tetrahedron. For $u \in \mathcal{M}^{m+1}_{\boldsymbol{a}+1}(\Omega)$, where $\boldsymbol{a}$ is given in (6.9)–(6.10), let u_I be its nodal interpolation on $\mathcal{T}_n$. Then for $0 \leq i < n$, we have*

$$|u - u_I|_{H^1(L_{v,i})} \leq Ch^m\|u\|_{\mathcal{M}^{m+1}_{\boldsymbol{a}+1}(L_{v,i})},$$

where $h = 2^{-n}$ and C is independent of n and u.

Proof For $(x, y, z) \in L_{v,i}$, let $(\hat{x}, \hat{y}, \hat{z}) \in L_{v,0}$ be its image under the dilation $\mathbf{B}_{v,i}$ (5.76). For a function v on $L_{v,i}$, we define $\hat{v}$ on $L_{v,0}$ by $\hat{v}(\hat{x}, \hat{y}, \hat{z}) := v(x, y, z)$. As part of $\mathcal{T}_n$, the triangulation on $L_{v,i}$ is mapped by $\mathbf{B}_{v,i}$ to a triangulation on $L_{v,0}$ with mesh size $O(2^{i-n})$. Then by the scaling argument and (5.77), we have

$$\begin{aligned}|u - u_I|^2_{H^1(L_{v,i})} &= \kappa^i|\hat{u} - \hat{u}_I|^2_{H^1(L_{v,0})} \leq C\kappa^i 2^{2m(i-n)}|\hat{u}|^2_{H^{m+1}(L_{v,0})}\\ &\leq C2^{2m(i-n)}\kappa^{2mi}|u|^2_{H^{m+1}(L_{v,i})} \leq C2^{2m(i-n)}\kappa^{2ai}\sum_{|\alpha|=m+1}\|r^{m-a}\partial^\alpha u\|^2_{L^2(L_{v,i})},\end{aligned}$$

where r and κ are defined as in Remark 5.5. Recall $\kappa = 2^{-m/a}$, where $a = a_{ec} \leq a_c$ or $a = a_e$ as in (6.6)–(6.7) depending on the type of $T_{(0)}$. Then by the definition of the weighted space and Proposition 6.1, we have

$$|u - u_I|^2_{H^1(L_{v,i})} \leq C2^{2m(i-n)}\kappa^{2ai} \sum_{|\alpha|=m+1} \|r^{m-a}\partial^\alpha u\|^2_{L^2(L_{v,i})} \leq Ch^{2m}\|u\|^2_{\mathcal{M}^{m+1}_{a+1}(L_{v,i})},$$

which completes the proof. □

Now, we give the error estimate on the entire initial tetrahedron $T_{(0)}$.

Corollary 6.1 *Let $T_{(0)} \in \mathcal{T}_0$ be either a v- or a v_e-tetrahedron. For $u \in \mathcal{M}^{m+1}_{\boldsymbol{a}+1}(\Omega)$, where $\boldsymbol{a}$ is given in (6.9)–(6.10), let u_I be its nodal interpolation on $\mathcal{T}_n$. Then*

$$|u - u_I|_{H^1(T_{(0)})} \leq Ch^m\|u\|_{\mathcal{M}^{m+1}_{\boldsymbol{a}+1}(T_{(0)})},$$

where $h = 2^{-n}$ and C is independent of n and u.

Proof By Lemma 6.2, it suffices to show the estimate for the last layer $L_{v,n}$. For $(x, y, z) \in L_{v,n}$, let $(\hat{x}, \hat{y}, \hat{z}) \in T_{(0)}$ be its image under the dilation $\mathbf{B}_{v,n}$. For a function v on $L_{v,n}$, we define $\hat{v}$ on $T_{(0)}$ by $\hat{v}(\hat{x}, \hat{y}, \hat{z}) := v(x, y, z)$. Suppose x_0 is the singular vertex of $T_{(0)}$. Now let χ be a smooth cut-off function on $T_{(0)}$ such that $\chi = 0$ in a neighborhood of x_0 and $= 1$ at every other node of $T_{(0)}$. Recall the distance function r from Remark 5.5. Thus, $r(\hat{x}, \hat{y}, \hat{z}) = \kappa^{-n}r(x, y, z)$. Since $\chi\hat{u} = 0$ in the neighborhood of x_0, we have

$$|\chi\hat{u}|^2_{H^{m+1}(T_{(0)})} \leq C \sum_{|\alpha|\leq m+1} \|r^{|\alpha|-1}\partial^\alpha \hat{u}\|^2_{L^2(T_{(0)})}.$$

Define $\hat{w} := \hat{u} - \chi\hat{u}$ and note that $(\chi\hat{u})_I = \hat{u}_I$. We have

$$\begin{aligned} |\hat{u} - \hat{u}_I|_{H^1(T_{(0)})} &= |\hat{w} + \chi\hat{u} - \hat{u}_I|_{H^1(T_{(0)})} \leq |\hat{w}|_{H^1(T_{(0)})} + |\chi\hat{u} - \hat{u}_I|_{H^1(T_{(0)})} \\ &= |\hat{w}|_{H^1(T_{(0)})} + |\chi\hat{u} - (\chi\hat{u})_I|_{H^1(T_{(0)})} \leq C\big(\|\hat{u}\|_{H^1(T_{(0)})} + |\chi\hat{u}|_{H^{m+1}(T_{(0)})}\big), \end{aligned} \quad (6.11)$$

where C depends on m and, through χ, the nodes in the triangulation. Then using (6.11), the scaling argument, $\kappa^{-n} \lesssim r^{-1}$ in $L_{v,n}$, and the definition of the weighted space, we have

$$\begin{aligned} |u - u_I|^2_{H^1(L_{v,n})} &= \kappa^n|\hat{u} - \hat{u}_I|^2_{H^1(T_{(0)})} \\ &\leq C\kappa^n\Big(\|\hat{u}\|^2_{H^1(T_{(0)})} + \sum_{|\alpha|\leq m+1} \|r^{|\alpha|-1}\partial^\alpha\hat{u}\|^2_{L^2(T_{(0)})}\Big) \\ &\leq C \sum_{|\alpha|\leq m+1} \|r^{|\alpha|-1}\partial^\alpha u\|^2_{L^2(L_{v,n})} \leq C\kappa^{2na}\|u\|^2_{\mathcal{M}^{m+1}_{a+1}(L_{v,n})} \\ &= C2^{-2mn}\|u\|^2_{\mathcal{M}^{m+1}_{a+1}(L_{v,n})} = Ch^{2m}\|u\|^2_{\mathcal{M}^{m+1}_{a+1}(L_{v,n})}. \end{aligned}$$

Then the desired estimate follows by summing up the estimates from different layers $L_{v,i}$, $0 \leq i \leq n$. □

6.1.2 Estimates on Initial e-Tetrahedra in $\mathcal{T}_0$

On an e-tetrahedron $T_{(0)} = \Delta^4 x_0x_1x_2x_3 \in \mathcal{T}_0$, the graded mesh (Algorithm 5.7) is anisotropic and needs special attention in the error analysis. Recall the mesh layer $L_{e,i}$ in Definition 5.4. Note that each tetrahedron $T_{(i+1)} \in \mathcal{T}_{i+1}$ that belongs to layer $L_{e,i}$ falls into either Case I or Case II of Lemma 5.9. Thus, there is a linear transformation $\mathbf{B}$ (either $\mathbf{B}_{e,i}$ or $\mathbf{B}_{i,k}$) that maps $T_{(i+1)}$ to an o-tetrahedron in either $\hat{\mathcal{T}}_1$ or $\hat{\mathcal{T}}_2$. Denote this o-tetrahedron by $\hat{T}_{(i+1)}$. It is clear that $\hat{T}_{(i+1)}$ belongs to a finite number of similarity classes determined by the o-tetrahedra in $\hat{\mathcal{T}}_1$ and $\hat{\mathcal{T}}_2$. Then for $(x, y, z) \in T_{(i+1)}$, we have

$$(\hat{x}, \hat{y}, \hat{z}) = \mathbf{B}(x, y, z) \in \hat{T}_{(i+1)}. \tag{6.12}$$

For a function v on $T_{(i+1)}$, we define $\hat{v}(\hat{x}, \hat{y}, \hat{z}) := v(x, y, z)$. In addition, we assume the singular edge x_0x_1 of $T_{(0)}$ is on the z-axis.

In the $i + 1$st refinement, $0 \le i < n$, when the layer $L_{e,i}$ is formed, it only contains tetrahedra in $\mathcal{T}_{i+1}$. To obtain the mesh $\mathcal{T}_n$, these tetrahedra in $L_{e,i}$ are further refined uniformly $n - i - 1$ times. In the following, we obtain a uniform interpolation error estimate for the mesh $\mathcal{T}_n$ in the layer $L_{e,i}$.

Theorem 6.2 *Let $T_{(0)} \in \mathcal{T}_0$ be an e-tetrahedron. For $u \in \mathcal{M}^{m+1}_{\boldsymbol{a}+1}(\Omega)$, where $\boldsymbol{a}$ is given in (6.9) – (6.10), let u_I be its nodal interpolation on $\mathcal{T}_n$. Then for $0 \le i < n$, we have*

$$|u - u_I|_{H^1(L_{e,i})} \le Ch^m \|u\|_{\mathcal{M}^{m+1}_{\boldsymbol{a}+1}(L_{e,i})},$$

where $h = 2^{-n}$, and C depends on $T_{(0)}$ and m, but not on i.

Proof Based on Algorithm 5.7, the layer $L_{e,i}$ is formed in the $i + 1$st refinement and is the union of tetrahedra in $\mathcal{T}_{i+1}$ between $P_{e,i}$ and $P_{e,i+1}$. Therefore, it suffices to verify the following interpolation error estimate on each tetrahedron $T_{(i+1)}$ such that $T_{(i+1)} \in \mathcal{T}_{i+1}$ and $T_{(i+1)} \subset L_{e,i}$,

$$|u - u_I|_{H^1(T_{(i+1)})} \le Ch^m \|u\|_{\mathcal{M}^{m+1}_{\boldsymbol{a}+1}(T_{(i+1)})}. \tag{6.13}$$

We show this estimate based on the type of $T_{(i+1)}$'s parent tetrahedron.

Case I: $T_{(i+1)}$'s parent is an e-tetrahedron in $\mathcal{T}_i$. Let $(x, y, z) \in T_{(i+1)}$ and $(\hat{x}, \hat{y}, \hat{z}) \in \hat{T}_{(i+1)}$ be as in (6.12). By the mapping in (5.85), we have

$$\begin{cases} dxdydz = 2^{-i}\kappa_e^{2i}\, d\hat{x}d\hat{y}d\hat{z}; \\ \partial_x v = (\kappa_e^{-i}\partial_{\hat{x}} + b_1\kappa_e^{-i}\partial_{\hat{z}})\hat{v}, \quad \partial_y v = (\kappa_e^{-i}\partial_{\hat{y}} + b_2\kappa_e^{-i}\partial_{\hat{z}})\hat{v}, \quad \partial_z v = 2^i\partial_{\hat{z}}\hat{v}; \\ \partial_{\hat{x}}\hat{v} = (\kappa_e^{i}\partial_x - b_1 2^{-i}\partial_z)v, \quad \partial_{\hat{y}}\hat{v} = (\kappa_e^{i}\partial_y - b_2 2^{-i}\partial_z)v, \quad \partial_{\hat{z}}\hat{v} = 2^{-i}\partial_z v. \end{cases} \tag{6.14}$$

Therefore, by Lemma 5.9, (6.14), the standard interpolation estimate on $\hat{T}_{(i+1)}$, (5.78), and (6.6), we have

$$\begin{aligned}
\|\partial_x(u-u_I)\|^2_{L^2(T_{(i+1)})} &\le C2^{-i}\Big(\|\partial_{\hat{x}}(\hat{u}-\hat{u}_I)\|^2_{L^2(\hat{T}_{(i+1)})} + \|\partial_{\hat{z}}(\hat{u}-\hat{u}_I)\|^2_{L^2(\hat{T}_{(i+1)})}\Big)\\
&\le C2^{-i}2^{2m(i-n)}|\hat{u}|^2_{H^{m+1}(\hat{T}_{(i+1)})}\\
&\le C2^{2m(i-n)}\sum_{|\alpha_\perp|+\alpha_3=m+1} 2^{-2i\alpha_3}\kappa_e^{2i(|\alpha_\perp|-1)}\|\partial^{\alpha_\perp}\partial_z^{\alpha_3}u\|^2_{L^2(T_{(i+1)})}\\
&\le C2^{2m(i-n)}\sum_{|\alpha_\perp|+\alpha_3=m+1} 2^{-2i\alpha_3}\|r_e^{|\alpha_\perp|-1}\partial^{\alpha_\perp}\partial_z^{\alpha_3}u\|^2_{L^2(T_{(i+1)})}\\
&\le C2^{2m(i-n)}\kappa_e^{2ia_e}\|u\|^2_{\mathcal{M}^{m+1}_{a+1}(T_{(i+1)})} \le Ch^{2m}\|u\|^2_{\mathcal{M}^{m+1}_{a+1}(T_{(i+1)})}.
\end{aligned}$$

A similar calculation for the derivative with respect to y gives

$$\|\partial_y(u-u_I)\|_{L^2(T_{(i+1)})} \le Ch^m\|u\|_{\mathcal{M}^{m+1}_{a+1}(T_{(i+1)})}.$$

In the z-direction, by Lemma 5.9, (6.14), the standard interpolation estimate, (5.78), and (6.6), we have

$$\begin{aligned}
\|\partial_z(u-u_I)\|^2_{L^2(T_{(i+1)})} &\le C2^i\kappa_e^{2i}\|\partial_{\hat{z}}(\hat{u}-\hat{u}_I)\|^2_{L^2(\hat{T}_{(i+1)})}\\
&\le C2^i\kappa_e^{2i}2^{2m(i-n)}|\hat{u}|^2_{H^{m+1}(\hat{T}_{(i+1)})}\\
&\le C2^{2m(i-n)}\sum_{|\alpha_\perp|+\alpha_3=m+1} 2^{2i}2^{-2i\alpha_3}\kappa_e^{2i|\alpha_\perp|}\|\partial^{\alpha_\perp}\partial_z^{\alpha_3}u\|^2_{L^2(T_{(i+1)})}\\
&\le C2^{2m(i-n)}\sum_{|\alpha_\perp|+\alpha_3=m+1} 2^{-2i\alpha_3}\|r_e^{|\alpha_\perp|-1}\partial^{\alpha_\perp}\partial_z^{\alpha_3}u\|^2_{L^2(T_{(i+1)})}\\
&\le C2^{2m(i-n)}\kappa_e^{2ia_e}\|u\|^2_{\mathcal{M}^{m+1}_{a+1}(T_{(i+1)})} \le Ch^{2m}\|u\|^2_{\mathcal{M}^{m+1}_{a+1}(T_{(i+1)})}.
\end{aligned}$$

Hence, we have completed the proof for (6.13).

Case II: $T_{(i+1)}$'s parent is a v_e-tetrahedron $T_{(i)} \in \mathcal{T}_i$. Let $T_{(k)} \in \mathcal{T}_k$, $1 \le k \le i$, be the v_e-tetrahedron, such that $T_{(i)} \subset T_{(k)}$ and $T_{(k)}$'s parent tetrahedron $T_{(k-1)} \in \mathcal{T}_{k-1}$ is an e-tetrahedron. Using the mapping (5.86), by (6.12), for $(x, y, z) \in T_{(i+1)}$ and $(\hat{x}, \hat{y}, \hat{z}) \in \hat{T}_{(i+1)}$, we have

$$\begin{cases}
dxdydz = 2^{1-k}\kappa_e^{3i-k-2}\,d\hat{x}d\hat{y}d\hat{z}, \quad \partial_z v = 2^{k-1}\kappa_e^{k-i}\partial_{\hat{z}}\hat{v}, \quad \partial_{\hat{z}}\hat{v} = 2^{1-k}\kappa_e^{i-k}\partial_z v;\\
\partial_x v = (\kappa_e^{1-i}\partial_{\hat{x}} + b_1\kappa_e^{1-i}\partial_{\hat{z}})\hat{v}, \quad \partial_y v = (\kappa_e^{1-i}\partial_{\hat{y}} + b_2\kappa_e^{1-i}\partial_{\hat{z}})\hat{v};\\
\partial_{\hat{x}}\hat{v} = (\kappa_e^{i-1}\partial_x - b_1 2^{1-k}\kappa_e^{i-k}\partial_z)v, \quad \partial_{\hat{y}}\hat{v} = (\kappa_e^{i-1}\partial_y - b_2 2^{1-k}\kappa_e^{i-k}\partial_z)v.
\end{cases} \tag{6.15}$$

Therefore, by Lemma 5.9, (6.15), the standard interpolation estimate, (5.78), and (6.6), we have

$$\begin{aligned}
\|\partial_x(u-u_I)\|^2_{L^2(T_{(i+1)})} &\le C2^{1-k}\kappa_e^{i-k}\Big(\|\partial_{\hat{x}}(\hat{u}-\hat{u}_I)\|^2_{L^2(\hat{T}_{(i+1)})}+\|\partial_{\hat{z}}(\hat{u}-\hat{u}_I)\|^2_{L^2(\hat{T}_{(i+1)})}\Big)\\
&\le C2^{1-k}\kappa_e^{i-k}2^{2m(i-n)}|\hat{u}|^2_{H^{m+1}(\hat{T}_{(i+1)})}\\
&\le C2^{2m(i-n)}\sum_{|\alpha_\perp|+\alpha_3=m+1}2^{2(1-k)\alpha_3}\kappa_e^{2(i-k)\alpha_3}\kappa_e^{(2i-2)(|\alpha_\perp|-1)}\|\partial^{\alpha_\perp}\partial_z^{\alpha_3}u\|^2_{L^2(T_{(i+1)})}\\
&\le C2^{2m(i-n)}\sum_{|\alpha_\perp|+\alpha_3=m+1}2^{2(1-k)\alpha_3}\|r_e^{|\alpha_\perp|-1}\partial^{\alpha_\perp}\partial_z^{\alpha_3}u\|^2_{L^2(T_{(i+1)})}\\
&\le C2^{2m(i-n)}\kappa_e^{2ia_e}\|u\|^2_{\mathcal{M}^{m+1}_{a+1}(T_{(i+1)})}\le Ch^{2m}\|u\|^2_{\mathcal{M}^{m+1}_{a+1}(T_{(i+1)})}.
\end{aligned}$$

A similar calculation for the derivative with respect to y gives

$$\|\partial_y(u-u_I)\|_{L^2(T_{(i+1)})}\le Ch^m\|u\|_{\mathcal{M}^{m+1}_{a+1}(T_{(i+1)})}.$$

In the z-direction, by Lemma 5.9, (6.15), the standard interpolation estimate, (5.78), and (6.6), we have

$$\begin{aligned}
\|\partial_z(u-u_I)\|^2_{L^2(T_{(i+1)})} &\le C(2^{1-k}\kappa_e^{i-k})\kappa_e^{2(i-1)}(2^{k-1}\kappa_e^{k-i})^2\|\partial_{\hat{z}}(\hat{u}-\hat{u}_I)\|^2_{L^2(\hat{T}_{(i+1)})}\\
&\le C(2^{1-k}\kappa_e^{i-k})\kappa_e^{2(i-1)}(2^{k-1}\kappa_e^{k-i})^2 2^{2m(i-n)}|\hat{u}|^2_{H^{m+1}(\hat{T}_{(i+1)})}\\
&\le C2^{2m(i-n)}\sum_{|\alpha_\perp|+\alpha_3=m+1}(2^{1-k}\kappa_e^{i-k})^{2(\alpha_3-1)}\kappa_e^{2|\alpha_\perp|(i-1)}\|\partial^{\alpha_\perp}\partial_z^{\alpha_3}u\|^2_{L^2(T_{(i+1)})}\\
&\le C2^{2m(i-n)}\sum_{|\alpha_\perp|+\alpha_3=m+1}(2^{1-k}\kappa_e^{i-k})^{2(\alpha_3-1)}\kappa_e^{2i-2|\alpha_\perp|}\|r_e^{|\alpha_\perp|-1}\partial^{\alpha_\perp}\partial_z^{\alpha_3}u\|^2_{L^2(T_{(i+1)})}\\
&\le C2^{2m(i-n)}\kappa_e^{2ia_e}\|u\|^2_{\mathcal{M}^{m+1}_{a+1}(T_{(i+1)})}\le Ch^{2m}\|u\|^2_{\mathcal{M}^{m+1}_{a+1}(T_{(i+1)})}.
\end{aligned}$$

This completes the proof for (6.13) of Case II.

Hence, the theorem is proved by summing up the estimates for all the tetrahedra $T_{(i+1)}$ in $L_{e,i}$. □

Remark 6.2 The main ingredients for the proof of Theorem 6.2 include the scaling argument based on the mappings in Lemma 5.9 and the standard interpolation estimates. Despite the slight difference in the interpolation error analysis for different directional derivatives, the calculations in the proof show that for the element $T_{(i+1)}$ away from the edge, we have

$$|u-u_I|_{H^1(T_{(i+1)})}\le C2^{m(i-n)}h_\perp^{-1}\sum_{|\alpha|=m+1}h_\perp^{|\alpha_\perp|}h_3^{\alpha_3}\|\partial^\alpha u\|_{L^2(T_{(i+1)})},$$

where $h_\perp$ is the size of the projection of the element in the xy-plane, and h_3 is the size of the element along the edge direction. In the proof, $h_\perp$ and h_3 are replaced by specific expressions in terms of the grading parameters, such that they can be transferred into the proper weight in the norm.

Then we extend the interpolation error estimate to the entire initial tetrahedron $T_{(0)}\in\mathcal{T}_0$.

Corollary 6.2 *Let $T_{(0)} \in \mathcal{T}_0$ be an e-tetrahedron. For $u \in \mathcal{M}^{m+1}_{a+1}(\Omega)$, where $\boldsymbol{a}$ is given in (6.9)–(6.10), let u_I be its nodal interpolation on $\mathcal{T}_n$. Then we have*

$$|u - u_I|_{H^1(T_{(0)})} \leq Ch^m \|u\|_{\mathcal{M}^{m+1}_{a+1}(T_{(0)})},$$

where $h = 2^{-n}$, and C depends on $T_{(0)}$ and m.

Proof By Theorem 6.2, it suffices to show the estimate for any tetrahedron $T_{(n)} \in \mathcal{T}_n$ in the last layer $L_{e,n}$. We derive the desired estimate in the following two cases.

Case I: $T_{(n)}$ is an e-tetrahedron. By Lemma 5.7, the mapping $\mathbf{B}_{e,n}$ translates $T_{(n)}$ into the reference tetrahedron $\hat{T}$. Consequently, it maps any point $(x, y, z) \in T_{(n)}$ to $(\hat{x}, \hat{y}, \hat{z}) \in \hat{T}$. For a function v on $T_{(n)}$, we define $\hat{v}$ on $\hat{T}$ by $\hat{v}(\hat{x}, \hat{y}, \hat{z}) := v(x, y, z)$. Now, let χ be a smooth cut-off function on $\hat{T}$ such that $\chi = 0$ in a neighborhood of the edge $\hat{e} := \hat{x}_0\hat{x}_1$ and $= 1$ at every other Lagrange node of $\hat{T}$. Let $r_{\hat{e}}$ be the distance to $\hat{e}$. Let $\hat{u}_I$ be the interpolation of $\hat{u}$ on the reference tetrahedron $\hat{T}$. Since $\chi\hat{u} = 0$ in the neighborhood of $\hat{e}$, $(\chi\hat{u})_I = \hat{u}_I$ and

$$|\chi\hat{u}|^2_{H^{m+1}(\hat{T})} \leq C \sum_{|\alpha_\perp|+\alpha_3 \leq m+1} \|r_{\hat{e}}^{|\alpha_\perp|-1} \partial^{\alpha_\perp} \partial^{\alpha_3}_{\hat{z}} \hat{u}\|^2_{L^2(\hat{T})}. \tag{6.16}$$

Define $\hat{w} := \hat{u} - \chi\hat{u}$. Then by the usual interpolation error estimate, we have

$$\begin{aligned} |\hat{u} - \hat{u}_I|_{H^1(\hat{T})} &= |\hat{w} + \chi\hat{u} - \hat{u}_I|_{H^1(\hat{T})} \leq |\hat{w}|_{H^1(\hat{T})} + |\chi\hat{u} - \hat{u}_I|_{H^1(\hat{T})} \\ &= |\hat{w}|_{H^1(\hat{T})} + |\chi\hat{u} - (\chi\hat{u})_I|_{H^1(\hat{T})} \leq C\big(\|\hat{u}\|_{H^1(\hat{T})} + |\chi\hat{u}|_{H^{m+1}(\hat{T})}\big), \end{aligned} \tag{6.17}$$

where C depends on m and, through χ, the nodes on $\hat{T}$. Then using the scaling argument based on (6.14), (6.17), (6.16), the relation $r_{\hat{e}}(\hat{x}, \hat{y}, \hat{z}) = \kappa_e^{-n} r_e(x, y, z)$, and (6.6), we have

$$\begin{aligned} \|\partial_x(u - u_I)\|^2_{L^2(T_{(n)})} &\leq C2^{-n}\Big(\|\partial_{\hat{x}}(\hat{u} - \hat{u}_I)\|^2_{L^2(\hat{T})} + \|\partial_{\hat{z}}(\hat{u} - \hat{u}_I)\|^2_{L^2(\hat{T})}\Big) \\ &\leq C2^{-n} \sum_{|\alpha_\perp|+\alpha_3 \leq m+1} \|r_{\hat{e}}^{|\alpha_\perp|-1} \partial^{\alpha_\perp} \partial^{\alpha_3}_{\hat{z}} \hat{u}\|^2_{L^2(\hat{T})} \\ &\leq C \sum_{|\alpha_\perp|+\alpha_3 \leq m+1} 2^{-2n\alpha_3} \|r_e^{|\alpha_\perp|-1} \partial^{\alpha_\perp} \partial^{\alpha_3}_z u\|^2_{L^2(T_{(n)})} \\ &\leq C \sum_{|\alpha_\perp|+\alpha_3 \leq m+1} 2^{-2n\alpha_3} \kappa_e^{2na_e} \|r_e^{|\alpha_\perp|-1-a_e} \partial^{\alpha_\perp} \partial^{\alpha_3}_z u\|^2_{L^2(T_{(n)})} \\ &\leq Ch^{2m} \|u\|^2_{\mathcal{M}^{m+1}_{a+1}(T_{(n)})}. \end{aligned}$$

A similar calculation for the derivative with respect to y gives

$$\|\partial_y(u - u_I)\|_{L^2(T_{(n)})} \leq Ch^m \|u\|_{\mathcal{M}^{m+1}_{a+1}(T_{(n)})}.$$

In the z-direction, using (6.17), (6.16), (6.14), and (6.6), we have

$$
\begin{aligned}
\|\partial_z(u-u_I)\|^2_{L^2(T_{(n)})} &= 2^n\kappa_e^{2n}\|\partial_{\hat z}(\hat u-\hat u_I)\|^2_{L^2(\hat T)}\\
&\le C2^n\kappa_e^{2n}\sum_{|\alpha_\perp|+\alpha_3\le m+1}\|r_{\hat e}^{|\alpha_\perp|-1}\partial^{\alpha_\perp}\partial_{\hat z}^{\alpha_3}\hat u\|^2_{L^2(\hat T)}\\
&\le C\sum_{|\alpha_\perp|+\alpha_3\le m+1}2^{-2n\alpha_3}\|r_e^{|\alpha_\perp|-1}\partial^{\alpha_\perp}\partial_z^{\alpha_3}u\|^2_{L^2(T_{(n)})}\\
&\le C\kappa_e^{2na_e}\|u\|^2_{\mathcal{M}^{m+1}_{a+1}(T_{(n)})}\le Ch^{2m}\|u\|^2_{\mathcal{M}^{m+1}_{a+1}(T_{(n)})}.
\end{aligned}
$$

Thus, we have proved the estimate for Case I.

Case II: $T_{(n)}$ is a v_e-tetrahedron. Let $T_{(k)}\in\mathcal{T}_k$, $1\le k\le n$, be the v_e-tetrahedron, such that $T_{(n)}\subset T_{(k)}$ and $T_{(k)}$'s parent tetrahedron $T_{(k-1)}\in\mathcal{T}_{k-1}$ is an e-tetrahedron. By Lemma 5.8, the mapping $\mathbf{B}_{n,k}$ translates $T_{(n)}$ into a v_e-tetrahedron in $\hat T_{(n)}\in\hat{\mathcal{T}}_1$. Thus, $\mathbf{B}_{n,k}$ maps every point $(x,y,z)\in T_{(n)}$ to $(\hat x,\hat y,\hat z)\in\hat T_{(n)}$. As in Case I, for a function v on $T_{(n)}$, we define $\hat v$ on $\hat T_{(n)}$ by $\hat v(\hat x,\hat y,\hat z):=v(x,y,z)$. Let χ be a smooth cut-off function on $\hat T_{(n)}$ such that $\chi=0$ in a neighborhood of the e-singular vertex on $\hat e:=\hat x_0\hat x_1$ and $=1$ at every other Lagrange node of $\hat T_{(n)}$. Recall the distance $r_{\hat e}$ to $\hat e$. Since $\chi\hat u=0$ in the neighborhood of the e-singular vertex, we have $(\chi\hat u)_I=\hat u_I$ on $\hat T_{(n)}$ and

$$
|\chi\hat u|^2_{H^{m+1}(\hat T_{(n)})}\le C\sum_{|\alpha_\perp|+\alpha_3\le m+1}\|r_{\hat e}^{|\alpha_\perp|-1}\partial^{\alpha_\perp}\partial_{\hat z}^{\alpha_3}\hat u\|^2_{L^2(\hat T_{(n)})}. \tag{6.18}
$$

Define $\hat w:=\hat u-\chi\hat u$. Then by the usual interpolation error estimate, we have

$$
\begin{aligned}
|\hat u-\hat u_I|_{H^1(\hat T_{(n)})} &= |\hat w+\chi\hat u-\hat u_I|_{H^1(\hat T_{(n)})}\le|\hat w|_{H^1(\hat T_{(n)})}+|\chi\hat u-\hat u_I|_{H^1(\hat T_{(n)})}\\
&=|\hat w|_{H^1(\hat T_{(n)})}+|\chi\hat u-(\chi\hat u)_I|_{H^1(\hat T_{(n)})}\le C\big(\|\hat u\|_{H^1(\hat T_{(n)})}+|\chi\hat u|_{H^{m+1}(\hat T_{(n)})}\big),
\end{aligned}\tag{6.19}
$$

where C depends on m and, through χ, the nodes on $\hat T_{(n)}$. In $L_{e,n}$, $r_e(x,y,z)=\kappa_e^{n-1}r_{\hat e}(\hat x,\hat y,\hat z)$. Therefore, by (6.15), (6.19), (6.18), and (6.6), we have

$$
\begin{aligned}
\|\partial_x(u-u_I)\|^2_{L^2(T_{(n)})} &\le C2^{1-k}\kappa_e^{n-k}\Big(\|\partial_{\hat x}(\hat u-\hat u_I)\|^2_{L^2(\hat T_{(n)})}+\|\partial_{\hat z}(\hat u-\hat u_I)\|^2_{L^2(\hat T_{(n)})}\Big)\\
&\le C2^{1-k}\kappa_e^{n-k}\sum_{|\alpha_\perp|+\alpha_3\le m+1}\|r_{\hat e}^{|\alpha_\perp|-1}\partial^{\alpha_\perp}\partial_{\hat z}^{\alpha_3}\hat u\|^2_{L^2(\hat T_{(n)})}\\
&\le C\sum_{|\alpha_\perp|+\alpha_3\le m+1}2^{2(1-k)\alpha_3}\kappa_e^{2(n-k)\alpha_3}\|r_e^{|\alpha_\perp|-1}\partial^{\alpha_\perp}\partial_z^{\alpha_3}u\|^2_{L^2(T_{(n)})}\\
&\le C\kappa_e^{2na_e}\|u\|^2_{\mathcal{M}^{m+1}_{a+1}(T_{(n)})}\le Ch^{2m}\|u\|^2_{\mathcal{M}^{m+1}_{a+1}(T_{(n)})}.
\end{aligned}
$$

A similar calculation for the derivative with respect to y gives

$$
\|\partial_y(u-u_I)\|_{L^2(T_{(n)})}\le Ch^m\|u\|_{\mathcal{M}^{m+1}_{a+1}(T_{(n)})}.
$$

In the z-direction, by (6.15), (6.19), (6.18), and (6.6), we have

$$\begin{aligned}\|\partial_z(u-u_I)\|^2_{L^2(T_{(n)})} &= (2^{1-k}\kappa_e^{n-k})\kappa_e^{2(n-1)}(2^{k-1}\kappa_e^{k-n})^2\|\partial_{\hat z}(\hat u-\hat u_I)\|^2_{L^2(\hat T_{(n)})}\\ &\le C(2^{1-k}\kappa_e^{n-k})\kappa_e^{2(n-1)}(2^{k-1}\kappa_e^{k-n})^2\sum_{|\alpha_\perp|+\alpha_3\le m+1}\|r_{\hat e}^{|\alpha_\perp|-1}\partial^{\alpha_\perp}\partial_{\hat z}^{\alpha_3}\hat u\|^2_{L^2(\hat T_{(n)})}\\ &\le C\sum_{|\alpha_\perp|+\alpha_3\le m+1}(2^{1-k}\kappa_e^{n-k})^{2\alpha_3}(2^{k-1}\kappa_e^{k})^2\|r_e^{|\alpha_\perp|-1}\partial^{\alpha_\perp}\partial_z^{\alpha_3}u\|^2_{L^2(T_{(n)})}\\ &\le C\kappa_e^{2na_e}\|u\|^2_{\mathcal{M}^{m+1}_{a+1}(T_{(n)})}\le Ch^{2m}\|u\|^2_{\mathcal{M}^{m+1}_{a+1}(T_{(n)})}.\end{aligned}$$

Thus, we have proved the estimate for Case II.

Hence, the corollary is proved by summing up the estimates in Theorem 6.2 and the estimates for all the tetrahedra $T_{(n)}$ in $L_{e,n}$. □

6.1.3 Estimates on Initial *ev*-Tetrahedra in $\mathcal{T}_0$

In this subsection, we denote by $T_{(0)} = \Delta^4 x_0x_1x_2x_3 \in \mathcal{T}_0$ an *ev*-tetrahedron, such that $x_0 = c \in C$ and x_0x_1 is on the edge $e \in \mathcal{E}$. We also assume that x_0x_1 is on the z-axis.

We proceed to present the interpolation error estimate in the mesh layers on $T_{(0)}$.

Theorem 6.3 *Let $T_{(0)} = \Delta^4 x_0x_1x_2x_3 \in \mathcal{T}_0$ be an ev-tetrahedron defined above. Let $L_{ev,i}$ be the mesh layer in Definition 5.7, $0 \le i < n$. Recall the parameters a_c, a_e, a_{sc}, and a_{ec} associated to κ_c, κ_e, κ_{sc}, and κ_{ec} in (6.5)–(6.8). Define*

$$a_C := (m+1)(1-a_c^{-1}a_{ec}) + a_{ec}. \tag{6.20}$$

Suppose

$$\sum_{|\alpha_\perp|+\alpha_3\le m+1}\|r_c^{\alpha_3-a_C+a_e}r_e^{|\alpha_\perp|-1-a_e}\partial^{\alpha_\perp}\partial_z^{\alpha_3}u\|^2_{L^2(T_{(0)})}<\infty.$$

Let u_I be the nodal interpolation on $\mathcal{T}_n$. Then we have

$$|u-u_I|^2_{H^1(L_{ev,i})}\le Ch^{2m}\sum_{|\alpha_\perp|+\alpha_3\le m+1}\|r_c^{\alpha_3-a_C+a_e}r_e^{|\alpha_\perp|-1-a_e}\partial^{\alpha_\perp}\partial_z^{\alpha_3}u\|^2_{L^2(L_{ev,i})},$$

where $h = 2^{-n}$, and C depends on $T_{(0)}$ and m.

Proof Let $T_{(i)} \subset T_{(0)}$ be the *ev*-tetrahedron in $\mathcal{T}_i$. By Definition 5.7, we have $L_{ev,i} = T_{(i)} \setminus T_{(i+1)}$. Then the mapping $\mathbf{B}_{ev,i}$ in (5.87) translates $L_{ev,i}$ into $\hat L$ (see Definition 5.8). For a point $(x, y, z) \in L_{ev,i}$, let $(\hat x, \hat y, \hat z) \in \hat L$ be its image under $\mathbf{B}_{ev,i}$. For a function v on $L_{ev,i}$, define the function $\hat v$ on $\hat L$ by $\hat v(\hat x, \hat y, \hat z) := v(x, y, z)$. Let $r_{\hat e}$ be the distance to $\hat x_0\hat x_1$ on the reference tetrahedron $\hat T$. Then it is clear that $r_e(x, y, z) = \kappa_{ec}^i r_{\hat e}(\hat x, \hat y, \hat z)$ on $L_{ev,i}$. Meanwhile, $\mathbf{B}_{ev,i}$ maps the triangulation $\mathcal{T}_n$ on $L_{ev,i}$ into a graded triangulation on $\hat L$ that is obtained after $n-i-1$ refinements of the initial mesh $\hat{\mathcal{L}}$. Note that the subsequent refinements on $\hat{\mathcal{L}}$ are anisotropic with the parameter κ_e toward $\hat{x}_0\hat{x}_1$ since $\hat{\mathcal{L}}$ does not contain ev- or v-tetrahedra.

Then by the mapping (5.87), the scaling argument, Corollary 6.2, (5.78), and (6.6), we have

$$\begin{aligned}
&\|\partial_x(u-u_I)\|^2_{L^2(L_{ev,i})} \leq C\kappa^i_{sc}\left(\|\partial_{\hat{x}}(\hat{u}-\hat{u}_I)\|^2_{L^2(\hat{\mathcal{L}})} + \|\partial_{\hat{z}}(\hat{u}-\hat{u}_I)\|^2_{L^2(\hat{\mathcal{L}})}\right)\\
&\leq C\kappa^i_{sc}2^{2m(i-n)} \sum_{|\alpha_\perp|+\alpha_3\leq m+1} \|r_{\hat{e}}^{|\alpha_\perp|-1-a_e}\partial^{\alpha_\perp}\partial^{\alpha_3}_{\hat{z}}\hat{u}\|^2_{L^2(\hat{\mathcal{L}})}\\
&\leq C2^{2m(i-n)} \sum_{|\alpha_\perp|+\alpha_3\leq m+1} \kappa_{sc}^{2i\alpha_3}\kappa_{ec}^{2ia_e}\|r_e^{|\alpha_\perp|-1-a_e}\partial^{\alpha_\perp}\partial^{\alpha_3}_z u\|^2_{L^2(L_{ev,i})}\\
&\leq C2^{-2mn} \sum_{|\alpha_\perp|+\alpha_3\leq m+1} 2^{2im}\kappa_{sc}^{2i\alpha_3}\kappa_{ec}^{2ia_e}\|r_e^{|\alpha_\perp|-1-a_e}\partial^{\alpha_\perp}\partial^{\alpha_3}_z u\|^2_{L^2(L_{ev,i})}.
\end{aligned}\tag{6.21}$$

Note that $\kappa^i_{ec} \lesssim r_c \lesssim \kappa^i_{sc}$ on $L_{ev,i}$, $a_c, a_e \geq a_{ec}$ (see (6.7)), and $a_C \geq a_c$. Then we consider all the possible cases below.

(I) ($\alpha_3 \leq a_c$). We have

$$\kappa_{sc}^{i(\alpha_3-a_c)} \lesssim r_c^{\alpha_3-a_c}.$$

Then by $a_c \geq a_{sc}$ and (6.8), we have

$$2^{im}\kappa_{sc}^{i\alpha_3}\kappa_{ec}^{ia_e} = \kappa_{sc}^{i(\alpha_3-a_{sc})}\kappa_{ec}^{ia_e} \leq \kappa_{sc}^{i(\alpha_3-a_c)}\kappa_{ec}^{ia_e} \lesssim r_c^{\alpha_3-a_c}\kappa_{ec}^{ia_e} \lesssim r_c^{\alpha_3-a_c+a_e}. \tag{6.22}$$

(II) ($(1-a_c^{-1}a_{ec})(m+1) < \alpha_3 \leq m+1$). Note that $0 < a_c \leq m$. Therefore, by (6.7), we have

$$\kappa_{sc}^{i\alpha_3} = 2^{-im\alpha_3/a_{sc}} \leq 2^{-im\alpha_3/a_c} \leq 2^{-im\alpha_3/a_{ec}+ima_C/a_{ec}-im} = \kappa_{ec}^{i(\alpha_3-a_C+a_{ec})}.$$

Note that $\alpha_3 - a_C + a_{ec} > 0$. Therefore,

$$2^{im}\kappa_{sc}^{i\alpha_3}\kappa_{ec}^{ia_e} \leq 2^{im}\kappa_{ec}^{i(\alpha_3-a_C+a_{ec})}\kappa_{ec}^{ia_e} \lesssim r_c^{\alpha_3-a_C+a_e}. \tag{6.23}$$

(III) ($a_c < \alpha_3 \leq (1-a_c^{-1}a_{ec})(m+1)$). If $a_{ec} = a_{sc} = a_c$, we have $(1-a_c^{-1}a_{ec})(m+1) = 0$, and therefore, such α_3 does not exist. Thus, we only need to consider the case $a_{ec} < a_c$. Note that $\alpha_3 - a_C + a_{ec} \leq 0$ and $a_{ec} \leq a_e$. Therefore, by (6.7), we have

$$2^{im}\kappa_{sc}^{i\alpha_3}\kappa_{ec}^{ia_e} = \kappa_{sc}^{i\alpha_3}\kappa_{ec}^{i(a_e-a_{ec})} \lesssim r_c^{\alpha_3-a_C+a_{ec}}r_c^{(a_e-a_{ec})} = r_c^{\alpha_3-a_C+a_e}. \tag{6.24}$$

Therefore, choosing a_C as in (6.20), by (6.21)–(6.24), we have shown that

$$\begin{aligned}
&\|\partial_x(u-u_I)\|^2_{L^2(L_{ev,i})}\\
&\leq C2^{-2mn}\sum_{|\alpha_\perp|+\alpha_3\leq m+1}\|r_c^{\alpha_3-a_C+a_e}r_e^{|\alpha_\perp|-1-a_e}\partial^{\alpha_\perp}\partial^{\alpha_3}_z u\|^2_{L^2(L_{ev,i})}.
\end{aligned}\tag{6.25}$$

In the y-direction, with a similar process, we obtain

$$\|\partial_y(u-u_I)\|^2_{L^2(L_{ev,i})} \leq C2^{-2mn} \sum_{|\alpha_\perp|+\alpha_3\leq m+1} \|r_c^{\alpha_3-a_C+a_e} r_e^{|\alpha_\perp|-1-a_e} \partial^{\alpha_\perp}\partial_z^{\alpha_3} u\|^2_{L^2(L_{ev,i})}. \quad (6.26)$$

In the z-direction, by the mapping (5.87), the scaling argument, and Corollary 6.2, we have

$$\begin{aligned}
\|\partial_z(u-u_I)\|^2_{L^2(L_{ev,i})} &= \kappa_{sc}^{-i}\kappa_{ec}^{2i}\|\partial_{\hat z}(\hat u-\hat u_I)\|^2_{L^2(\hat L)} \\
&\leq C2^{2m(i-n)}\kappa_{sc}^{-i}\kappa_{ec}^{2i} \sum_{|\alpha_\perp|+\alpha_3\leq m+1} \|r_{\hat e}^{|\alpha_\perp|-1-a_e}\partial^{\alpha_\perp}\partial_{\hat z}^{\alpha_3}\hat u\|^2_{L^2(\hat L)} \\
&\leq C2^{-2mn} \sum_{|\alpha_\perp|+\alpha_3\leq m+1} 2^{2im}\kappa_{sc}^{2i(\alpha_3-1)}\kappa_{ec}^{2i(1+a_e)}\|r_e^{|\alpha_\perp|-1-a_e}\partial^{\alpha_\perp}\partial_z^{\alpha_3}u\|^2_{L^2(L_{ev,i})} \\
&\leq C2^{-2mn} \sum_{|\alpha_\perp|+\alpha_3\leq m+1} 2^{2im}\kappa_{sc}^{2i\alpha_3}\kappa_{ec}^{2ia_e}\|r_e^{|\alpha_\perp|-1-a_e}\partial^{\alpha_\perp}\partial_z^{\alpha_3}u\|^2_{L^2(L_{ev,i})} \\
&\leq C2^{-2mn} \sum_{|\alpha_\perp|+\alpha_3\leq m+1} \|r_c^{\alpha_3-a_C+a_e} r_e^{|\alpha_\perp|-1-a_e}\partial^{\alpha_\perp}\partial_z^{\alpha_3}u\|^2_{L^2(L_{ev,i})}, \quad (6.27)
\end{aligned}$$

where the last inequality follows from the analysis in (6.22)–(6.24).

Hence, the proof is completed by the estimates in (6.25)–(6.27). □

Then we are ready to derive the interpolation error estimate on the entire *ev*-tetrahedron $T_{(0)}$.

Corollary 6.3 *Let $T_{(0)} \in \mathcal{T}_0$ be an ev-tetrahedron as in Theorem 6.3. Recall a_C from (6.20). Suppose*

$$\sum_{|\alpha_\perp|+\alpha_3\leq m+1} \|r_c^{\alpha_3-a_C+a_e} r_e^{|\alpha_\perp|-1-a_e}\partial^{\alpha_\perp}\partial_z^{\alpha_3}u\|^2_{L^2(T_{(0)})} < \infty.$$

Let u_I be the nodal interpolation on $\mathcal{T}_n$. Then we have

$$|u-u_I|^2_{H^1(T_{(0)})} \leq Ch^{2m} \sum_{|\alpha_\perp|+\alpha_3\leq m+1} \|r_c^{\alpha_3-a_C+a_e} r_e^{|\alpha_\perp|-1-a_e}\partial^{\alpha_\perp}\partial_z^{\alpha_3}u\|^2_{L^2(T_{(0)})},$$

where $h = 2^{-n}$, and C depends on $T_{(0)}$ and m.

Proof By Theorem 6.3, it suffices to show

$$|u-u_I|^2_{H^1(L_{ev,n})} \leq C2^{-2mn} \sum_{|\alpha_\perp|+\alpha_3\leq m+1} \|r_c^{\alpha_3-a_C+a_e} r_e^{|\alpha_\perp|-1-a_e}\partial^{\alpha_\perp}\partial_z^{\alpha_3}u\|^2_{L^2(L_{ev,n})}.$$

By Lemma 5.11, $\mathbf{B}_{ev,n}(L_{ev,n}) = \hat T$. For $(x, y, z) \in L_{ev,n}$, let $(\hat x, \hat y, \hat z) \in \hat T$ be its image under $\mathbf{B}_{ev,n}$. For a function v on $L_{ev,n}$, we define $\hat v$ on $\hat T$ by $\hat v(\hat x, \hat y, \hat z) := v(x, y, z)$. Let χ be a smooth cut-off function on $\hat T$ such that $\chi = 0$ in a neighborhood of the edge $\hat e := \hat x_0\hat x_1$ and $= 1$ at every other node of $\hat T$. Let $r_{\hat c}$ be the distance from $(\hat x, \hat y, \hat z)$ to $\hat x_0$. Then by (5.87), we have

$$\kappa_{ec}^n r_{\hat{c}}(\hat{x}, \hat{y}, \hat{z}) \lesssim r_c(x, y, z) \lesssim \kappa_{sc}^n r_{\hat{c}}(\hat{x}, \hat{y}, \hat{z}), \tag{6.28}$$

and $\kappa_{ec}^n r_{\hat{e}}(\hat{x}, \hat{y}, \hat{z}) = r_e(x, y, z)$. Let $\hat{u}_I$ be the interpolation of $\hat{u}$ on the reference tetrahedron $\hat{T}$. Since $\chi\hat{u} = 0$ in the neighborhood of $\hat{e}$, $(\chi\hat{u})_I = \hat{u}_I$ and

$$|\chi\hat{u}|^2_{H^{m+1}(\hat{T})} \le C \sum_{|\alpha_\perp|+\alpha_3 \le m+1} \|r_{\hat{e}}^{|\alpha_\perp|-1-a_e} r_{\hat{c}}^{\alpha_3-a_C+a_e} \partial^{\alpha_\perp} \partial_{\hat{z}}^{\alpha_3} \hat{u}\|^2_{L^2(\hat{T})}. \tag{6.29}$$

Note that by (6.20), $a_C \ge a_{ec}$. Define $\hat{w} := \hat{u} - \chi\hat{u}$. Then by the usual interpolation error estimate, $r_{\hat{e}} \lesssim r_{\hat{c}}$, and (6.29), we have

$$\begin{aligned}
|\hat{u} - \hat{u}_I|_{H^1(\hat{T})} &= |\hat{w} + \chi\hat{u} - \hat{u}_I|_{H^1(\hat{T})} \le |\hat{w}|_{H^1(\hat{T})} + |\chi\hat{u} - \hat{u}_I|_{H^1(\hat{T})} \\
&= |\hat{w}|_{H^1(\hat{T})} + |\chi\hat{u} - (\chi\hat{u})_I|_{H^1(\hat{T})} \le C\big(\|\hat{u}\|_{H^1(\hat{T})} + |\chi\hat{u}|_{H^{m+1}(\hat{T})}\big), \\
&\le C\Bigg(\sum_{|\alpha_\perp|+\alpha_3 \le m+1} \|r_{\hat{e}}^{|\alpha_\perp|-1-a_e} r_{\hat{c}}^{\alpha_3-a_C+a_e} \partial^{\alpha_\perp} \partial_{\hat{z}}^{\alpha_3} \hat{u}\|^2_{L^2(\hat{T})}\Bigg)^{1/2},
\end{aligned} \tag{6.30}$$

where C depends on m and, through χ, the nodes on $\hat{T}$. Then using (6.30), the scaling argument based on (5.87), and the relation $r_{\hat{e}}(\hat{x}, \hat{y}, \hat{z}) = \kappa_{ec}^{-n} r_e(x, y, z)$, we have

$$\begin{aligned}
&\|\partial_x(u - u_I)\|^2_{L^2(L_{ev,n})} \le C\kappa_{sc}^n\Big(\|\partial_{\hat{x}}(\hat{u} - \hat{u}_I)\|^2_{L^2(\hat{T})} + \|\partial_{\hat{z}}(\hat{u} - \hat{u}_I)\|^2_{L^2(\hat{T})}\Big) \\
&\le C\kappa_{sc}^n \sum_{|\alpha_\perp|+\alpha_3 \le m+1} \|r_{\hat{c}}^{\alpha_3-a_C+a_e} r_{\hat{e}}^{|\alpha_\perp|-1-a_e} \partial^{\alpha_\perp} \partial_{\hat{z}}^{\alpha_3} \hat{u}\|^2_{L^2(\hat{T})} \\
&\le C \sum_{|\alpha_\perp|+\alpha_3 \le m+1} \kappa_{sc}^{2n\alpha_3} \kappa_{ec}^{2na_e} \|r_{\hat{c}}^{\alpha_3-a_C+a_e} r_e^{|\alpha_\perp|-1-a_e} \partial^{\alpha_\perp} \partial_z^{\alpha_3} u\|^2_{L^2(L_{ev,n})} \\
&\le Ch^{2m} \sum_{|\alpha_\perp|+\alpha_3 \le m+1} (2^m \kappa_{sc}^{\alpha_3} \kappa_{ec}^{a_e})^{2n} \|r_{\hat{c}}^{\alpha_3-a_C+a_e} r_e^{|\alpha_\perp|-1-a_e} \partial^{\alpha_\perp} \partial_z^{\alpha_3} u\|^2_{L^2(L_{ev,n})}.
\end{aligned} \tag{6.31}$$

Then we consider the following cases:
(I) ($\alpha_3 \le a_c$). By (6.8), (6.28), $a_C \ge a_c \ge a_{sc}$, and $\alpha_3 - a_c \le 0$, we have

$$\begin{aligned}
&2^{nm} \kappa_{sc}^{n\alpha_3} \kappa_{ec}^{na_e} r_{\hat{c}}^{\alpha_3-a_C+a_e} = \kappa_{sc}^{n(\alpha_3-a_{sc})} r_{\hat{c}}^{\alpha_3-a_c} \kappa_{ec}^{na_e} r_{\hat{c}}^{a_e} r_{\hat{c}}^{a_c-a_C} \\
&\le \kappa_{sc}^{n(\alpha_3-a_c)} r_{\hat{c}}^{\alpha_3-a_c} \kappa_{ec}^{na_e} r_{\hat{c}}^{a_e} r_{\hat{c}}^{a_c-a_C} \lesssim r_c^{\alpha_3-a_c+a_e} r_{\hat{c}}^{a_c-a_C} \lesssim r_c^{\alpha_3-a_C+a_e}.
\end{aligned} \tag{6.32}$$

(II) ($(1 - a_c^{-1} a_{ec})(m + 1) < \alpha_3 \le m + 1$). Following the calculation in (6.23), by (6.7) and (6.28), we have

$$\begin{aligned}
2^{nm} \kappa_{sc}^{n\alpha_3} \kappa_{ec}^{na_e} r_{\hat{c}}^{\alpha_3-a_C+a_e} &\le 2^{nm} \kappa_{ec}^{n(\alpha_3-a_C+a_{ec})} \kappa_{ec}^{na_e} r_{\hat{c}}^{\alpha_3-a_C+a_e} \\
&= \kappa_{ec}^{n(\alpha_3-a_C)} \kappa_{ec}^{na_e} r_{\hat{c}}^{\alpha_3-a_C+a_e} \lesssim r_c^{\alpha_3-a_C+a_e}.
\end{aligned} \tag{6.33}$$

(III) ($a_c < \alpha_3 \le (1 - a_c^{-1} a_{ec})(m + 1)$). If $a_{ec} = a_c$, we have $(1 - a_c^{-1} a_{ec})(m + 1) = 0$, and therefore such α_3 does not exist. Thus, we only need to consider the case $a_{ec} < a_c$. Note that $\alpha_3 - a_C + a_{ec} \le 0$ and $a_{ec} \le a_e$. Therefore, by (6.7) and (6.28), we have

$$2^{nm}\kappa_{sc}^{n\alpha_3}\kappa_{ec}^{na_e}r_{\hat{c}}^{\alpha_3-a_C+a_e} = \kappa_{sc}^{n\alpha_3}\kappa_{ec}^{n(a_e-a_{ec})}r_{\hat{c}}^{\alpha_3-a_C+a_e}$$
$$\le \kappa_{sc}^{n(\alpha_3-a_C+a_{ec})}r_{\hat{c}}^{\alpha_3-a_C+a_{ec}}\kappa_{ec}^{n(a_e-a_{ec})}r_{\hat{c}}^{a_e-a_{ec}} \lesssim r_c^{\alpha_3-a_C+a_e}. \quad (6.34)$$

Therefore, by (6.31)–(6.34), we conclude

$$\|\partial_x(u-u_I)\|^2_{L^2(L_{ev,n})}$$
$$\le Ch^{2m}\sum_{|\alpha_\perp|+\alpha_3\le m+1}\|r_c^{\alpha_3-a_C+a_e}r_e^{|\alpha_\perp|-1-a_e}\partial^{\alpha_\perp}\partial_z^{\alpha_3}u\|^2_{L^2(L_{ev,n})}. \quad (6.35)$$

A similar error estimate in the y-direction leads to

$$\|\partial_y(u-u_I)\|^2_{L^2(L_{ev,n})}$$
$$\le Ch^{2m}\sum_{|\alpha_\perp|+\alpha_3\le m+1}\|r_c^{\alpha_3-a_C+a_e}r_e^{|\alpha_\perp|-1-a_e}\partial^{\alpha_\perp}\partial_z^{\alpha_3}u\|^2_{L^2(L_{ev,n})}. \quad (6.36)$$

In the z-direction, using (6.30), $\kappa_{sc} \ge \kappa_{ec}$, the scaling argument based on (5.87), (6.6), and (6.32)–(6.34), we have

$$\|\partial_z(u-u_I)\|^2_{L^2(L_{ev,n})} = \kappa_{sc}^{-n}\kappa_{ec}^{2n}\|\partial_{\hat{z}}(\hat{u}-\hat{u}_I)\|^2_{L^2(\hat{T})}$$
$$\le C\kappa_{sc}^{-n}\kappa_{ec}^{2n}\sum_{|\alpha_\perp|+\alpha_3\le m+1}\|r_{\hat{c}}^{\alpha_3-a_C+a_e}r_{\hat{e}}^{|\alpha_\perp|-1-a_e}\partial^{\alpha_\perp}\partial_{\hat{z}}^{\alpha_3}\hat{u}\|^2_{L^2(\hat{T})}$$
$$\le C\sum_{|\alpha_\perp|+\alpha_3\le m+1}\kappa_{sc}^{2n(\alpha_3-1)}\kappa_{ec}^{2n(1+a_e)}\|r_{\hat{c}}^{\alpha_3-a_C+a_e}r_e^{|\alpha_\perp|-1-a_e}\partial^{\alpha_\perp}\partial_z^{\alpha_3}u\|^2_{L^2(L_{ev,n})}$$
$$\le C2^{-2mn}\sum_{|\alpha_\perp|+\alpha_3\le m+1}2^{2nm}\kappa_{sc}^{2n\alpha_3}\kappa_{ec}^{2na_e}\|r_{\hat{c}}^{\alpha_3-a_C+a_e}r_e^{|\alpha_\perp|-1-a_e}\partial^{\alpha_\perp}\partial_z^{\alpha_3}u\|^2_{L^2(L_{ev,n})}$$
$$\le Ch^{2m}\sum_{|\alpha_\perp|+\alpha_3\le m+1}\|r_c^{\alpha_3-a_C+a_e}r_e^{|\alpha_\perp|-1-a_e}\partial^{\alpha_\perp}\partial_z^{\alpha_3}u\|^2_{L^2(L_{ev,n})}. \quad (6.37)$$

Then the proof is completed by (6.35)–(6.37). □

Consequently, we have the finite element error analysis for the anisotropic mesh.

Theorem 6.4 *Recall* $\boldsymbol{a}$ *in (6.9)–(6.10). Let* $\mathcal{T}_n$ *be the triangulation defined in Algorithm 5.7 with grading parameters given in (6.5)–(6.8). Suppose the solution of the Dirichlet problem (6.1)* $u \in \mathcal{M}^{m+1}_{\mu+1}(\Omega)$ *for* $\boldsymbol{0} \le \boldsymbol{\mu} < \boldsymbol{\eta}$, *where* $\boldsymbol{\eta}$ *is given in (5.5). Let* u_I *be its nodal interpolation on* $\mathcal{T}_n$. *Then we have*

$$|u-u_I|_{H^1(\Omega)} \le Ch^m\|u\|_{\mathcal{M}^{m+1}_{\mu+1}(\Omega)},$$

where $h = 2^{-n}$. *Consequently, for the finite element solution* u_n *(6.2), we have*

$$|u-u_n|_{H^1(\Omega)} \le C\dim(S_n)^{-m/3}\|u\|_{\mathcal{M}^{m+1}_{\mu+1}(\Omega)}, \quad (6.38)$$

where $\dim(S_n)$ *is the dimension of the finite element space associated with* $\mathcal{T}_n$. *In both estimates, the constant* C *depends on* $\mathcal{T}_0$ *and* m, *but not on* n.

Proof Recall $\boldsymbol{0} < \boldsymbol{a} \leq \boldsymbol{\mu}$. The first inequality is the consequence of the definition of the weighted space $\mathcal{M}^m_{\boldsymbol{\mu}}$ and the local interpolation error estimates on different initial tetrahedra: the o-tetrahedra (Lemma 6.1), the v- or v_e-tetrahedra (Corollary 6.1), the e-tetrahedra (Corollary 6.2), and the ev-tetrahedra (Corollary 6.3).

Note that for each refinement, a tetrahedron is decomposed into 8 child tetrahedra. Therefore, the dimension of the finite element space $\dim(S_n) \sim 2^{3n}$. Thus, the second inequality follows from the best approximation property (6.3) and $h \sim \dim(S_n)^{-1/3}$.□

Remark 6.3 Using (6.20), one obtains

$$a_c(a_C - a_c) = (m + 1 - a_c)(a_c - a_{ec}).$$

Since $0 < a_{ec} \leq a_c \leq m$, it can be seen that for ev-tetrahedra, $a_c \leq a_C$. Therefore, by (6.10), for a given regularity index μ_c in the weighted space, the optimal value of a_c is usually smaller than that of μ_c. Intuitively, the additional regularity represented by the difference in the regularity index $\mu_c - a_c$ is needed to compensate for the lack of the maximum angle condition in the mesh when $a_{sc} > a_{ec}$. In the special case when $a_c = a_{ec}$, we have $a_C = a_c$, and the new ev-tetrahedra generated in each refinement will satisfy the maximum angle condition. Nevertheless, the conditions in (6.9) and (6.10) give rise to an optimal range of the grading parameter, for which the proposed finite element solution converges in the optimal rate (6.38) when $u \in \mathcal{M}^{m+1}_{\boldsymbol{\mu}+\boldsymbol{1}}(\Omega)$. In particular, according to (5.6), $u \in \mathcal{M}^{m+1}_{\boldsymbol{\mu}+\boldsymbol{1}}(\Omega)$ when the given data $f \in \mathcal{M}^{m+1}_{\boldsymbol{\mu}-\boldsymbol{1}}(\Omega)$.

6.2 Error Estimates for $u \in \mathcal{H}^2_\gamma(\Omega)$ (Rough Given Data)

We consider again the Dirichlet problem (6.1) ($\overline{\Gamma_D} = \partial\Omega$) but with rough given data f. In particular, recall the spaces $L^2_{\mu^*}(\Omega)$, $\mathcal{H}^2_\gamma(\Omega)$, and the parameters γ_e and γ_c in Corollary 5.2. In this section, suppose $f \in L^2_{\mu^*}(\Omega)$. Then the solution $u \in \mathcal{H}^2_\gamma(\Omega)$ (see(5.36)). We here include finite element error analysis results that ensure the graded mesh (Algorithm 5.7) can also be applied when f merely belongs to the weighted L^2 space. This shall extend the application of the anisotropic finite element method to broader practical computations.

Algorithm 6.5 (Anisotropic Algorithm for $\boldsymbol{f \in L^2_{\mu^*}(\Omega)}$) For the Dirichlet problem (6.1) with $f \in L^2_{\mu^*}(\Omega)$, we choose the anisotropic mesh $\mathcal{T}_n$ (Algorithm 5.7) with the grading parameters (6.5)–(6.8) such that

$$1 - \mu_e^* \leq a_e \leq \mu_e^* \quad \text{if } \omega_e > \pi; \qquad a_e = 1 \quad \text{if } \omega_e \leq \pi; \tag{6.39}$$

$$a_C := 2 + a_{ec} - \frac{2a_{ec}}{a_c} \leq \gamma_c, \tag{6.40}$$

where ω_e is the opening angle at the edge e. We use the linear ($m = 1$) finite element space $S_n \subset H_0^1(\Omega)$ associated with the mesh $\mathcal{T}_n$ in the numerical approximation. Then let $u_n \in S_n$ be the finite element solution defined in (6.2). □

Remark 6.4 In Algorithm 6.5, based on (6.40), we have $a_C \geq a_c \geq a_{ec}$, with the equal sign being taken when $a_c = a_{ec}$. By choosing a_e close to μ_e^* and a_c small, it is clear that the conditions (6.39) and (6.40) lead to a non-empty set. Note further that if c is a regular vertex ($\gamma_c = 1$), the condition (6.40) becomes $a_c \leq \frac{2a_{ec}}{1+a_{ec}}$. Hence, the choice $a_c = 1$ leads to $a_{ec} = 1$. In other words, the choice $a_c = 1$ is only possible if every edge $e \in \mathcal{E}_c$ satisfies $\omega_e \leq \pi$; otherwise, the simplest choice is to take $a_c \leq a_e$ for all $e \in \mathcal{E}_c$, which leads to $a_c = a_{ec}$.

The finite element error analysis for $|u - u_n|_{H^1(\Omega)}$ follows along similar lines as in the last section by considering the interpolation error estimates on initial tetrahedra of different types. Therefore, we only state the results and leave the proof to the readers. We also refer the readers to [85] for more details on the analysis.

Theorem 6.6 *Under the assumption in Lemma 5.1, for $f \in L_{\mu^*}(\Omega)$ defined in (5.8), the finite element approximation (Algorithm 6.5) of (6.1) achieves the optimal rate of convergence. Namely,*

$$|u - u_n|_{H^1(\Omega)} \leq Ch\|u\|_{\mathcal{H}^2_{\gamma}(\Omega)} \leq C\dim(S_n)^{-1/3}\|f\|_{L^2_{\mu^*}(\Omega)},$$

where $h := 2^{-n}$, $dim(S_n)$ is the dimension of the finite element space associated with $\mathcal{T}_n$, and the constant C depends on $\mathcal{T}_0$, but not on n.

6.3 Error Estimates for Mixed Boundary Conditions

In this section, we consider (6.1) with the mixed boundary condition and $\Gamma_D \neq \emptyset$, which includes the case where Neumann boundary conditions are allowed on adjacent faces of the boundary. Recall the parameters λ_e, τ_c, and σ_c in (5.37) and Corollary 5.4. Suppose $f \in H^{\sigma}(\Omega)$ for some $\sigma \in [0, 1)$. Consequently, the solution u satisfies the regularity estimates in $\mathcal{V}_e^o$ (Theorem 5.5) and in $\mathcal{V}_c$ (Theorem 5.6 and Corollary 5.4). The error analysis results in this section shall extend the application of the anisotropic finite element method to elliptic problems with mixed boundary conditions.

Algorithm 6.7 (Anisotropic Algorithm for Mixed Boundary Condition) For the aforementioned problem with the mixed boundary condition, we choose the anisotropic mesh $\mathcal{T}_n$ (Algorithm 5.7) with the grading parameters (6.5)–(6.8) such that $a_c \leq a_e$ for any $e \in \mathcal{E}_c$ and

$$1 - \sigma \leq a_e < \lambda_e \text{ if } e \text{ is singular}; \qquad a_e = 1 \text{ if } e \text{ is regular}; \tag{6.41}$$

$$a_c < \tau_c + 1/2 \text{ if } c \text{ is singular}; \qquad a_c = 1 \text{ if } c \text{ and all } e \in \mathcal{E}_c \text{ are regular}. \tag{6.42}$$

See (5.37)–(5.38) for definitions of singular edges and vertices. We use the linear ($m = 1$) finite element space S_n associated with the mesh $\mathcal{T}_n$ in the numerical approximation. Then let $u_n \in S_n$ be the finite element solution defined in (6.2). □

Remark 6.5 For any $c \in \mathcal{C}$, recall $\kappa_{ec} := \min_{e\in\mathcal{E}_c}\{\kappa_c, \kappa_e\}$ in Algorithm 5.7. Based on the parameter selections in Algorithm 6.7, it is clear that for any $c \in \mathcal{C}$, $\kappa_{ec} = \kappa_{sc} = \kappa_c$. Note that a_e has a lower bound $1 - \sigma$ in (6.41). The condition (5.39) $\sigma > \lambda_{\mathcal{E}} \geq \max_{e\in\mathcal{E}}\{1 - \lambda_e\}$ ensures the set given in (6.41) is not empty. For $0 < a_e < 1$, it is clear that refinements for an e- or ev-tetrahedron lead to anisotropic meshes toward the edge that do not preserve the maximum angle condition.

The finite element error analysis for $|u - u_n|_{H^1(\Omega)}$ again follows along similar lines as in Sect. 6.1. Therefore, we only state the results and leave the proof to the readers. We refer the readers to [86] for more details on the analysis.

Theorem 6.8 *For the anisotropic finite element method in Algorithm 6.7 solving (6.1) with $f \in H^\sigma(\Omega)$ and $\sigma \in [0, 1)$ satisfying (5.39), $\sigma \neq \lambda_{e,k} - 1$, for all $k \in \mathbb{N}$, $e \in \mathcal{E}$ (see Theorem 5.5), and $\sigma \neq \tau_{c,k} - \frac{1}{2}$ for all $k \in \mathbb{N}^*$, $c \in \mathcal{C}$ (see Theorem 5.6), we have*

$$\|u - u_n\|_{H^1(\Omega)} \leq Ch,$$

where $h = 2^{-n}$, and C depends on the initial triangulation $\mathcal{T}_0$ and f, but not on n.

6.4 Numerical Illustrations

In this section, we demonstrate the anisotropic finite element method (Algorithm 5.7) solving (6.1) in various situations: 1. $u \in \mathcal{M}^2_{\mu+1}(\Omega)$ ($f \in \mathcal{M}^2_{\mu-1}(\Omega)$); 2. $u \in \mathcal{H}^2_\gamma(\Omega)$ ($f \in L^2_{\mu^*}(\Omega)$); 3. when the mixed boundary condition is imposed. In these examples, the linear ($m = 1$) finite element method is used on graded meshes in two typical polyhedral domains: the prism domain and the Fichera domain. In addition, the numerical convergence rate $\mathcal{R}$ (3.2) is reported for each case, which confirms the optimal convergence when the grading parameters are chosen within the ranges in Algorithms 6.1, 6.5, and 6.7. Recall that the trend $\mathcal{R} \to 1$ as the level j of refinements increases implies the optimal convergence is achieved.

6.4.1 The Case of $u \in \mathcal{M}^2_{\mu+1}(\Omega)$

Assume $\overline{\Gamma_D} = \partial\Omega$ and let $f = 1$. We demonstrate the method in two examples.

Example 6.1 (Prism Domain) Let T be the triangle with vertices $(0, 0)$, $(1, 0)$, and $(0.5, 0.5)$. Let $\Omega := \big((0, 1)^2 \setminus T\big) \times (0, 1)$ be the prism domain. See Fig. 6.1 for the domain and the initial mesh. According to the regularity results in Sect. 5.1, the solution $u \in H^2$ in any sub-region of Ω that is away from the edge e where the opening angle is $3\pi/2$. Therefore, a quasi-uniform mesh in such a region will yield a first-order (optimal) convergence for the interpolation error. In the neighborhood of the edge e, by (5.5) and Table 1 in [46], we have

$$f \in \mathcal{M}^2_{\mu-1}, \quad u \in \mathcal{M}^2_{\mu+1}, \qquad \text{for } \mu_e < \eta_e = 2/3 \text{ and } \mu_c < \eta_c = 13/6.$$

Then by (6.9) and (6.10), a sufficient condition to attain the optimal convergence rate for the finite element solution is that the mesh parameters satisfy $a_e < 2/3$ and $a_C < 13/6$.

For linear ($m = 1$) elements, using (6.20), $0 < a_{ec} \leq a_c$, $0 < a_{ec} \leq a_e$, and the fact $a_c, a_e \in (0, 1]$, we have

$$a_C = 2 - a_c^{-1}(2 - a_c)a_{ec} \leq 2 - a_{ec} < 2 < 13/6.$$

Therefore, the condition $a_C < 13/6$ for the vertex c is always satisfied for all the feasible values of a_c and a_e, which implies that the vertex c shall not affect the convergence rate on the anisotropic mesh. Hence, to improve the convergence rate, we only need to implement special edge refinement based on the value of a_e. To see the impact of the grading parameter for e, we choose the parameters such that

$$0 < a_e \leq 1 \qquad \text{and} \qquad a_c = 1.$$

Then based on Theorem 6.4 and (6.9), in order to recover the optimal convergence rate for the finite element solution, it suffices to choose $0 < a_e < 2/3$, namely, $0 < \kappa_e = 2^{-1/a_e} < 0.353$. Recall that for $\kappa_c = 2^{-1/a_c} = 0.5$ and $\kappa_{ec} = \kappa_e < 0.5$, the resulting mesh is graded toward the edge e without special refinement for the vertex c. See Fig. 6.1 for such graded meshes when $\kappa_e = 0.2$.

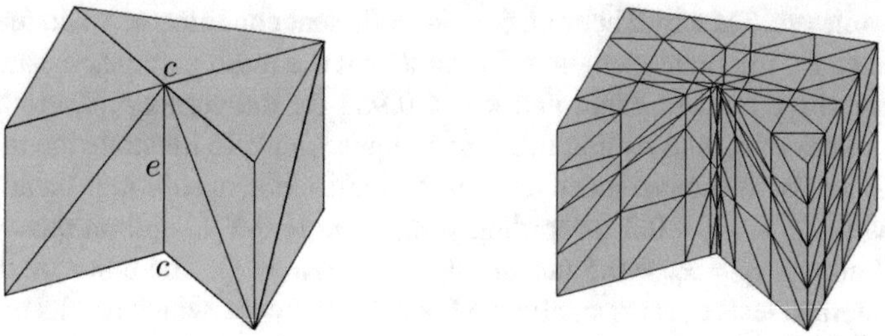

Fig. 6.1 The prism domain: the initial triangulation (left) and the mesh after two graded refinements toward the singular edge e ($\kappa_e = 0.2$).

In Table 6.3, we display the convergence rates of the finite element solution on these anisotropic meshes associated with different values of the grading parameter κ_e. Here, j is the level of refinements. It is clear from the table that the first-order convergence rate is obtained for $\kappa_e = 0.1, 0.2, 0.3 < 0.353$, while we lose the optimal

$j\backslash\mathcal{R}$	$\kappa_e = 0.1$	$\kappa_e = 0.2$	$\kappa_e = 0.3$	$\kappa_e = 0.4$	$\kappa_e = 0.5$
3	0.75	0.79	0.82	0.84	0.83
4	0.91	0.93	0.94	0.93	0.90
5	0.97	0.98	0.98	0.96	0.91
6	0.99	0.99	0.99	0.97	0.89
7	1.00	1.00	1.00	0.97	0.86

Table 6.1 Convergence rates $\mathcal{R}$ (3.2) for the prism domain (Fig. 6.1), $u \in \mathcal{M}^2_{\mu+1}(\Omega)$.

convergence if $\kappa_e = 0.4, 0.5$, both larger than the critical value 0.353. Note that when $\kappa_e = 0.4$, that is $0.353 < \kappa_e < 0.5$, this choice still leads to an anisotropic mesh graded toward the singular edge. However, such grading is insufficient to resolve the edge singularity in the solution and hence does not lead to the optimal rate of convergence. These results are in agreement with the theoretical estimates. We also point out that on quasi-uniform meshes ($\kappa_e = 0.5$), the theoretical convergence rate is about $h^{0.66}$. We see in Table 6.1 that the numerical rate for $\kappa_e = 0.5$ is decreasing and shall approach the theoretical rate 0.66 as j increases.

Example 6.2 (Fichera Domain) Let $D_0 = [0.5, 1) \times (0, 0.5] \times [0.5, 1)$, and let the domain $\Omega := (0, 1)^3 \setminus D_0$. See Fig. 6.2 for the domain and the initial mesh. Then Ω is featured with the Fichera corner at the vertex c and three adjacent edges e with the opening angle $3\pi/2$. For any sub-region away from these three edges, the solution $u \in H^2$, and therefore, a quasi-uniform mesh will lead to the optimal convergence rate for the interpolation error. In the neighborhood of the three edges, including the Fichera corner, by (5.5), (5.6), and Table 1 in [46], we have

$$f \in \mathcal{M}^2_{\mu-1}, \quad u \in \mathcal{M}^2_{\mu+1}, \qquad \text{for } \mu_e < \eta_e = 2/3 \text{ and } \mu_c < \eta_c \approx 0.954.$$

For the endpoints of the three edges (marked as e), which are not at the Fichera corner, the upper bound of the regularity index is $13/6$. For the same reason as in Example 6.1, these vertices shall not affect the convergence rate with feasible mesh parameters. Then by Theorem 6.4, the sufficient condition to attain the optimal convergence for the finite element solution is that the mesh parameters satisfy $a_e < 2/3$ for the three marked edges and $a_C < 0.954$ for the vertex c. There are many possible values of a_e and a_c that fulfill this requirement. To illustrate the method, in Table 6.2, we list the convergence rates of the finite element solutions on anisotropic meshes with $\kappa_e = \kappa_c = 0.3$ (accordingly, $a_e = a_c \approx 0.576$) and on quasi-uniform meshes, namely, $\kappa_e = \kappa_c = 0.5$ (accordingly, $a_e = a_c = 1$). The finest mesh ($j = 8$) in the numerical tests consists of about 5.9×10^8 tetrahedra, which results in a system of over 10^8 linear equations.

In the case $\kappa_e = \kappa_c = 0.3$, by (6.20), we have $a_e \approx 0.576 < 2/3$ and $a_C = a_c \approx 0.576 < 0.954$. Therefore, by Theorem 6.4, (6.9), and (6.10), we expect to obtain the first-order optimal convergence rate in the finite element approximation. For the quasi-uniform mesh ($\kappa_e = \kappa_c = 0.5$), since the solution is not globally in H^2, by (6.4), we expect a sub-optimal convergence rate. We observe that the numerical results in Table 6.4 validate this theoretical prediction.

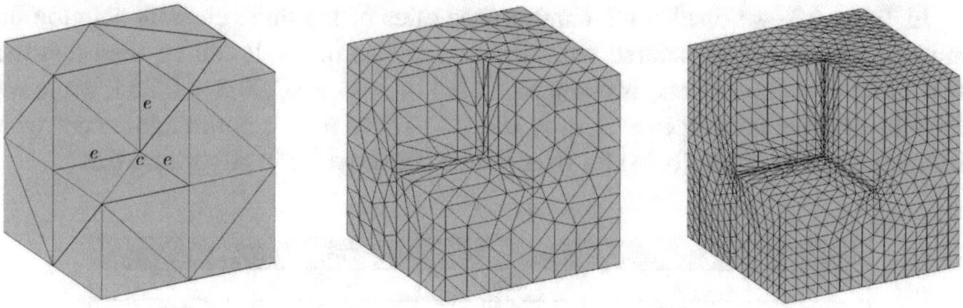

Fig. 6.2 The Fichera corner (left–right): the initial mesh, mesh after two refinements, mesh after three refinements ($\kappa_e = \kappa_c = 0.3$).

$j\backslash\mathcal{R}$	$\kappa_c = 0.3 \quad \kappa_e = 0.3$	$\kappa_c = 0.5 \quad \kappa_e = 0.5$
3	0.84	0.82
4	0.94	0.86
5	0.97	0.86
6	0.99	0.83
7	0.99	0.80

Table 6.2 Convergence rates $\mathcal{R}$ (3.2) for the Fichera corner (Fig. 6.2), $u \in \mathcal{M}^2_{\mu+1}(\Omega)$.

6.4.2 The Case of Rough Given Data

Assume $\overline{\Gamma_D} = \partial\Omega$. We demonstrate the method in two examples using the same polyhedral domains in Examples 6.1 and 6.2, respectively.

Example 6.3 (Prism Domain) Recall the edge e is on the z-axis (see Fig. 6.1). Let $f = |z - 0.5|^{0.1}$ in (6.1). Then we have $f \notin H^1(\Omega)$ and $f \in L^2_{\mu^*}$ near the edge e for any

$$1/2 \leq \mu_e^* < \eta_e = \lambda_e = 2/3. \tag{6.43}$$

Meanwhile, the solution is in H^2 in the sub-region of Ω that is away from the edge e. Therefore, a quasi-uniform mesh in this sub-region shall give the first-order convergence in the interpolation error. In the neighborhood of e, based on the regularity estimates in Corollary 5.2 and Algorithm 6.5, the condition (6.43) leads to the optimal range of the grading parameter:

$$1/3 \leq a_e < 2/3, \quad \text{namely,} \quad 0.125 \leq \kappa_e = 2^{-1/a_e} < 0.353. \tag{6.44}$$

In a sufficiently small neighborhood of the endpoints c of e, as discussed in Example 6.1, we have $f \in \mathcal{M}^2_{\mu-1}$, for $0 < \mu_e < 2/3$ and $0 < \mu_c < 13/6$, and therefore $u \in \mathcal{M}^2_{\mu+1}$. This implies that for any feasible $a_c, a_e \in (0, 1]$, these vertices c shall not affect the convergence rate of the numerical solution. Consequently, to achieve the optimal convergence rate, it is sufficient to implement special edge refinement based on the value of κ_e in (6.44).

In Table 6.3, we display the convergence rates of the finite element solution on anisotropic meshes associated with different values of κ_e. It can be seen that the first-order convergence rate is obtained for $0.125 \leq \kappa_e = 0.2, 0.3 < 0.353$, while we lose the optimal convergence if $\kappa_e = 0.4, 0.5$. These results are in agreement with the sufficient condition (6.44) for the optimal convergence in Algorithm 6.5.

$j\backslash\mathcal{R}$	$\kappa_e = 0.2$	$\kappa_e = 0.3$	$\kappa_e = 0.4$	$\kappa_e = 0.5$
3	0.77	0.81	0.83	0.82
4	0.92	0.93	0.93	0.90
5	0.97	0.98	0.96	0.91
6	0.99	0.99	0.97	0.89

Table 6.3 Convergence rates $\mathcal{R}$ (3.2) for the prism domain (Fig. 6.1), $f \in L^2_{\mu^*}$.

Example 6.4 (Fichera Domain) The singular edges are the three edges e joining at the Fichera corner c (Fig. 6.2). Let $\Omega_0 := (0, 0.5) \times (0.5, 1) \times (0, 0.5)$. We set in (6.1)

$$f = \begin{cases} 0 & \text{in } \Omega_0, \\ |x - 0.75|^{0.1} + |y - 0.25|^{0.1} + |z - 0.75|^{0.1} & \text{in } \Omega \setminus \Omega_0. \end{cases}$$

It is clear that $f \notin H^1(\Omega)$ and $f \in L^2_{\mu^*}$ for any $1/2 \leq \mu_e^* < 2/3$ in the neighborhood of the singular edges. Meanwhile, the solution is in H^2 in any sub-region away from the singular edges, where we shall use the quasi-uniform mesh in the numerical approximation. As discussed in Example 6.2, the endpoints of the singular edges that are not at the Fichera corner c shall not affect the convergence rate for any feasible parameters $a_e, a_c \in (0, 1]$.

In the neighborhood of the Fichera corner c, by Corollary 5.2 and Table 1 in [46], the solution satisfies

$$u \in \mathcal{H}^2_\gamma \quad \text{for } \mu_e^* \leq \gamma_e < \lambda_e = 2/3 \text{ and } \gamma_c < \eta_c \approx 0.954.$$

Then by Algorithm 6.5, the sufficient condition to attain the optimal convergence rate for the finite element solution is that the mesh parameters give rise to

$$1/3 \leq a_e < 2/3 \text{ for the three singular edges} \quad \text{and} \quad a_C < 0.954 \text{ for } c. \tag{6.45}$$

There are many possible values of a_e and a_c that fulfill this requirement. In Table 6.4, we list the convergence rates of the finite element solutions on meshes with $\kappa_e = \kappa_c = 0.3$ (accordingly, $a_e = a_c \approx 0.576$) and on quasi-uniform meshes $\kappa_e = \kappa_c = 0.5$ (accordingly, $a_e = a_c = 1$). These numerical results are aligned with the predication for the optimal range of the grading parameters (6.45).

$j\backslash\mathcal{R}$	$\kappa_c = 0.3 \quad \kappa_e = 0.3$	$\kappa_c = 0.5 \quad \kappa_e = 0.5$
3	0.83	0.81
4	0.93	0.87
5	0.97	0.87
6	0.99	0.84

Table 6.4 Convergence rates $\mathcal{R}$ (3.2) for the Fichera corner (Fig. 6.2), $f \in L^2_{\mu^*}$.

6.4.3 The Case of Mixed Boundary Conditions

We now demonstrate the convergence of the anisotropic finite element method (Algorithm 6.7) solving (6.1) with mixed boundary conditions.

Example 6.5 (Prism Domain) Let Ω_1 be the square with vertices at $(1, 0)$, $(0, 1)$, $(-1, 0)$, $(0, -1)$, and let T be the triangle with vertices at $(0, 0)$, $(-1, 0)$, $(0, -1)$. Define the prism domain $\Omega := (\Omega_1 \setminus T) \times (0, 1)$. See Fig. 6.3 for the domain and the initial mesh. For a point $(x, y, z) \in \Omega$, denote by (r, θ) the polar coordinates of its projection in the xy-plane ($r(x, y, z) = r(x, y)$ and $\theta(x, y, z) = \theta(x, y)$). In the first numerical example, we are especially interested in the performance of the numerical method when the Neumann condition is imposed on both adjacent faces of the singular edge.

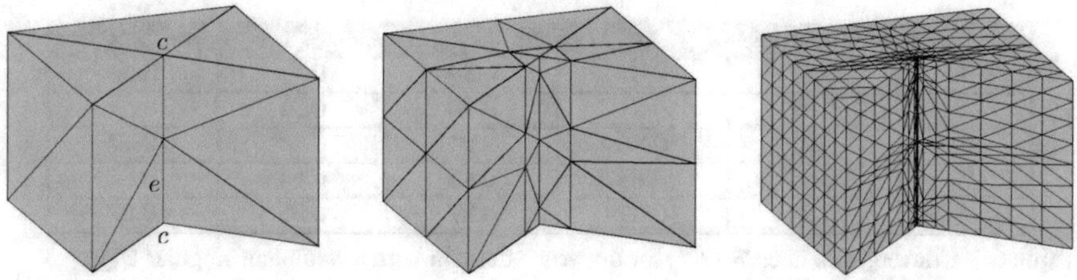

Fig. 6.3 The prism domain (left–right): the initial mesh, mesh after one refinement, mesh after three refinements ($\kappa_e = \kappa_c = 0.2$).

Consider the following elliptic equation with the mixed boundary condition:

$$\begin{cases} -\Delta u = 1 & \text{in } \Omega, \\ u = r^{2/3} \cos \frac{2(\theta + \frac{\pi}{2})}{3} & \text{on } \Gamma_D, \\ \partial_n u = 0 & \text{on } \Gamma_N. \end{cases} \tag{6.46}$$

We choose Γ_N such that the Neumann condition is imposed on the two faces adjacent to the edge e and the Dirichlet condition is on all the other faces (including the top and bottom faces). See Fig. 6.4. Thus, e is a Neumann edge, and other edges are either Dirichlet edges or DN edges (see the description in (5.37)).

According to (5.37) (see also [46]), the edge e is the only singular edge with $\lambda_e = \frac{2}{3}$, and the two vertices c (its two endpoints) satisfy $\tau_c + \frac{1}{2} > \lambda_e$. Other vertices and edges of Ω are regular. In addition, the right hand side function in (6.46) belongs

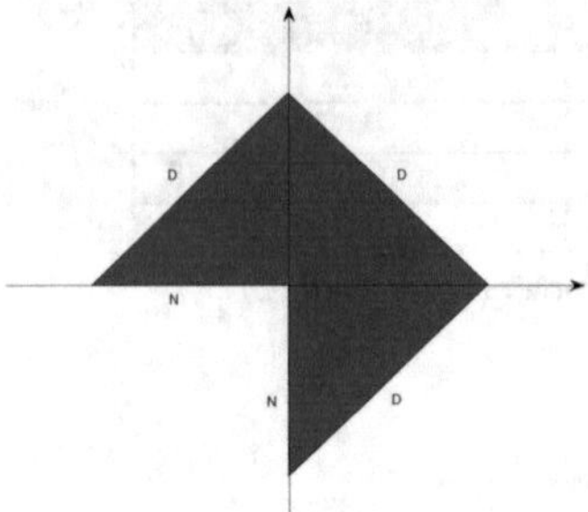

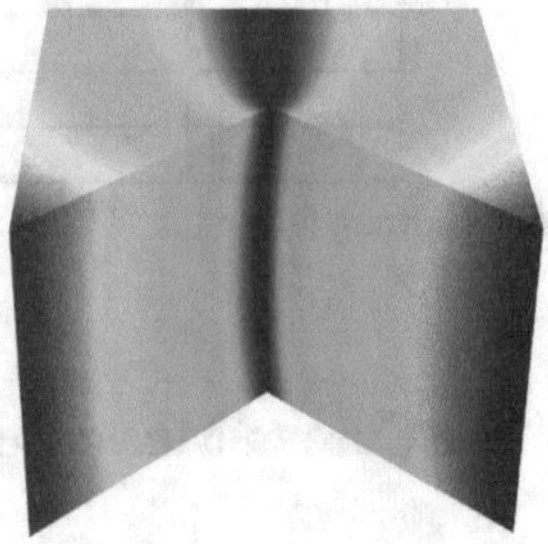

Fig. 6.4 The Neumann singular edge: the top view of Ω, top Dirichlet face marked in blue (left); the absolute value of the numerical solution (right).

to $H^\sigma(\Omega)$ for any $\sigma \in [0, 1)$. In fact, the solution $u \in H^{1+s}(\Omega)$ for $s < \frac{2}{3}$. Therefore, based on the conditions in (6.41) and (6.42), and by Theorem 6.8, it is sufficient to obtain the optimal convergence rate in the finite element method if we choose $a_e \in (0, \frac{2}{3})$ for the edge e and $a_c \in (0, a_e]$ for its two endpoints. This gives rise to the following optimal range of the grading parameters: $\kappa_c \leq \kappa_e = 2^{-1/a_e} < 2^{-3/2} \approx 0.353$ near e, and quasi-uniform meshes (the associated grading parameters being $\kappa_e, \kappa_c = 0.5$) for all the other edges and vertices.

$j \backslash \mathcal{R}$	$\kappa = 0.1$	$\kappa = 0.2$	$\kappa = 0.3$	$\kappa = 0.4$	$\kappa = 0.5$
4	0.81	0.83	0.84	0.77	0.64
5	0.89	0.91	0.90	0.81	0.66
6	0.95	0.96	0.94	0.83	0.67
7	0.98	0.98	0.96	0.84	0.67
8	0.99	0.99	0.97	0.85	0.67

Table 6.5 Convergence rates $\mathcal{R}$ (3.2) for the prism domain with a Neumann singular edge.

In Table 6.5, we display the convergence rates of the finite element solution on graded meshes. In all these meshes, we choose $\kappa = \kappa_e = \kappa_c$ for the singular edge e and the two endpoints c (Fig. 6.3). We see that the above theoretical predictions are confirmed by the numbers in the table. Namely, $\kappa = \kappa_e = \kappa_c = 0.1, 0.2, 0.3 < 0.353$ leads to optimal finite element methods. In addition, the convergence rate for $\kappa = 0.5$ is also very close to the theoretical bound $\frac{2}{3}$ as discussed above.

In the second numerical example, we test the anisotropic algorithm in the prism domain that has a DN singular edge. In particular, we solve (6.1) with $f = 1$, impose the Neumann condition $\partial_n u = 0$ on one face adjacent to the edge e, and impose the Dirichlet condition $u = 0$ on all the other faces (see Fig. 6.5). Thus, e is the singular edge surrounded by faces with mixed boundary conditions.

According to (5.37), e is the only singular edge with $\lambda_e = \frac{1}{3}$. Note that $f = 1$ belongs to $H^\sigma(\Omega)$ for any $\sigma \in [0, 1)$, and the solution $u \in H^{1+s}(\Omega)$ for $s < \frac{1}{3}$. Based on Algorithm 6.7, it is sufficient to obtain the optimal convergence rate in the

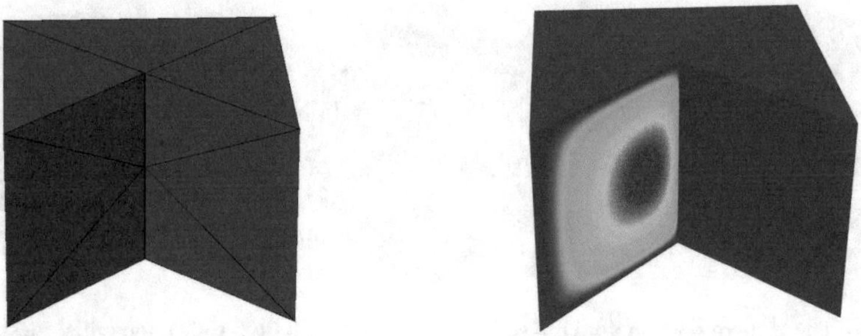

Fig. 6.5 The DN singular edge: Neumann face marked in red and Dirichlet face in blue (left); the numerical solution (right).

finite element method if we choose $a_e \in (0, \frac{1}{3})$ for the edge e and $a_c \in (0, a_e]$ for its two endpoints. Hence, the optimal range of the grading parameters is: $\kappa_c \le \kappa_e = 2^{-1/a_e} < 2^{-3} = 0.125$ near e, and quasi-uniform meshes for all the other edges and vertices.

$j\backslash\mathcal{R}$	$\kappa = 0.1$	$\kappa = 0.2$	$\kappa = 0.3$	$\kappa = 0.4$	$\kappa = 0.5$
4	0.72	0.73	0.72	0.70	0.68
5	0.87	0.86	0.81	0.72	0.64
6	0.93	0.90	0.80	0.66	0.54
7	0.96	0.90	0.76	0.58	0.45
8	0.97	0.89	0.70	0.52	0.39

Table 6.6 Convergence rates $\mathcal{R}$ (3.2) for the prism domain with a DN singular edge.

We summarize the numerical convergence rates in Table 6.6, which again verifies the theoretical prediction. Namely, when the grading parameter $\kappa = \kappa_e = \kappa_c < 0.125$, the optimal convergence rate is achieved, while it is not the case for $\kappa > 0.125$.

Example 6.6 (Fichera Domain) Consider the same domain Ω as in Example 6.2. See Fig. 6.6 for the domain and the initial mesh. Let $f = 1$ in (6.1). We impose the Neumann condition $\partial_n u = 0$ on the two faces that touch the center vertex c of Ω and impose the Dirichlet condition $u = 0$ on all the other faces.

According to (5.37), there are three singular edges e (the edges touching the center vertex), and the endpoints c of these singular edges are possible singular vertices. For the Neumann singular edge, we have $\lambda_e = \frac{2}{3}$, and for the two DN singular edges, we have $\lambda_e = \frac{1}{3}$. In fact, the solution satisfies the global regularity $u \in H^{1+s}(\Omega)$ for $s < \frac{1}{3}$. As described in Algorithm 6.7, we use the graded mesh toward the singular edges and their endpoints. To simplify the presentation, we choose the same parameter κ_e for all the singular edges and the same parameter $\kappa_c = \kappa_e$ for all their endpoints. See Fig. 6.6 for the case $\kappa_e = \kappa_c = 0.2$. Recall that for any possible singular vertex c, $\tau_c + \frac{1}{2} > \frac{1}{2}$. Hence, based on (6.41) and (6.42), it is sufficient to choose $\kappa_c = \kappa_e < 2^{-3} = 0.125$ for all the singular edges and their endpoints, in order to obtain the optimal convergence in the numerical approximation.

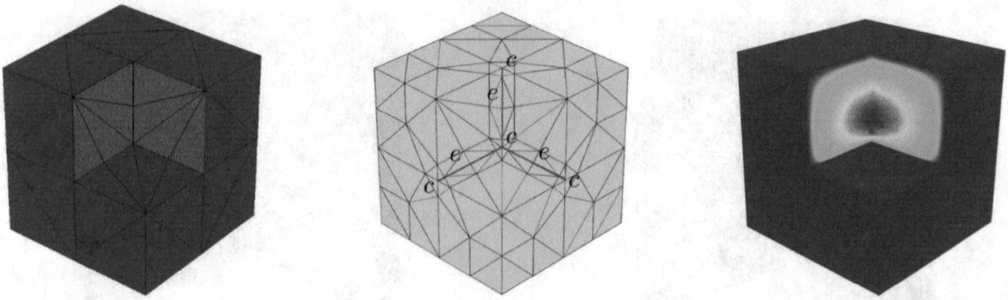

Fig. 6.6 The Fichera domain with mixed boundary conditions (left–right): the initial mesh, Neumann faces marked in red, and Dirichlet faces in blue; the mesh after one refinement, $\kappa_e = \kappa_c = 0.2$; the numerical solution.

The numerical convergence rates $\mathcal{R}$ (3.2) are listed in Table 6.7. As predicted by the theory, for $\kappa = \kappa_c = \kappa_e > 0.125$, the convergence is not optimal, while for $\kappa = 0.1 < 0.125$, the numbers are increasing toward the optimal rate. We stopped at level 7 because the resources needed to extend the calculation to the next level of refinement have exceeded our computing capability. For instance, the refinement to the next level will produce more than 5 billion tetrahedra and 1 billion nodes. Nevertheless, according to Table 6.7, there is a notable improvement in convergence rates using the appropriate graded meshes ($\kappa < 0.125$) compared with other graded meshes ($\kappa > 0.125$), and it is reasonable to expect the rates for $\kappa = 0.1$ will converge to 1 when the asymptotic region is reached with further refinements.

$j \backslash \mathcal{R}$	$\kappa = 0.1$	$\kappa = 0.2$	$\kappa = 0.3$	$\kappa = 0.4$	$\kappa = 0.5$
3	0.58	0.63	0.65	0.65	0.63
4	0.78	0.81	0.80	0.76	0.70
5	0.85	0.88	0.83	0.75	0.65
6	0.88	0.89	0.81	0.68	0.54
7	0.91	0.88	0.75	0.59	0.45

Table 6.7 Convergence rates $\mathcal{R}$ (3.2) for the Fichera domain with mixed boundary conditions.

Remark 6.6 In Chaps. 4–6, we have addressed some of the most important theoretical and numerical features regarding various graded finite element methods based on the general mesh algorithm (Algorithm 3.9). These algorithms are effective to approximate singular solutions of elliptic equations from the nonsmoothness of the domain and from the change of boundary conditions. In addition, these methods have been applied to other mathematical models with singularities, such as the Schrödinger equation with the inverse square potential [87, 88], the axisymmetric Poisson and Stokes equations [77, 80], the elliptic equations with Dirac sources [7, 89], the transmission problem [84], the acoustic fluid–structure interaction problem [36], the singular optimal control problems [11], and the problem with the Ventcel boundary condition [102]. For other theoretical developments of these graded finite element methods, we mention the error analysis in non-energy norms [12, 52, 81], the multigrid methods on 2D graded meshes [32, 34, 119] and on 3D anisotropic meshes for edge refinement [14], and the condition number estimates for 3D anisotropic meshes [6, 83]. Meanwhile, despite the progress made in recent years, the regularity theory for 2D and 3D singular problems is still an evolving field. Many critical questions on finite element methods for 3D anisotropic singular problems, such as novel mesh algorithms, optimal error analysis, sharp regularity estimates, efficient numerical implementations, and fast numerical solvers, remain open. In this book, we organize the content to systematically present the basic ideas and results on graded finite element methods for singular problems, in hope that it serves as a useful resource for readers who are interested in this topic.

References

1. R. Adams. *Sobolev Spaces*, volume 65 of *Pure and Applied Mathematics*. Academic Press, New York-London, 1975.
2. S. Agmon. *Lectures on Elliptic Boundary Value Problems*. Prepared for publication by B. Frank Jones, Jr. with the assistance of George W. Batten, Jr. Van Nostrand Mathematical Studies, No. 2. D. Van Nostrand Co., Inc., Princeton, N.J.-Toronto-London, 1965.
3. M. Ainsworth and J. T. Oden. *A Posteriori Error Estimation in Finite Element Analysis*. Pure and Applied Mathematics (New York). Wiley-Interscience [John Wiley & Sons], New York, 2000.
4. B. Ammann, A. D. Ionescu, and V. Nistor. Sobolev spaces on Lie manifolds and regularity for polyhedral domains. *Doc. Math.*, 11:161–206, 2006.
5. B. Ammann and V. Nistor. Weighted Sobolev spaces and regularity for polyhedral domains. *Comput. Methods Appl. Mech. Engrg.*, 196(37–40):3650–3659, 2007.
6. T. Apel. *Anisotropic Finite Elements: Local Estimates and Applications*. Advances in Numerical Mathematics. B. G. Teubner, Stuttgart, 1999.
7. T. Apel, O. Benedix, D. Sirch, and B. Vexler. A priori mesh grading for an elliptic problem with Dirac right-hand side. *SIAM J. Numer. Anal.*, 49(3):992–1005, 2011.
8. T. Apel and B. Heinrich. Mesh refinement and windowing near edges for some elliptic problem. *SIAM J. Numer. Anal.*, 31(3):695–708, 1994.
9. T. Apel and S. Nicaise. The finite element method with anisotropic mesh grading for elliptic problems in domains with corners and edges. *Math. Methods Appl. Sci.*, 21(6):519–549, 1998.
10. T. Apel, S. Nicaise, and J. Schöberl. Finite element methods with anisotropic meshes near edges. In *Finite element methods (Jyväskylä, 2000)*, volume 15 of *GAKUTO Internat. Ser. Math. Sci. Appl.*, pages 1–8. Gakkōtosho, Tokyo, 2001.

H. Li, *Graded Finite Element Methods for Elliptic Problems in Nonsmooth Domains*, Surveys and Tutorials in the Applied Mathematical Sciences 10,
https://doi.org/10.1007/978-3-031-05821-9

11. T. Apel, J. Pfefferer, and A. Rösch. Finite element error estimates for Neumann boundary control problems on graded meshes. *Comput. Optim. Appl.*, 52(1):3–28, 2012.
12. T. Apel, A. Rösch, and D. Sirch. L^∞-error estimates on graded meshes with application to optimal control. *SIAM J. Control Optim.*, 48(3):1771–1796, 2009.
13. T. Apel, A.-M. Sändig, and J. Whiteman. Graded mesh refinement and error estimates for finite element solutions of elliptic boundary value problems in non-smooth domains. *Math. Methods Appl. Sci.*, 19(1):63–85, 1996.
14. T. Apel and J. Schöberl. Multigrid methods for anisotropic edge refinement. *SIAM J. Numer. Anal.*, 40(5):1993–2006 (electronic), 2002.
15. D. Arnold, A. Mukherjee, and L. Pouly. Locally adapted tetrahedral meshes using bisection. *SIAM J. Sci. Comput.*, 22(2):431–448 (electronic), 2000.
16. D. N. Arnold, D. Boffi, and R. S. Falk. Approximation by quadrilateral finite elements. *Math. Comp.*, 71(239):909–922 (electronic), 2002.
17. I. Babuška and A. Aziz. On the angle condition in the finite element method. *SIAM J. Numer. Anal.*, 13(2):214–226, 1976.
18. I. Babuška, R. Kellogg, and J. Pitkäranta. Direct and inverse error estimates for finite elements with mesh refinements. *Numer. Math.*, 33(4):447–471, 1979.
19. I. Babuška, T. von Petersdorff, and B. Andersson. Numerical treatment of vertex singularities and intensity factors for mixed boundary value problems for the Laplace equation in $\mathbf{R}^3$. *SIAM J. Numer. Anal.*, 31(5):1265–1288, 1994.
20. I. Babuška and A. K. Aziz. Survey lectures on the mathematical foundations of the finite element method. In *The mathematical foundations of the finite element method with applications to partial differential equations (Proc. Sympos., Univ. Maryland, Baltimore, Md., 1972)*, pages 1–359, 1972. With the collaboration of G. Fix and R. B. Kellogg.
21. C. Bacuta, J. Bramble, and J. Xu. Regularity estimates for elliptic boundary value problems in Besov spaces. *Math. Comp.*, 72:1577–1595, 2003.
22. C. Bacuta, J. H. Bramble, and J. Xu. Regularity estimates for elliptic boundary value problems with smooth data on polygonal domains. *J. Numer. Math.*, 11(2):75–94, 2003.
23. C. Bacuta, H. Li, and V. Nistor. Anisotropic graded meshes and quasi-optimal rates of convergence for the FEM on polyhedral domains in 3D. In *CCOMAS 2012 - European Congress on Computational Methods in Applied Sciences and Engineering*, e-Book Full Papers, pages 9003–9014. Proceedings of European Congress on Computational Methods in Applied Sciences and Engineering, 2012.
24. C. Bacuta, H. Li, and V. Nistor. Differential operators on domains with conical points: precise uniform regularity estimates. *Rev. Roumaine Math. Pures Appl.*, 62:383–411, 2017.

25. C. Bacuta, V. Nistor, and L. Zikatanov. Improving the rate of convergence of 'high order finite elements' on polygons and domains with cusps. *Numer. Math.*, 100(2):165–184, 2005.
26. C. Bacuta, V. Nistor, and L. Zikatanov. Improving the rate of convergence of high-order finite elements on polyhedra. II. Mesh refinements and interpolation. *Numer. Funct. Anal. Optim.*, 28(7–8):775–824, 2007.
27. J. Bergh and J. Löfström. *Interpolation Spaces. An introduction*. Springer-Verlag, Berlin, 1976. Grundlehren der Mathematischen Wissenschaften, No. 223.
28. C. Bernardi, M. Dauge, and Y. Maday. Polynomials in the Sobolev world. 2007.
29. J. Bey. Tetrahedral grid refinement. *Computing*, 55(4):355–378, 1995.
30. P. Binev, W. Dahmen, and R. DeVore. Adaptive finite element methods with convergence rates. *Numer. Math.*, 97(2):219–268, 2004.
31. H. Blum and M. Dobrowolski. On finite element methods for elliptic equations on domains with corners. *Computing*, 28(1):53–63, 1982.
32. J. Brannick, H. Li, and L. Zikatanov. Uniform convergence of the multigrid V-cycle on graded meshes for corner singularities. *Numer. Linear Algebra Appl.*, 15(2–3):291–306, 2008.
33. S. Brenner. Multigrid methods for the computation of singular solutions and stress intensity factors. I. Corner singularities. *Math. Comp.*, 68(226):559–583, 1999.
34. S. Brenner, J. Cui, and L.-Y. Sung. Multigrid methods for the symmetric interior penalty method on graded meshes. *Numer. Linear Algebra Appl.*, 16(6):481–501, 2009.
35. S. Brenner and L. Scott. *The Mathematical Theory of Finite Element Methods*, volume 15 of *Texts in Applied Mathematics*. Springer-Verlag, New York, second edition, 2002.
36. S. C. Brenner, A. l. Çeşmelioğlu, J. Cui, and L.-Y. Sung. A nonconforming finite element method for an acoustic fluid-structure interaction problem. *Comput. Methods Appl. Math.*, 18(3):383–406, 2018.
37. F. Brezzi and M. Fortin. *Mixed and Hybrid Finite Element Methods*. Springer-Verlag, New York, 1991.
38. A. Buffa, M. Costabel, and M. Dauge. Anisotropic regularity results for Laplace and Maxwell operators in a polyhedron. *C. R. Math. Acad. Sci. Paris*, 336(7):565–570, 2003.
39. Z. Cai and S. Kim. A finite element method using singular functions for the Poisson equation: corner singularities. *SIAM J. Numer. Anal.*, 39(1):286–299 (electronic), 2001.
40. C. Carstensen and S. Bartels. Each averaging technique yields reliable a posteriori error control in FEM on unstructured grids. I. Low order conforming, nonconforming, and mixed FEM. *Math. Comp.*, 71(239):945–969 (electronic), 2002.

41. C. Carstensen and J. Hu. A unifying theory of a posteriori error control for nonconforming finite element methods. *Numer. Math.*, 107(3):473–502, 2007.
42. Z. Chen and S. Dai. On the efficiency of adaptive finite element methods for elliptic problems with discontinuous coefficients. *SIAM J. Sci. Comput.*, 24(2):443–462, 2002.
43. P. Ciarlet. *The Finite Element Method for Elliptic Problems*, volume 4 of *Studies in Mathematics and Its Applications*. North-Holland, Amsterdam, 1978.
44. M. Costabel. Boundary integral operators on curved polygons. *Ann. Mat. Pura Appl. (4)*, 133:305–326, 1983.
45. M. Costabel and M. Dauge. Weighted regularization of maxwell equations in polyehdral domains. a rehabilitation of nodal finite elements. *Numerische Mathematik*, 93:239–277, 2002.
46. M. Costabel, M. Dauge, and S. Nicaise. Weighted analytic regularity in polyhedra. *Comput. Math. Appl.*, 67(4):807–817, 2014.
47. M. Costabel, M. Dauge, and C. Schwab. Exponential convergence of *hp*-FEM for Maxwell equations with weighted regularization in polygonal domains. *Math. Models Methods Appl. Sci.*, 15(4):575–622, 2005.
48. A. Craig, J. Zhu, and O. Zienkiewicz. A posteriori error estimation, adaptive mesh refinement and multigrid methods using hierarchical finite element bases. In *The mathematics of finite elements and applications, V (Uxbridge, 1984)*, pages 587–594. Academic Press, London, 1985.
49. M. Dauge. *Elliptic Boundary Value Problems on Corner Domains*, volume 1341 of *Lecture Notes in Mathematics*. Springer-Verlag, Berlin, 1988.
50. C. De Coster and S. Nicaise. Singular behavior of the solution of the Helmholtz equation in weighted L^p-Sobolev spaces. *Adv. Differential Equations*, 16(1–2):165–198, 2011.
51. A. Demlow. A posteriori error estimation and adaptivity. Lecture Notes.
52. A. Demlow, D. Leykekhman, A. Schatz, and L. Wahlbin. Best approximation property in the W^1_∞ norm for finite element methods on graded meshes. *Math. Comp.*, 81(278):743–764, 2012.
53. A. Demlow and R. Stevenson. Convergence and quasi-optimality of an adaptive finite element method for controlling L_2 errors. *Numer. Math.*, 117(2):185–218, 2011.
54. T. Dupont and L. Scott. Polynomial approximation of functions in Sobolev spaces. *Math. Comp.*, 34(150):441–463, 1980.
55. A. Ern and J. Guermond. *Theory and Practice of Finite Elements*, volume 159 of *Applied Mathematical Sciences*. Springer-Verlag, New York, 2004.
56. L. Evans. *Partial Differential Equations*, volume 19 of *Graduate Studies in Mathematics*. AMS, Rhode Island, 1998.
57. R. Fritzsch. *Optimale Finite-Elemente-Approximationen für Funktionen mit Singularitäten*. 1990. Thesis (Ph.D.)–TU Dresden.

58. D. Gilbarg and N. Trudinger. *Elliptic Partial Differential Equations of Second Order*. Springer-Verlag, Berlin, 1977. Grundlehren der Mathematischen Wissenschaften, Vol. 224.
59. V. Girault and P. Raviart. *Finite Element Methods for Navier-Stokes Equations*, volume 5 of *Springer Series in Computational Mathematics*. Springer-Verlag, Berlin, 1986. Theory and algorithms.
60. P. Grisvard. Théorèmes de traces relatifs à un polyèdre. *C. R. Acad. Sci. Paris Sér. A*, 278:1581–1583, 1974.
61. P. Grisvard. *Elliptic Problems in Nonsmooth Domains*, volume 24 of *Monographs and Studies in Mathematics*. Pitman (Advanced Publishing Program), Boston, MA, 1985.
62. P. Grisvard. Problèmes aux limites dans les polygones. Mode d'emploi. *EDF Bull. Direction Études Rech. Sér. C Math. Inform.*, (1):3, 21–59, 1986.
63. P. Grisvard. Edge behavior of the solution of an elliptic problem. *Math. Nachr.*, 132:281–299, 1987.
64. P. Grisvard. Singularités en elasticité. *Arch. Rational Mech. Anal.*, 107(2):157–180, 1989.
65. P. Grisvard. *Singularities in Boundary Value Problems*, volume 22 of *Research Notes in Applied Mathematics*. Springer-Verlag, New York, 1992.
66. B. Guo and I. Babuška. Regularity of the solutions for elliptic problems on nonsmooth domains in $\mathbf{R}^3$. I. Countably normed spaces on polyhedral domains. *Proc. Roy. Soc. Edinburgh Sect. A*, 127(1):77–126, 1997.
67. E. Hunsicker, H. Li, V. Nistor, and V. Uski. Analysis of Schrödinger operators with inverse square potentials I: regularity results in 3D. *Bull. Math. Soc. Sci. Math. Roumanie (N.S.)*, 55(103)(2):157–178, 2012.
68. E. Hunsicker, H. Li, V. Nistor, and V. Uski. Analysis of Schrödinger operators with inverse square potentials II: FEM and approximation of eigenfunctions in the periodic case. *Numer. Methods Partial Differential Equations*, 30(4):1130–1151, 2014.
69. C. Johnson. *Numerical Solution of Partial Differential Equations by the Finite Element Method*. Dover Publications, Inc., Mineola, NY, 2009. Reprint of the 1987 edition.
70. R. B. Kellogg. On the Poisson equation with intersecting interfaces. *Applicable Anal.*, 4:101–129, 1974/75.
71. V. Kondrat′ev. Boundary value problems for elliptic equations in domains with conical or angular points. *Trudy Moskov. Mat. Obšč.*, 16:209–292, 1967.
72. V. Kozlov, V. Maz'ya, and J. Rossmann. *Elliptic Boundary Value Problems in Domains with Point Singularities*, volume 52 of *Mathematical Surveys and Monographs*. American Mathematical Society, Providence, RI, 1997.
73. V. Kozlov, V. Maz'ya, and J. Rossmann. *Spectral Problems Associated with Corner Singularities of Solutions to Elliptic Equations*, volume 85 of *Mathematical Surveys and Monographs*. American Mathematical Society, Providence, RI, 2001.
74. M. Křížek. On the maximum angle condition for linear tetrahedral elements. *SIAM J. Numer. Anal.*, 29(2):513–520, 1992.

75. A. Kufner. *Weighted Sobolev Spaces*. John Wiley & Sons, 1984.
76. R. Lauter and V. Nistor. Analysis of geometric operators on open manifolds: a groupoid approach. In N. Landsman, M. Pflaum, and M. Schlichenmaier, editors, *Quantization of Singular Symplectic Quotients*, volume 198 of *Progress in Mathematics*, pages 181–229. Birkhäuser, Basel - Boston - Berlin, 2001.
77. Y.-J. Lee and H. Li. Axisymmetric Stokes equations in polygonal domains: regularity and finite element approximations. *Comput. Math. Appl.*, 64(11):3500–3521, 2012.
78. D. Leguillon and E. Sánchez-Palencia. *Computation of Singular Solutions in Elliptic Problems and Elasticity*. John Wiley & Sons, Ltd., Chichester; Masson, Paris, 1987.
79. H. Li. A-priori analysis and the finite element method for a class of degenerate elliptic equations. *Math. Comp.*, 78:713–737, 2009.
80. H. Li. Regularity and multigrid analysis for Laplace-type axisymmetric equations. *Math. Comp.*, 84(293):1113–1144, 2015.
81. H. Li. The W_p^1 stability of the Ritz projection on graded meshes. *Math. Comp.*, 86(303):49–74, 2017.
82. H. Li. An anisotropic finite element method on polyhedral domains: interpolation error analysis. *Math. Comp.*, 87(312):1567–1600, 2018.
83. H. Li and X. Lu. Condition numbers of finite element methods on a class of anisotropic meshes. *Appl. Numer. Math.*, 158:22–43, 2020.
84. H. Li, A. Mazzucato, and V. Nistor. Analysis of the finite element method for transmission/mixed boundary value problems on general polygonal domains. *Electron. Trans. Numer. Anal.*, 37:41–69, 2010.
85. H. Li and S. Nicaise. Regularity and a priori error analysis on anisotropic meshes of a Dirichlet problem in polyhedral domains. *Numer. Math.*, 139(1):47–92, 2018.
86. H. Li and S. Nicaise. A priori analysis of an anisotropic finite element method for elliptic equations in polyhedral domains. *Comput. Methods Appl. Math.*, 21(1):145–177, 2021.
87. H. Li and V. Nistor. Analysis of a modified Schrödinger operator in 2D: regularity, index, and FEM. *J. Comput. Appl. Math.*, 224(1):320–338, 2009.
88. H. Li and J. S. Ovall. A posteriori error estimation of hierarchical type for the Schrödinger operator with inverse square potential. *Numer. Math.*, 128(4):707–740, 2014.
89. H. Li, X. Wan, P. Yin, and L. Zhao. Regularity and finite element approximation for two-dimensional elliptic equations with line Dirac sources. *J. Comput. Appl. Math.*, 393:113518, 16, 2021.
90. H. Li and Q. Zhang. Optimal quadrilateral finite elements on polygonal domains. *J. Sci. Comput.*, 70(1):60–84, 2017.
91. J.-L. Lions and E. Magenes. *Non-Homogeneous Boundary Value Problems and Applications. Vol. I*. Springer-Verlag, New York, 1972. Translated from the French by P. Kenneth, Die Grundlehren der mathematischen Wissenschaften, Band 181.

92. J. Lubuma and S. Nicaise. Dirichlet problems in polyhedral domains. II. Approximation by FEM and BEM. *J. Comput. Appl. Math.*, 61(1):13–27, 1995.
93. V. Maz'ya and J. Rossmann. *Elliptic Equations in Polyhedral Domains*, volume 162 of *Mathematical Surveys and Monographs*. American Mathematical Society, Providence, RI, 2010.
94. A. L. Mazzucato and V. Nistor. Well-posedness and regularity for the elasticity equation with mixed boundary conditions on polyhedral domains and domains with cracks. *Arch. Ration. Mech. Anal.*, 195(1):25–73, 2010.
95. P. Monk. *Finite Element Methods for Maxwell's Equations*. Numerical Mathematics and Scientific Computation. Oxford University Press, New York, 2003.
96. P. Morin, R. Nochetto, and K. Siebert. Convergence of adaptive finite element methods. *SIAM Rev.*, 44(4):631–658 (electronic) (2003), 2002. Revised reprint of "Data oscillation and convergence of adaptive FEM" [SIAM J. Numer. Anal. **38** (2000), no. 2, 466–488 (electronic); MR1770058 (2001g:65157)].
97. S. Moroianu and V. Nistor. Index and homology of pseudodifferential operators i. manifolds with boundary. Proceedings of OAT Conference, 2008.
98. S. Nazarov and B. Plamenevsky. *Elliptic Problems in Domains with Piecewise Smooth Boundaries*, volume 13 of *de Gruyter Expositions in Mathematics*. Walter de Gruyter & Co., Berlin, 1994.
99. S. A. Nazarov, G. Sweers, and A. Stylianou. Paradoxes in problems on bending of polygonal plates with a hinged/supported edge. In *Doklady Physics*, volume 56, pages 439–443. Springer, 2011.
100. J. Nečas. *Les Méthodes Directes En Théorie Des Équations Elliptiques*. Masson et Cie, Éditeurs, Paris, 1967.
101. S. Nicaise. *Polygonal Interface Problems*. Lang, Peter Publishing, Incorporated, 1993.
102. S. Nicaise, H. Li, and A. Mazzucato. Regularity and a priori error analysis of a Ventcel problem in polyhedral domains. *Math. Methods Appl. Sci.*, 2016.
103. L. Oganesjan, V. Rivkind, and L. Ruhovec. Variational-difference methods for the solution of elliptic equations. I. *Differencial′nye Uravnenija i Primenen.—Trudy Sem. Processy Differentsial′nye Uravneniya i ikh Primenenie*, pages 3–389, 391, 1973.
104. G. Raugel. Résolution numérique par une méthode d'éléments finis du problème de Dirichlet pour le laplacien dans un polygone. *C. R. Acad. Sci. Paris Sér. A-B*, 286(18):A791–A794, 1978.
105. W. Rudin. *Functional Analysis*. International Series in Pure and Applied Mathematics. McGraw-Hill, Inc., New York, second edition, 1991.
106. D. Schötzau, C. Schwab, and T. P. Wihler. *hp*-dGFEM for second-order mixed elliptic problems in polyhedra. *Math. Comp.*, 85(299):1051–1083, 2016.
107. E. Schrohe. Spectral invariance, ellipticity, and the Fredholm property for pseudodifferential operators on weighted Sobolev spaces. *Ann. Global Anal. Geom.*, 10(3):237–254, 1992.

108. C. Schwab. *P- And Hp- Finite Element Methods: Theory and Applications in Solid and Fluid Mechanics*. Oxford University Press, 1999.
109. E. Stein. *Singular Integrals and Differentiability Properties of Functions*. Princeton Mathematical Series, No. 30. Princeton University Press, Princeton, N.J., 1970.
110. W. Strang and G. Fix. *An Analysis of the Finite Element Method*. Wellesley Cambridge Pr, 1973.
111. B. Szabó and I. Babuška. Computation of the Amplitude of Stress Singular Terms for Cracks and Reentrant Corners. In *Fracture Mechanics: Nineteenth Symposium, ed. T. Cruse (West Conshohocken, PA: ASTM International, 1988)*, pages 101–124.
112. M. Taylor. *Partial Differential Equations I, Basic theory*, volume 115 of *Applied Mathematical Sciences*. Springer-Verlag, New York, 1995.
113. H. Triebel. *Interpolation Theory, Function Spaces, Differential Operators*. North-Holland Pub. Co., 1978.
114. R. Verfürth. *A Review of A Posteriori Error Estimation and Adaptive Mesh-Refinement Techniques*. Wiley-Teubner, 1996.
115. R. Verfürth. *A posteriori error estimation techniques for finite element methods*. Numerical Mathematics and Scientific Computation. Oxford University Press, Oxford, 2013.
116. L. Wahlbin. On the sharpness of certain local estimates for H^1 projections into finite element spaces: influence of a re-entrant corner. *Math. Comp.*, 42(165):1–8, 1984.
117. J. Xu and T. Tang, editors. *Adaptive Computations: Theory and Algorithms*. Science Press, 2007.
118. K. Yosida. *Functional Analysis*, volume 123 of *A Series of Comprehensive Studies in Mathematics*. Springer-Verlag, New York, fifth edition, 1978.
119. H. Yserentant. The convergence of multilevel methods for solving finite-element equations in the presence of singularities. *Math. Comp.*, 47(176):399–409, 1986.
120. S. Zhang. On the nested refinement of quadrilateral and hexahedral finite elements and the affine approximation. *Numer. Math.*, 98:559–579, 2004.
121. O. Zienkiewicz, R. Taylor, and J. Zhu. *Finite Element Method - Its Basis and Fundamentals*. Elsevier, 2005.

Index

H. Li, *Graded Finite Element Methods for Elliptic Problems in Nonsmooth Domains*, Surveys and Tutorials in the Applied Mathematical Sciences 10,
https://doi.org/10.1007/978-3-031-05821-9

Printed in the United States
by Baker & Taylor Publisher Services